Science and technology

AN INTRODUCTION TO THE LITERATURE

Science and technology

AN INTRODUCTION TO THE LITERATURE

DENIS GROGAN

Head of the Department of Bibliographical Studies
College of Librarianship Wales
(University of Wales School of Librarianship
and Information Studies)

FOURTH EDITION

CLIVE BINGLEY LONDON

First edition published 1970
Second edition published 1973
Third edition published 1976
This fourth edition first published 1982
by Clive Bingley Ltd, 16 Pembridge Road, London W11 3HL
Set in 10 on 11 point Press Roman by Allset
Printed and bound in the UK by Redwood Burn Ltd, Trowbridge, Wilts
Copyright © Denis Joseph Grogan 1982
All rights reserved
ISBN: 0-85157-315-0 (hardback)
ISBN: 0-85157-340-1 (paperback)

British Library Cataloguing in Publication Data

Grogan, Denis
  Science and technology: an introduction to
  the literature.—4th ed.
  1. Science—Bibliography
  2. Technology—Bibliography
  I. Title
  016.5   Z7401

  ISBN 0-85157-315-0
  ISBN 0-85157-340-1 (Pbk)

CONTENTS

OCT 1987

INTRODUCTION

This is a book written primarily for students, not practitioners. For the toiler actually in the field of scientific and technical information there are a host of excellent guides, as chapter 2 will indicate: this work is for the *would-be* practitioner, who has reached an understanding of the general sources of information such as encyclopedias and yearbooks and bibliographies, but now wishes to move on to the sources of scientific and technical information. Ultimately, in the field (or even in the classroom or seminar), he may well be required to concentrate his attention even more narrowly on a specific subject area within science or technology. It is hoped that this text will not only initiate him into the overall structure of the literature of science and technology, but will also prepare the way for just that kind of detailed study of constituent subject parts.

Of course, many experienced scientific and technical information workers were obliged to commence their study of the literature of their subject at the deep end, so to speak, inasmuch as they had to learn on the job, working with the literature and its users in an appropriate specialized library. And for a librarian to master the literature of a particular subject, such daily use is still the best and probably the only satisfactory method. But this haphazard approach is not only unscientific, it is inefficient, dependent as it is on the librarian gaining his acquaintance with the literature in a sequence determined by the random demands of enquirers. It is true that the requisite familiarity with the various documentary sources of information can be attained this way, but it takes far longer than it need.

The information workers of tomorrow, on the other hand, are given the opportunity to embark on the study of the literature in a more systematic fashion. As the students of today, not only do they commence with a basic course in general reference sources before progressing to scientific and technical literature, but they are encouraged to investigate the general structure of that literature before turning to more specific areas within the field. As novice practitioners in the library such students will of course still be faced with the need to come to grips with the detailed literature of their subject, but the hope is

7

that they will be able to accomplish this more methodically and rapidly by virtue of their theoretical grasp of the overall pattern of the literature of science and technology. Perhaps even more importantly, such students should be particularly well placed to meet the challenge that will increasingly confront the future practitioner in all fields: rapid obsolescence of current knowledge and the advent of new physical formats as a result of accelerating technical advance.

It is hoped that there will be at least some practising librarians or information workers for whom this book will have an interest: those for example thoroughly familiar with their own special area, but who might appreciate the opportunity to take a more general view, or those learning on the job but seeking a more structured approach, or those contemplating or already embarked upon a change of subject field, particularly from the humanities to the sciences. Here I have been fortunate enough to be able to draw on my own experience over several years in supervising the day-to-day work of public service staff in a very large scientific and technical library.

The needs of the reader without formal scientific training have been kept especially in mind. A survey carried out in 1965 indicated that over a third of the staff employed in scientific and technical information work in the United Kingdom fell into this category, and the 'swing from science' in the schools and universities suggests that this proportion will increase. To take but one measure: though full-time students taking university courses in the UK in the decade of the 1970s increased by almost a quarter, the proportion studying science and technology (including medicine and agriculture) actually decreased by over 3%. For students (and indeed practitioners) lacking a scientific background it is of particular importance that they have the opportunity of examining the general structure of the literature of science and technology before exploring an area in depth.

It is popularly thought that the way a librarian studies the literature of a subject is to memorize as many authors and titles as possible. Ever since the library catalogue was invented, this has never been really necessary, but even in these days of vast and instantaneous computer memories, many (including myself) have been reluctant to shed their admiration for the traditional reference librarian's encyclopedic knowledge of his stock, and with Goldsmith's village rustics we still marvel 'that one small head could carry all he knew'.

However, a guide to the literature for a student demands a radically different approach from a guide to the literature for the user in the library. The latter is searching for information, and wants his guide to be comprehensive; the former is striving to understand the underlying pattern of the literature and needs a guide that demonstrates this by pointing to particular selected examples worthy of closer study. The

8

object of attention is the general rather than the particular, or more precisely, it is the particular not for its own sake but only insofar as it represents an instance of the general. The zoology student dissects the dogfish not because it is the most important or even the most common of the fishes but because it serves particularly well as a representative of one of the group's two main types.

The wide selection of worthy publications enumerating and describing those scientific and technical reference books, bibliographies and data bases that librarians and others find useful makes it unnecessary here even to attempt to list basic titles. This is a teaching tool, and as the essence of teaching is selection, attention is concentrated on *types* of literature, and individual titles are normally listed merely as manifestations of those types. The student should of course endeavour to examine and study them closely as representatives of their class, but since they are merely examples chosen from many possible alternatives, he might gain an even more valuable insight into this particular aspect of scientific and technical communication if he identifies for himself in the library collections or information systems to which he has access similar instances of each type, and devotes his attention to these. Where a guide such as this has to endeavour to be comprehensive is in demonstrating all the *types* of scientific and technical literature a student might meet. Of course, once in the field, his task will be to build on this largely theoretical framework by identifying within each type no longer simply representative titles, but *all* those works within his chosen area which might be of value to their potential users in his library.

It would be superfluous in this book to rehearse the features of many of the titles quoted as examples. This has already been far better done in *Walford's guide to reference material: volume I, Science and technology* (Library Association, fourth edition 1980) and E P Sheehy *Guide to reference books* (Chicago, American Library Association, ninth edition 1976) and *Supplement* (1980), greatly strengthened in the pure and applied sciences chapters and now reinforced, despite the word 'books' in its title, with a separate chapter on computer-readable data bases. However, both of these guides concentrate on reference works, whereas much of the actual literature used by the scientist and technologist is non-reference material such as monographs and textbooks, as well as a whole range of non-book materials such as periodicals, patents, research reports, etc. Obviously, comment in the text on examples quoted from these categories is more appropriate, as it is with certain other groups of reference material such as British directories (covered only selectively by Walford) and reviews of progress (excluded by Sheehy).

It should be emphasized that this is a guide to the *literature*, and it is

9

a mistake to assume that this is synonymous with information. It is an even graver error to underestimate the importance of non-documentary information in science and technology: a number of surveys have shown that 'live' sources (eg consultation with colleagues, attendance at professional meetings, etc) play a large part in communication. After all, in the last analysis, it is people who are the ultimate source of all knowledge. When he was a professor at Princeton President Woodrow Wilson used to advise his students 'I would never read a book if it were possible to talk half an hour with the man who wrote it'. A generation or more later J Robert Oppenheimer is reported to have said 'If you really want to communicate, send a man'. And quite apart from any other reason, the pace of development is such that in many fields any information that has got into print has almost certainly been overtaken by events. It has been rightly said that the main disadvantage of the literature is that it is history and not news. Nevertheless, it is equally foolish to sell short the printed word: this is to throw away cumulative human experience. In Carlyle's words: 'All that mankind has done, thought, gained or been – it is lying, as in magic preservation, in the pages of books'. Science and technology in particular are dependent on the printed word as a means of communication. Surveys have shown a surprising amount of duplicated research and general wasted effort due to disregard of the published literature. In a recent ten-year period 950 new antibiotics were reported in the literature: 250 of them were duplicate discoveries. Some years ago it was reported in Germany that 'Approximately two thirds of all patent applications have to be rejected because the alleged inventions fail to meet the novelty criterion and because the applicants were unaware that the problem they worked on had already been solved and the solution published.' But the lesson is slowly being learned: increasingly, for example, governments are recognizing that the primary scientific and technical literature is one of the few non-consumable national resources. R T Bottle reminds us that 'Measured in terms of the man-years taken to produce it, the chemical literature is the most expensive tool available to the chemist.' The vital role played by the literature has been highlighted in recent years by the mushrooming computerized information systems. These, of course, are 'literature-based', even though the data they store is encoded in digital form. As such it is quite incomprehensible and both input and output must ultimately take the form of alphanumeric characters.

The reader will not find here any advice on how to use libraries and library catalogues, or any account of classification schemes. No attempt will be made to instruct him in the answering of reference enquiries or in the compilation of bibliographies. He will not be taught how to carry out a literature search or construct a retrieval profile or interrogate a computerized data base. It is true that such assistance is given in

10

many guides to the literature, and they are useful and indeed indispensable accomplishments of the 'compleat' librarian. This book, however, is written in the belief that personal service to the users of our libraries can best be improved by a more sophisticated approach to the literature by the librarian, allied to a more refined awareness of the user's needs. A much deeper understanding is required of the types of literature and of the special role of each type in the network of scientific communication. This has become an even more essential requirement with the advent of computerized data bases as part of the literature. Compared to the conventional sources in printed form, individual data bases are much more difficult to evaluate with regard to coverage, arrangement, indexing, accuracy, reliability, up-to-dateness, and all the other criteria of judgement that a librarian uses in studying an information resource. Their true role, too, *vis-à-vis* the printed literature, has still to be assessed. Of course the literature is only part of the pattern, but it is that part peculiarly within the librarian's domain and he should understand it fully. Merely to match subjects is to operate far below the optimum level of service. To provide 'something on' the topic the user is interested in is not enough. In their striving to satisfy user needs librarians have much to learn here from the scientific and technical publishers, commercial and otherwise: they do not just produce 'books' and 'periodicals' on a subject. They (and their authors) direct their productions at particular groups of consumers, personal and institutional: although somewhat similar to the casual glance, textbooks are really quite different from monographs, research journals quite different from technical journals.

The student will find that practically all of the examples chosen are in the English language: he will appreciate that this is a mere matter of convenience and that in many subjects there are vital works available only in other tongues. Again, for reasons of convenience, if no place of publication is cited with the publisher's name, it means that London appears in the imprint. Medicine has not been rigorously excluded: where a point can best be illustrated by a medical example it has been used. The lists of further reading appended to the chapters are deliberately highly selective, being confined to items thought to be of real value to the student, not too inaccessible, and capable of being read during a course of study. Although I have drawn extensively on the writings of others no attempt has been made to document every reference and quotation in the text: the reader's attention will not be distracted by what a *Times* reviewer once called 'the perpetual patter of tiny footnotes'; such excess of bibliographical scruple is out of place in a textbook for students.

It has been assumed throughout that the user of this book is familiar with *general* reference and bibliographical sources and the relevant

11

terminology. He will not find, therefore, his attention drawn in the chapter on biographical sources to *Who's who* or the *Dictionary of national biography*, even though both works contain their share of scientists and technologists. *The British union-catalogue of periodicals* is not mentioned among the lists of periodicals for the same reason, although it contains the locations of thousands of scientific journals. And so on.

I owe a great debt to the authors of those classic guides to the literature of the various sciences and technologies, some of which are noted in chapter 2. They have been constantly consulted over the years during my own exploration of the field. It should be added, however, that practically all the works mentioned in the text I have personally examined, and in most cases made use of.

In the introduction I have written for each successive revision of this work I have had occasion to remark that such has been the pace of change in the literature of science and technology, even in the few years since the previous edition, that scarcely a page of the earlier text has remained unrevised − a minor instance, in fact, of that rapid obsolescence of current knowledge about which the student was warned in the introduction to the first edition. If anything this pace has accelerated over the last five years. To select only three of the more striking changes: by Act of Parliament the British patent system has been transformed; the quite amazing capacity of the videodisc is already being exploited for the storage of scientific and technical information; and the lethal combination of world recession and soaring inflation has placed the scientific journal in great jeopardy. Indeed, in the area where we have seen the greatest advances made, the computerization of abstracting and indexing services, the multiplication not only of data bases but also of producers, suppliers, vendors, hosts, brokers and intermediaries is rapidly nearing the point where it will no longer be possible to see the wood for the trees.

And while the scientists and technologists continue to be responsible for the proliferation of their literature, and indeed for new forms of literature, the systematic bibliographers, many of them librarians, attempt to keep pace as they have been striving to do since the sixteenth century.

I have been increasingly encouraged in my own strivings by the adoption of this work as a textbook in universities over five continents, which would suggest that its particular approach to its subject still fulfils a real need. I have been especially heartened by the many readers who have written from all over the world to express their appreciation, and I would like to take this opportunity to thank them.

Aberystwyth, August 1981   D J GROGAN

Chapter 1

THE LITERATURE

Einstein believed that 'The whole of science is nothing more than a refinement of everyday thinking'. The way this refinement has been achieved has been through the discovery and perfection of the experimental method — possibly the greatest contribution science has made to human progress. Indeed, several writers have claimed that science is no more than this method — the scientific method as it is sometimes called. Karl Pearson, the founder of the twentieth century science of statistics, claimed that 'the unity of all science consists alone in its method, not in its material'. This method holds good for all the sciences, and the technologies also, and is of course widely applied in other disciplines.

The implications for the literature of science and technology are so far-reaching that it is essential for the student librarian to grasp the elements of the method. The first step a scientist (or technologist) takes towards solving a problem is to collect all the information that may have a bearing on the question: this is the *observation* stage. He then formulates a tentative theory as to how such facts are to be interpreted: this is the *hypothesis* stage. He then designs and carries out a series of controlled tests to try to confirm his working hypothesis: this is the *experimental* stage. If findings of the experiments prove his theory correct he formulates his answer to the problem: this is the *conclusion* stage. Of course, it frequently happens that the working hypothesis does not stand up under experiment; T H Huxley called this 'the great tragedy of science — the slaying of a beautiful hypothesis by an ugly fact.' When this occurs the scientist must go back as often as necessary until he achieves a hypothesis that not only accounts for all the observed facts but can be confirmed by controlled experiment. This is the classic *inductive* theory of scientific method 'still taught to every generation of students'.

J H Poincaré, the French genius who dominated the world of mathematics in the late nineteenth century, once wrote: 'Science is built of facts, the way a house is built of bricks; but an accumulation of facts is no more a science than a pile of bricks is a house'. These facts, deriving from observation and experiment, have first to be communicated to the

13

scientific community and then consciously integrated into the structure of knowledge. As John Gray and Brian Perry have recently reminded us, 'Science would not be science without scientific communication'.

Then, when a later scientist comes along, wishing to advance knowledge in his field, it is clearly vital for him first to discover what has already been achieved. He turns therefore to those records of observations and experiments left by his predecessors, the 'long avenues of carefully reasoned logical thought', to the literature, in fact. And when he in his turn adds to the archive by communicating his findings in a book or article he indicates the extent of his indebtedness by the references to previous work that he cites in his bibliography. As the Royal Society has proclaimed, 'Science rests on its published record', and it is this characteristic that allows Max Gluckman, the anthropologist, to define a science as 'any discipline in which a fool of this generation can go beyond the point reached by the genius of the last generation'. A century earlier Ernest Renan, the French philosopher, said much the same thing: 'The simplest schoolboy is now familiar with truths for which Archimedes would have sacrificed his life'. Even the great Sir Isaac Newton acknowledged: 'If I have seen further than most men it is by standing on the shoulders of giants'.

It should be pointed out that the classic description of scientific method just outlined is not universally accepted. Though the accretive, proven-fact view is still the most widely held conception of science, it has not remained without challenge from some philosophers of science, such as Thomas Kuhn, and especially Karl Popper with his 'searchlight' analogy of the human mind and his theory of conjectures and refutations. We must not forget the role of imagination and even intuition in science: J D Bernal pointed out that 'It is characteristic of science that the full explanations are often seized in their essence by the percipient scientist long in advance of any possible proof'.

The primary sources
The original reports of scientific and technical investigations make up the bulk of what is known as the *primary* literature. Some of these records may be largely observational (eg reports of scientific expeditions), or descriptive (eg some trade literature), or theoretical (eg much mathematics and physics), but most of them are accounts of experiments with findings and conclusions. A piece of research is not regarded as complete until the results are made available publicly, and it is a basic principle of scientific investigation that sufficient detail should be given to enable the work described to be repeated (and therefore double-checked) by any competent investigator. Before being added to the corpus of science all advances have to be critically assessed by a

14

worker's peers. It was the irreproducibility of the experimental results published by Dr William Summerlin, an American immunologist, that led to his unmasking in 1974 in what has been described as the 'medical Watergate'. He was said to have inked black patches on white mice to simulate successful transplants.

These contributions then represent new knowledge (or at least new interpretations of old knowledge) and constitute the latest available information. They are published in a variety of forms:

1 Periodicals (many of these are solely devoted to reporting original work)
2 Research reports
3 Conference proceedings
4 Reports of scientific expeditions
5 Official publications
6 Patents
7 Standards
8 Trade literature
9 Theses and dissertations.

They also form the archive or permanent record of the progress of science, available to all, whenever they should wish to see it.

Many of course remain unpublished, and outside the mainstream of scientific progress, but do occasionally become accessible later in their original form, and are often consulted for their historical interest, eg

1 Laboratory notebooks, diaries, memoranda, etc
2 Internal research reports, minutes of meetings, company files, etc
3 Correspondence, personal files, etc.

By its very nature the primary literature is widely scattered, disconnected, and unorganized. It records information as yet unassimilated to the body of scientific and technical knowledge. Although of vital importance, it is difficult to locate and to apply, and over a period there has therefore grown up a second tier of more accessible information sources.

Secondary sources
These are compiled from the primary sources and are arranged according to some definite plan. They represent 'worked-over' knowledge rather than new knowledge, and they organize the primary literature in more convenient form. By their nature they are often more widely available than the primary sources, and in many cases more self-sufficient:

1 Periodicals (a number of these specialize in interpreting and commenting on developments reported in the primary literature)
2 Indexing and abstracting services

3 Reviews of progress
4 Reference books, eg
    a encyclopedias
    b dictionaries
    c handbooks
    d tables
    e formularies
5 Treatises
6 Monographs
7 Textbooks.

In addition to repackaging the information from the primary literature many of these have the further useful function of guiding the
worker to the original documents. In other words, they serve not only
as repositories of digested facts, but as bibliographical keys to the
primary sources.

Tertiary sources
It is possible to distinguish a less well-defined group of sources the main
function of which is to aid the searcher in using the primary and
secondary sources. They are unusual in that most of them do not
carry 'subject' knowledge at all:
1 Directories and yearbooks
2 Bibliographies, eg
    a lists of books
    b location lists of periodicals
    c lists of indexing and abstracting services
3 Guides to 'the literature'
4 Lists of research in progress
5 Guides to libraries and sources of information
6 Guides to organizations.

Non-documentary sources
All the sources so far listed, primary, secondary and tertiary alike, take
the form of physical documents. But neither information nor its communication require such material embodiment.

Not least because talking and listening are more congenial than
reading and writing, such 'paperless' sources form a substantial part of
the communication system in some disciplines within science and technology, as investigations by the Centre for Research in User Studies at
Sheffield University have shown: 'personal communication is one of
the most important means of transmitting information'. This is particularly the case during the innovation process: a classic example can be

16

studied in the remarkable case history by J D Watson *The double helix: a personal account of the discovery of the structure of DNA* (Weidenfeld and Nicolson, 1968). A major investigation of physicists found that personal communication, at around nine hours per week, far outweighed all the other sources of information put together, a total of less than four hours. One classic United States survey found that more than half of all scientific communication is 'informal', prior to formal publication. More recently the Physics Information Review Committee confirmed this unequivocally: 'Informal communication is the means used for "first disclosure" of any sort, usually to close colleagues within a subject area or company. This can take place months, or even years, before publication'. It is clear that oral sources provide something that the others do not (and perhaps cannot):

1 Formal, eg
    a government departments, central and local
    b research organizations
    c learned and professional societies
    d industry, private and public
    e universities and colleges
    f consultants

2 Informal, eg
    a discussions with colleagues, visitors, etc.
    b 'corridor meetings' at conferences, etc.
    c casual conversations, social gatherings, etc.
    d telephone calls, which with direct dialling are increasingly replacing written correspondence.

Such oral communication is often more concentrated, more exclusive, more up-to-date, and more within the control of the actual participants. It can therefore be tailored to match the hearer, with the bonus of instantaneous feedback. The fact that it is so widespread has led Robert Fairthorne to conclude: 'This means that at any time the bulk of scientific knowledge is not yet recorded.' It is a fact that a lot of practical research 'know-how' could never be written down, such as special techniques, hints on the use of apparatus, or warnings about pitfalls in the application of certain methods. We are told that 'Every good laboratory has its wise man, who can tell you the answer to a practical question that may never have been discussed at all in the official literature'. Studies in Sweden have shown that young scientists commonly know fewer fellow-workers than their senior colleagues, and so they are warned how important it is to build up a personal contact network as soon as possible.

The price paid of course for the flexibility and convenience of these word-of-mouth sources is their varying reliability. Constraints of time and place impose limits on the actual amount of such communication,

17

and the kind of information that can be conveyed orally or by gesture is also restricted. In 1979 the Agriculture Information Review Committee stressed that 'Great emphasis is placed on personal relationships and communication', but went on to warn, 'but in many cases these informal links prove to be less than adequate'. There is a greater risk of inaccuracy too with oral sources, compared to documentary sources, which by definition are recorded and therefore more likely to have been subject to the process of peer review. In many cases they serve merely as pointers to the documentary (primary or secondary) sources, and the following up of a 'personal contact' often leads ultimately to the printed page. A 1978 study for the National Science Foundation found that 18% of the 'readings' of journal articles were as a result of personal referral by somebody else.

The chill wind of commercial competition is of course one curb on this free and friendly interchange of non-documentary information among industrial scientists and technologists, for it is not uncommon for them to be forbidden by their contracts from communicating proprietory information or ideas outside their companies, but even academic scientists occasionally find themselves competing with one another to be the first to present their findings in print. A J Meadows, Professor of Astronomy at the University of Leicester, reminds us that 'One of the first lessons that a scientist learns is that, if he wishes to gain the rewards that science offers, he must outpace his fellow-scientists in reaching a result'. Sir William Osler, the great Canadian physician, warned many years ago: 'In science the credit goes to the man who convinces the world, not to the man to whom the idea first occurs'. And obviously publication in a scientific journal furnishes 'a much surer basis for a priority claim than private correspondence or oral communications', or even the former practice of depositing a sealed letter, carefully dated, with the secretary of a learned society. A US survey of two thousand scientists revealed that two-thirds of them had been anticipated at least once by a colleague in the publication of their research results; some had suffered this frustration as many as ten times or more. As one of the harsh facts of scientific life, such rivalry can clearly hinder interpersonal communication, and secretive behaviour among scientists is not unknown.

Perhaps these oral sources and the literature are best regarded as complementary rather than contending, with reading and listening and talking and writing all strands in the web of scientific communication, without which science would not be science.

Technology and the literature
Apart of course from unpublished and oral sources, all that have been listed so far, primary, secondary and tertiary, are the products of a

18

particular technology, that of printing. But within the last few generations we have seen the advent of new technologies, notably photography, telecommunications, electronics and computers; more significantly, within the last few years we have witnessed a confluence of these technologies. It is this that is changing the communication of information in a manner at least as radical as the invention of the printing press more than five hundred years ago.

It does not yet seem that these innovations have added new *categories* of information sources, with the possible exception of cinefilm and video. Rather have they extended the range of formats in which the standard sources are recorded and consulted: microforms by and large are versions of books and periodicals in alternative physical form; audiocassettes used as information sources are usually recorded versions of oral communication; data bases are machine-readable versions of printed abstracting and indexing services. And so on.

At the moment it is convenient to regard these alternative technological formats as the equivalents or supplements of their printed counterparts, and it is still premature to see them as total replacements. But it is worth noting the prediction of Lewis Branscomb, Chief Scientist at IBM: 'Documents will become only occasional by-products of information access, not the primary embodiment of it'. This kind of prophetic dream sees the literary record of the future existing only as digital data encoded on magnetic tape, disk, or other form of storage. Consultation of this record, however, since it can only be assimilated by the user through the age-old process of reading, would still require words projected on a visual display unit, but printout on paper or computer-generated microfilm would become an optional extra. The literature, in other words, would have been stood on its head.

Use of the literature
The demands of the experimental method being what they are, it is obvious that scientists and technologists are in constant need of information and the average worker spends an estimated 20% of his time searching for it. Surveys show that industrial scientists and technologists give to it a rather larger slice of their day, academic scientists a somewhat smaller share. Now there are at least three ways to find out a fact that you need: you can work it out for yourself, you can ask someone who knows, or you can look it up. We tend to avoid the first where we can, even if it is quite feasible, such as measuring the boiling point of alcohol or the weight of a seagull. The second way, by asking someone, is often the easiest, and for certain kinds of information (such as methods) an important channel of scientific and technical communication, as we have seen. It is the third way, however, that concerns

us here, 'looking it up' by consulting the literature.

A worker may consult all these types of sources for a host of reasons, but in general he uses literature only when he finds it essential for some particular facet of his task. It is easier to understand the pattern of use (and indeed the pattern of the information sources) if one groups various kinds of consultation into a series of 'approaches', as has been demonstrated by Melvin J Voigt.

a *The current approach* arises from the need to keep up, to know what other workers in the field are doing (or about to do). Surveys of both scientists and technologists have shown that the most important source of such information is a fairly small core of primary periodicals, systematically scanned. Of great importance too are personal contacts with colleagues, with visitors, and at conferences. Secondary sources such as abstracts and indexes, if used at all, are regarded as guides to the primary sources.

Outside a worker's specific field of interest, he often tries to keep up to date in a broader area, knowledge of which is necessary to give depth and meaning to his special subject. Here it is not necessary for him to know his material so intimately or thoroughly, nor so vital to be absolutely up to the minute, and he can place more reliance on secondary sources. One estimate is that he will scan perhaps three thousand articles a year, and read about three hundred of them closely, although there are very great variations between individuals. Abstracts are systematically perused as substitutes for reading the original papers. Reviews of progress are of particular value, and monographs (often located via book reviews in periodicals) are found useful for updating, despite their lateness in appearing.

b *The everyday approach* arises in the course of daily work, regularly and frequently, usually in the form of a need for some specific piece of information vital for further progress. Unlike information needed in the current approach, it can usually be found in a number of places, and to the user the source is of little concern provided it is reliable. Almost invariably the one nearest to hand is chosen, and the emphasis is on finding what is wanted as soon as possible. This approach is the one scientists and technologists most frequently use.

Probably because it is easiest, consulting other people is the most common method: it is possibly more important here than all the printed sources put together, though it should be pointed out again that such consultation often leads to the literature, particularly in those cases where *numerical* data is required. As a 1972 ASLIB study showed, 'When we come to look at the sources from which chemists seek data, it is notable that published sources are more important than personal ones — to a much greater extent than is usually found in "information" use surveys. This dependence on the documentary record is particularly

20

marked in the case of physico-chemical properties. This importance of formal documentation for the provision of numerical data must therefore be stressed'. It is of course the secondary sources that are most commonly used, reference books especially. Most of them have been specifically compiled and arranged to furnish just such everyday information, eg encyclopedias, dictionaries, handbooks, data banks, tables, etc, and they usually provide what the searcher wants without making him look beyond the book in hand. Bibliographical sources are less important for this everyday approach: not only do they take longer to peruse but they usually provide more on a topic than is wanted. And in many cases they do not provide actual subject data, merely indicating references. Because of their unorganized state, primary sources are used as the last resort, unless the searcher knows exactly where to look.

A distinct variation of the everyday approach might be called *the background approach*. This too arises in the course of daily work, but in the form of a demand not for a specific piece of information, nor even for 'something on' a topic, but for an introductory or outline account sufficient to enable the worker to understand a subject new or unfamiliar to him. This has been called the 'orientation' or 'brush up on a field' approach, but J D Bernal's 'pilot information' is as expressive a description as any. According to Bernal this need is 'fairly infrequent in fundamental research but extremely common in applied research where projects are often embarked on in fields quite unfamiliar to the research teams'. This approach is common too with teachers (and students) who often do not have the time to digest original material. And it is a regular recourse for the new graduate when he discovers the need to span the gap between his undergraduate course and the research front. Again, books are an obvious resort, as surveys have shown, but the secondary sources most sought after for this purpose are reviews of progress: a recent 'state-of-the-art' article on the required topic is usually the ideal answer.

c *The exhaustive approach* usually arises when work begins on a new investigation, and involves a check through all the relevant information on a given subject. It is called for less frequently than the current or everyday approaches, but is vitally important, and often urgent. Some studies have suggested that on occasion such a search is used less as a guide on how to proceed and more as confirmation that the path being followed is the right one. 'Exhaustive' is almost always a relative term: E J Crane, for many years Director and Editor of *Chemical abstracts*, has written 'it is necessary to learn to recognize the point beyond which it is not reasonable to go because of the improbability of finding something of sufficient value to compensate for the time, effort, and expense of proceeding further'.

This approach is dependent very largely on printed sources, and bibliographical tools such as indexes and abstracts are all-important, but it must not be forgotten that literature references are included in many other secondary sources, eg reviews of progress, treatises, monographs, encyclopedias, etc. In certain fields where computerized bibliographic data bases are now available the burden of exhaustive searching has been greatly eased by the machine, though of course a manual search is still the only method for an investigator who wishes to go back more than a dozen years or so. Identifying the primary sources is of course the object of the literature search, but once they have been located they can then themselves be used in the search by following the references invariably cited at the end of every original article. This, picturesquely called the 'snowball method', is a very common way of proceeding. Indeed, for the scientists surveyed for the 1981 Royal Society report 'tracing back from the reference list of a recent relevant paper' was the most favoured method of all, ahead of using reviews, abstract journals or on-line searching.

d *The browsing approach*, by definition unplanned, and apparently inefficient, is nevertheless a fruitful path to information, definitely part of the scientific and technical communication system, and could well be added to Voigt's three approaches. Here, deliberately or otherwise, but without any specific topic in mind, workers seek information outside their predefined areas of interest, the aim (or result) being to extend the boundaries of that interest. This has been well described as seeking 'contact with the unknown, using chance as an informal kind of random sampling'. For the browser, this process 'brings into view a wide variety of knowledge with just enough depth to allow him to judge whether any one item would probably merit more study'. Surveys have shown that printed sources play a major 'triggering' role in stimulating new ideas and interests, or suggesting a new means of assault on a current problem, and one study at a major university library in the US showed that 18% of the books borrowed by scientists were items they did not go there specifically to find. A more recent study for the National Science Foundation discovered that 40% of the 'readings' of scientific and technical journals were of articles identified by browsing. Indeed, browsing through a comparatively small number of quality journals is the method most commonly used by a scientist on the look out for interesting new research. Engineers in particular, it would seem, browse regularly, especially in journals, and are always discovering useful items they were not looking for. A recent British Library study of the use of medical literature found that almost half of the hospital doctors surveyed browsed in the literature weekly or monthly, and over half of doctors in general practice had browsed within the previous four weeks. It is interesting to note the
22

complaint regularly heard from scientists and technologists that their load of required reading leaves them little time to browse: they obviously feel they are missing something. J M Ziman, Professor of Theoretical Physics at the University of Bristol, writing in 1980, tells us that 'modern methods of selective retrieval and dissemination have not yet adequately replaced the traditional practice of "browsing"'.

User studies

Much of what we know about the use of the literature of science and technology is derived from 'user studies', of which we now have a very large number. It is not possible here even to summarize the findings, and in any case it is unwise to draw conclusions as to the *value* of different types of literature solely on the basis of surveys of *use*. In its literal sense use means no more than the physical examination of an item; the value of that item may be substantial or slight; indeed it may have a negative value, inasmuch as its examination may turn out to have been a waste of time. Surveys of *opinion* have their own weaknesses, necessarily subjective as they are. Interpretation of such studies needs considerable care, but it would seem that there is a surprising uniformity of literature practices within particular subject fields. Between disciplines, however, there are marked variations: pure scientists for example scan about ten journals on a regular basis; industrial scientists and technologists read less. A 1966 survey of British engineers revealed that a fifth of them read no professional journal regularly. Engineers leaf through a journal page by page, pure scientists study the title page first. Of all scientists it is the chemists who have the reputation of being the most literature-conscious. Zoologists depend a great deal on the literature of other subject fields, whereas botanists do not. Distribution of reprints is most common in biology, whereas the distribution of 'preprints' (see chapter 9) is especially characteristic of physics. Even within a particular discipline it is often possible to chart variations according to type of activity: in engineering, predictably perhaps, 'pure' or 'academic' literature (primary journals, research reports, reviews of progress, etc) is used predominantly by those in research and teaching; 'applied' or 'industrial' literature (handbooks, standards, trade journals, etc) is used mainly by those whose work is concerned with design, maintenance, and testing. Government scientists make far more use of research reports than university scientists. One study has shown that creative chemists use the literature more than less creative chemists.

Variations between countries have also been detected: it would seem that American workers, for instance, make more use of personal contacts than do their British colleagues. The great differences between the Soviet Bloc and the West in the organization of science and its publication have a marked effect on usage patterns. In communist countries geology plays a much more significant role in the economy than in the West, with over a hundred thousand geologists in the Soviet Union and a corresponding imbalance in the literature and its use. One striking effect is that the literature of geology has the lowest proportion of English-language publications and the highest proportion in the Slavonic languages of any of the major sciences.

These variations in use can be matched by a whole range of other measures: in terms of articles written, for example, biologists are much more productive than engineers. Nevertheless, some investigators have recently cast doubt on the validity of many of the findings of user studies, and have dismissed some of the more naive comparisons between groups of users and different subject disciplines, based as some of them undoubtedly are on inadequate methods. So many discrepancies can be found between user studies that many experienced investigators would now agree with J M Brittain that cross-survey comparisons are almost impossible. After a detailed examination of many hundreds of studies the Centre for Research in User Studies concluded that 'the field is weak in concept definition and theorizing'. Recent researches have tended to show that the variations of information-seeking behaviour relate less to the subject of study than to the role being performed by the user at a particular stage in the search (or research) process.

A number of authoritative claims made recently would appear to bear this out. In 1978 the Biological Information Review Committee gave contrasting examples from its own subject field: 'The taxonomist values carefully prepared abstracts and indexes produced by secondary services, placing less emphasis on receiving information immediately. The biochemist, however, is mainly interested in current awareness, pays major attention to the source of the information, and is less interested in abstracts'. Indeed, the Committee went on to assert that 'assessing the information requirements of biologists as a body is usually unhelpful'. It is difficult to think of a group of biologists in a better position to know.

The members of the Chemical Information Review Committee reporting in 1978 also laid stress on the variation between different types of chemist. The academic chemist is conservative in his attitude towards the form taken by the primary literature and is concerned to maintain its standard. He is wary of change. He makes little use of patents, or of mechanized methods of information retrieval. The industrial chemist's need for access to the primary literature is as great as the academic chemist's, but he contributes to it less himself and is less concerned about its survival in its present form. His interest in patents is greater, understandably so inasmuch as his rep'itation depends on his ability to innovate as well as on his publications. He is much more likely to accept modern advances in information processing.

24

By way of contrast the parallel Agriculture Information Review Committee concluded in 1979 that 'much . . . remains obscure about the information-gathering habits of different categories of UK agriculturalists'.

Neglect of the literature
Paradoxically, the most striking finding of many of these studies is that scientists and technologists do not use the literature enough. The Chief Scientist of IBM has spoken of 'the yawning chasm . . . between what some people have learned yet others have not yet put to use'. In an ASLIB survey published in 1964, 647 research scientists reported 43 instances of finding literature, too late, which showed that their own research duplicated other work, and 245 instances of finding literature too late for the information to have full value. In a 1966 survey what was then the National Lending Library for Science and Technology found that only a small proportion of 2,355 scientists knew or made use of existing translation services and indexes. A pioneering survey of industrial technologists reported in 1959 that for information on a technical problem only 22% of the sample would make straight for the library or some other source of literature. The consequences can be expensive: the European Community spends more than £5,000 million per year on research and development, and it has been estimated that '15% of this research and development is either duplication pure and simple, or unproductive, as it fails to take account of the results already available, through ignorance of their existence'.

The seminal studies of T J Allen at the Massachusetts Institute of Technology in the 1960s revealed the intriguing process by which many practising engineers acquire their information without doing any reading at all. The sources they rely on are informal and depend on personal contact, in particular with what he called 'technological gatekeepers', individual colleagues who are themselves thoroughly familiar with the literature and who have acquired, often unconsciously, a bridging role in the information flow, as a result of their interest and skill in collecting information and distributing it to their colleagues.

Any observer of the information-seeking habits of scientists and technologists soon encounters the Principle of Least Effort, which as enunciated, for example, by G K Zipf of Harvard in 1949 governs every individual's entire behaviour. As applied to our field it was best expressed by C N Mooers: 'An information retrieval system will tend not to be used whenever it is more painful and troublesome for a customer to have information than for him not to have it'. There is of course nothing new here; over two hundred years ago the great Dr Johnson proclaimed: 'If it rained knowledge, I'd hold out my hand; but I would not give myself the trouble to go in quest of it'. And indeed, it does seem to be true that scientists are no exception: they do
25

not like to work hard or travel to any extent for their information even at the risk of losing some. As the Centre for Research in User Studies puts its, 'Accessibility is a key factor in determining the use made of an information source'. This is one of the reasons that personal communication is so popular with scientists. They will often prefer to use a source they know to be inferior or inadequate because it is more accessible. Indeed, as A J Meadows points out, many scientists prefer to avoid acquiring information at all: 'They are interested in the research process itself rather than in the results of other people's work.' P J Vinken, himself a medical scientist, has told us that 'The consumer of scientific literature is generally-speaking arrogant, conservative, lazy and ignorant'. Some, indeed, will suppress information received so that they can continue to behave as though they had never received it. This is perhaps less a criticism of scientists and technologists than a general comment on human nature. Part of the problem, of course, as the report of the Committee on Scientific and Technical Communication makes clear, is that 'Users of scientific and technical information were trained and work today under conditions that offer far larger rewards for doing "new" work than for finding, through careful literature search, the results of work already done.'

Recent studies of the role of information in the process of invention and innovation have shown that many workers feel that 'literature surveys impede the triggering and lubricating function of information ... Researchers relying on surveys in the earliest phase of innovation were viewed as less productive and lacking sufficient expertise to work in the field'. Frederic Banting, winner of the Nobel prize for his discovery of insulin, confessed later that if he had been aware in advance of all the complications as manifested in the vast literature on the subject he would never have had the heart to continue his own investigations. We have the testimony of Albert Szent-Gyorgyi, Hungarian biochemist and Nobel laureate, to indicate that there is room for improvement here. He writes: 'I could find more knowledge new to me in an hour's time spent in the library than I would find at my workbench in a month or a year'. To set against this we have the oft-quoted warning of Lord Rayleigh, Cambridge physicist and also a Nobel prizeman: 'rediscovery in the library may be a more difficult and uncertain process than the first discovery in the laboratory'. Indeed, Enrico Fermi, Italian nuclear physicist and another Nobel prize-winner, preferred to derive equations himself rather than look them up. He used to wager with his colleagues that he could work them out more quickly than they could find them in the literature. Making a more philosophical point, which is not surprising, Bertrand Russell, yet another Nobel prize-winner (though for literature not science), claimed that 'all advances in non-Euclidean geometry had been made in

26

ignorance of the previous literature, and even because of that ignorance'.

Proliferation of the literature
One undoubted reason for this reluctance to make the best use of the literature is the vast increase in the amount published over the last generation or so. Of course this is a phenomenon not confined to science and technology, but it is seen at its most acute here, where the growth has been exponential, ie the rate of increase itself has increased more and more rapidly with the passage of time. In the great period of expansion stretching from the Second World War to today the scientific literature has been doubling in bulk every fifteen years. This is seen at its most obvious in the number of new journals founded each year, and perhaps even more strikingly in the frequent splitting of established titles into two, three, or even more new journals. Many scientists are also of the opinion that this increase in bulk has been accompanied by a serious deterioration in quality, though reliable evidence is hard to find.

There are dangers in exaggerating the effect of this 'information explosion', and in any case the graph has shown a marked levelling off during the late 1970s and early 1980s. The growth of journals has no more than kept pace with the numbers of scientists and technologists, and is not therefore an unexpected phenomenon. The rate of growth too varies greatly among different disciplines, with some increasing within fifteen years by a factor of ten. Nevertheless, for the *individual* scientist, any growth in the current literature inevitably means that he can only keep in touch with an ever diminishing fraction of the whole, with serious consequences for the fragmentation of science. But as J M Ziman reminds us, such proliferation is not a pathological sign; it is an essentially natural process that we have to learn to cope with. Encouragingly he adds that 'the growth of scientific and technical literature in recent years has not been as rapid as the development of new and economical means for classifying and retrieving it in detail. Despite the apparently relentless growth in the bulk of scientific information over the past 10 or 20 years, the information contained by an archive today is probably more complete, better organized, and more easily accessible than formerly'.

The librarian and the literature
Whatever the reasons for a scientist's 'information-phobia', as A J Meadows has called it, whether ignorance, or deliberate preference for other channels of communication, or fear of being overwhelmed, or simple disinclination, there is clearly a role here for an intermediary,

a guide, and even a teacher. In a paper read to an ASLIB Annual Conference, G A Somerfield of what was then the Office for Scientific and Technical Information, a scientist himself, had this to say: 'Perhaps the most important factor restricting the development of better information services is the conservatism of the average user of scientific information and a general lack of understanding of exactly how information is used ... considerable education of the potential customers is required before they will appreciate the virtues of the different services offered to them'. A H Rubenstein, an engineer by profession, has pointed out that much of the observed behaviour of scientists using the literature and many of the preferences they express 'are a result of the media that have been available historically and have little to say about optimal and ideal media and forms of information transfer. Much work needs to be done in educating researchers about the currently and potentially available new forms of information exchange ... When they are acquainted with the possible range and form of media, researchers' preferences and behaviour may change radically'. It is to play at least part of this educative role that the librarian strives to fit himself.

Of course his approach to the literature does differ from the scientist's or technologist's approach. Only rarely would he attempt to master the primary literature. His stock-in-trade is the secondary literature, in particular the bibliographical and reference tools. As for the tertiary sources, these are peculiarly the librarian's concern, many of them of course having been compiled by librarians for their colleagues. The librarian's particular strength, however, lies in his grasp through systematic study of the underlying pattern discernible in the literature, and his familiarity with the variety in which it manifests itself, combined with an understanding of the relationships of these forms with one another, their comparative reliability, and their varying uses. This bibliographical expertise is given a human face by its integration with the librarian's professional commitment to serve, informed by a close knowledge of his users' needs.

The social role of the literature
It would be a mistake to regard the literature of science and technology merely as that part of the world's writings with a scientific or technical content, for it is different in nature from the literature of other subjects. Its cumulative character has already been referred to: this can be contrasted with the literature of the humanities where a contribution depends for its value far more on the individual writer. Even if nothing further is ever discovered about Alexander the Great, for instance, individual historians and biographers to the end of time will continue
28

to offer their own interpretation of him — and their work will be counted original. D J de Solla Price and others have also demonstrated, for example, that there are significant bibliometric differences between science and non-science in their patterns of referencing and citation. If one examines, for example, the citations in research publications on chemistry or physics, one finds that no more than 10% refer to books; in sociology, the proportion is more than half. In the humanities the references appended to a research paper will be markedly fewer than in the sciences, and they will be much more concerned with the older literature.

But the literature of science has a further vital role beyond the transfer of information and the maintenance of the archive, for one of the characteristics of science is its inbuilt system of self-criticism and reward. It is 'This sociological peculiarity', according to D H de Solla Price, 'which distinguishes scientific scholarship from all else'. Since every discovery, every new idea, has to be set out in the literature, it is immediately subject to critical scrutiny by the writer's colleagues. Non-documentary communication will not suffice: a permanent public record is required. This valuable, though often forgotten, social function of the literature combining quality control and the recognition of priority within the scientific community is one of the reasons for the continuing pre-eminence of the learned journal, the traditional arena for the public unveiling of a scientific discovery, and the chief medium through which a scientist builds his professional reputation (see chapter 9). It should be added however that the scientist in industry contends with a different set of pressures. He may of course publish in the open literature like his academic colleagues, but his primary consideration is the welfare of his firm or his industry, which inclines him towards secrecy rather than disclosure. He seeks his recognition from colleagues and superiors in his company, who base their assessment on his worth to them, his problem-solving ability, his success at innovation, his skill as a manager, etc. His 'publications' are more often internal company reports or perhaps that specially protected form of disclosure, the patent specification.

The forms of the literature
The 1968-70 UNISIST study on a world science information system (see chapter 10) has stated that 'Almost all countries and almost all fields of science modify only to a minor degree the fundamental institutions of scientific communications: journals, critical reviews, data compendia, libraries, etc. One result of this circumstance is that information specialists have come to talk a common technical language, regardless of the field of science they may represent'. This is comforting

to the librarian, but should not tempt him into complacency. The forms that the literature has taken and which are listed above are not to be regarded as immutable or eternal. They have no prescriptive right to remain as methods of scientific and technical communication. Even today we can discern overlapping and merging of categories: between dictionaries and encyclopedias, for example, or between yearbooks and directories. According to M J Thomson of the Science Reference Library 'there is no clear dividing line between trade directories and trade catalogues'. And within categories functions and uses change: because of their great increase in size some abstracting services are now found less useful than they used to be for the current approach, although remaining invaluable for the exhaustive approach. It could be argued too that some of the sources listed above as primary should be secondary, and vice versa. Even non-documentary channels of communication have been known to change: a number of the documentary sources listed have themselves grown out of word-of-mouth originals, eg periodicals, theses, textbooks. It should be remembered too that the reasons information is packaged in certain ways may not always be altruistic: scientific information is also an industry and decisions may be commercial rather than strictly scientific.

Old categories may be revived: citation indexes were unknown to the current generation of scientists and technologists when *Science citation index* appeaared in 1961, yet *Physics abstracts* had included citations to particular authors in its name index from 1898 to 1924. What we know now as KWIC/KWOC indexes were developed by H P Luhn at IBM in the 1950s, yet what was clearly the identical form of index had been used for the printed catalogue of the Manchester Reference Library in 1864. Even microforms, virtually unknown in libraries before the 1930s, were used for storing and transmitting documentary information, via pigeon, at the siege of Paris during the Franco-Prussian war of 1870-71.

As we have seen earlier new formats too evolve: microfiche, punched cards, magnetic tape, videodisc. These often represent advances in technology which we borrow from other fields and apply to our own purposes. We read too that 'Problem-solving, interdisciplinary science has created an insatiable demand for the "packaging" of scientific information in new forms'. These forms are 'less permanent; problems have a tendency to be solved, and the group redeployed'.

A further hazard in the path of the student striving to understand the structure and forms of the literature is the imprecise way descriptive terms are used in the titles of some scientific and technical publications. J N Friend *Textbook of inorganic chemistry* (Griffin, 1914-37) is a work of twenty-four volumes: obviously this is a treatise rather than a textbook in the usually accepted sense of that term. J Gomez

Dictionary of symptoms: a medical dictionary . . . (Centaur, 1967) is not arranged in the alphabetical order that most users expect of a dictionary, but it is set out systematically. In fact, the author of the contributed foreword does comment: 'I feel the term "dictionary" may be misleading'. R G White *Handbook of ultraviolet methods* (New York, Plenum, 1965) is not a handbook in any ordinary sense, but a bibliography. As the introduction says: 'This volume consists of more than 1,600 references . . . arranged alphabetically by senior author'.

Library users when asking for help use the terms 'standard' and 'specification' interchangeably, as they do the words 'thesis' and 'dissertation'. We find 'preprints', so-called, which are never subsequently printed. The word 'data' is used in two quite different senses in 'data bank' and 'data base'. And any librarian who has worked with a collection of report literature (see chapter 14) is certain to have encountered at least one report bearing on its front cover the legend: 'This document is not a report.'

In the succeeding chapters the types of scientific and technical literature are described and illustrated. Since this is a textbook the types of literature are discussed in the order in which it is suggested they should be studied. This is not the evolutionary order outlined above, nor is it necessarily the order the practising scientist would find most useful.

Further reading

Ciba Foundation *Communication in science* (Churchill, 1967).

S Passman *Scientific and technological communication* (Pergamon, 1969).

Committee on Scientific and Technical Communication (SATCOM) *Scientific and technical communication: a pressing national problem and recommendations for its solution* (Washington, National Academy of Sciences, 1969).

C E Nelson and D K Pollock *Communication among scientists and engineers* (Lexington, Mass, Heath, 1970).

J Gray and B Perry *Scientific information* (Oxford University Press, 1972).

J Hills *A review of the literature on primary communications in science and technology* (ASLIB, 1972).

A J Meadows *Communication in science* (Butterworths, 1974).

D W King *Statistical indicators of scientific and technical communication, 1960-1980: volume 1 A summary report* (Rockville, Md, King Research, 1976).

G Ford *User studies: an introductory guide and select bibliography*

(University of Sheffield Centre for Research in User Studies, 1977) [BL R&D report no 5375].

A K J Singleton *The communication system of physics: final report of the Physics Information Review Committee* (British Library Research and Development Department, 1977) [Report no 5386].

T D Crane *Final report of the Biological Information Review Committee* (British Library Research and Development Department, 1978) [Report no 5438].

J F B Rowland *Information transfer and use in chemistry: final report of the Chemical Information Review Committee* (British Library Research and Development Department, 1978) [Report no 5385].

G M Craig *Information systems in UK agriculture: final report of the Agriculture Information Review Committee* (British Library Research and Development Department, 1979) [Report no 5469].

K Subramanyan 'Scientific literature' *Encyclopedia of library and information science* 26 1979 376-548.

J M Ziman 'The proliferation of scientific literature: a natural process' *Science* 208 1980 369-71.

Royal Society Scientific Information Committee *A study of the scientific information system in the United Kingdom* (1981) [BL R&D report no 5626].

Chapter 2

GUIDES TO THE LITERATURE

The very first work a student should attempt to locate when commencing his study of the literature of a particular subject is a guide to the literature, if one exists. Fortunately, over the last few years these have greatly increased in number and there are few major disciplines in science and technology that still lack some such guide. One that has stood for over forty years and can still serve as a classic example is M G Mellon *Chemical publications: their nature and use* (McGraw-Hill, fourth edition 1965). Guides which attempt to cover the whole field such as E J Lasworth *Reference sources in science and technology* (Metuchen, NJ, Scarecrow, 1972) are far less common than those devoted to a single subject area like physics, or nuclear energy, or radioactive isotopes. An established major discipline like chemistry may indeed have a dozen such guides.

They are not exactly subject bibliographies in the ordinary sense, because they usually go beyond the normal limits of enumerative bibliography in including not merely lists of references but discussions of the functions and uses of the various types of literature. Some go further: for instance, S A J Parsons *How to find out about engineering* (Oxford, Pergamon, 1972) has two chapters on 'Careers in engineering' and 'Education and training for careers in engineering'; the final chapter of A Antony *Guide to basic information sources in chemistry* (New York, Wiley, 1979) comprises eleven case studies of actual information searches, using the literature; R T Bottle and H V Wyatt *The use of biological literature* (Butterworths, second edition 1971) has a 14-page chapter 'Libraries and their use'; the guide by Mellon mentioned above has a 78-page chapter on library 'problems', including 2,500 specific assignments for setting to individual students.

Like many other tertiary sources, they are surprisingly little known by the scientist or technologist, and it is the librarian or information worker serving them in whose hands they are most often found. R H Whitford *Physics literature: a reference manual* (Metuchen, NJ, Scarecrow, second edition 1968) is described by its author as 'a borderline contribution between the fields of physics and library science'. Many, indeed, nave been compiled by individual librarians: L J Anthony

33

Sources of information on atomic energy (Pergamon, 1966) draws largely on the author's experience with the United Kingdom Atomic Energy Authority, and is a much expanded version of the pamphlet of the same title first issued by the Atomic Energy Research Establishment Library at Harwell in 1956. A Pritchard *A guide to computer literature* (Bingley, second edition 1972) is the outcome of a research project carried out at the North-Western Polytechnic School of Librarianship. Library associations too have produced guides: E B Gibson and E W Tapia *Guide to metallurgical information* (New York, Special Libraries Association, second edition 1965); and so have the scientists' own professional societies, particularly those with special sections concerned with literature problems: L R A Melton *An introductory guide to information sources in physics* (Institute of Physics, 1978). One interesting example deriving from a 1949 work published by ASLIB is *A guide to sources of information in the textile industry* (Manchester, Textile Institute, second edition 1974): the first edition of 1970 was a joint publication by ASLIB and the Textile Institute.

In some cases (as in the earlier editions of the work by Anthony, just mentioned) it has been the libraries that have produced these guides, primarily for the users of their own collections. A 1979 survey by ASLIB of over five hundred such brief printed UK library guides revealed a number of shortcomings, including some confusion of purpose and little monitoring of use. Evidence of considerable overlap was found: nine university libraries had produced guides to the literature of physics; eight polytechnic libraries had produced guides to electrical engineering. Little sign of cooperation was evident, yet it is the case that where such collections are substantial the published guides have a value far beyond the walls of a particular library: J E Wild *Patents: a brief guide to the patents collection in the Technical Library* (Manchester Public Libraries, second edition 1966) describes one of the largest collections in Britain outside the Patent Office collection itself.

The majority of guides are aimed at a wide public — students, teachers, research workers, practising scientists and technologists, as well as information workers and librarians. A number, however, are produced specially for librarians, such as H R Malinowsky and J M Richardson *Science and engineering literature: a guide* (Littleton, Colo, Libraries Unlimited, third edition 1980). Of these, several (such as the present work) are particularly for student librarians, and not uncommonly are for use in particular university or college courses in the literature of science and technology, or are at least based on such courses. F B Jenkins *Science reference sources* (MIT Press, fifth edition 1969), for instance, is 'for use in the course in science reference service at the University of Illinois Graduate School of Library Science'.

There is an increasing number of literature courses not for library

34

students but for science and technology students, and we now have examples of guides that have stemmed from these. R J P Carey *Finding and using technical information* (Arnold, 1966) and R C Smith and M Reid *Guide to the literature of the life sciences* (Minneapolis, Burgess, eighth edition 1972) are both in the form of a textbook for just such a course, drawing largely on the teaching experience of the authors, the first in Britain, the second in the United States. E Mount *Guide to basic information sources in engineering* (Wiley, 1976) is 'based on a one-semester course in technical literature . . . given for several years at the School of Engineering and Applied Science at Columbia University'. R T Bottle *The use of chemical literature* (Butterworths, third edition 1979) started life as an edited version of actual lectures delivered at Liverpool College of Technology. Comparatively little known are the series of guides for Open University students compiled by members of the library staff, eg *Information sources in chemistry, Literature and information sources in computer science.*

Practising scientists and technologists too have their courses: S Herner *A brief guide to sources of scientific and technical information* (Washington, Information Resources Press, 1969) is a recapitulation of the content of a one-and-a-half day course for United States Federal scientists and engineers. An interesting by-product of this course is the companion volume S Herner and J Moody *Exhibits of sources of scientific and technical information* (Washington, Information Resources Press, 1971), a series of almost two hundred facsimile pages reproduced from the published sources treated on the course.

It is probably true to say, however, that the majority of separately published guides are normal commercial book-trade productions. E J Crane and others *A guide to the literature of chemistry* (Wiley, second edition 1957) claims that its 1927 edition was the 'first comprehensive book to appear in its field'. Another very well-known work is N G Parke *Guide to the literature of mathematics and physics* (New York, Dover, second edition 1959). Indeed, with the increase in demand for such guides, particularly from information workers, and the public urgings of the Library Association, ASLIB and the Fédération Internationale de Documentation, a number of publishers have seen the advantage in bringing out such titles in a series. In the United States in 1963 Interscience produced the first of its 'Guides to information sources in science and technology': B M Fry and F E Mohrhardt *A guide to information sources in space science and technology*, a work of close on six hundred pages, even though only current material (1958 to 1962) was included. In Britain, Pergamon's 'International series of monographs in library and information science' includes among its titles guides such as E R Yescombe *Sources of information on the rubber, plastics and allied industries* (1968), now updated by another publisher

35

as E R Yescombe *Plastics and rubber: world sources of information* (Applied Science, 1976). Pergamon is also responsible for a particularly well-known series, all entitled *How to find out . . .* , in the 'Commonwealth and international library'. A now extensive series is Butterworths' 'Information sources for research and development', each with the title *Use of . . . literature.* The name of the series was later changed to the more indicative 'Guide to information sources' and the titles of the most recent individual volumes to *Information sources in . . . .* The United Nations Industrial Development Organization also has a 'Guide to information sources' series, with over three dozen titles published to date, eg *Information sources on the petrochemical industry* (New York, 1978).

Not all such guides are separately published in book or pamphlet form. Several hundred have appeared as articles in periodicals, either in the scientific and technical press or in the professional library journals: examples are S C Traugher 'Searching entomological literature' *Bulletin of the Entomological Society of America* 20 1974 303-15; H M Woodburn 'Retrieval and use of the literature of organic chemistry' *Journal of chemical education* 49 1972 689-96; R J Ardern 'Countryside literature: a literature review' *ASLIB proceedings* 25 1973 274-87. Reference books too, both those for librarians and those for the scientists and technologists themselves, occasionally contain a literature guide as a separate article or chapter: two good examples are the 56-page article 'Geological literature' in *Encyclopedia of library and information science* (New York, Dekker, 1968-) and the 30-page article 'Literature of chemistry and chemical technology' in *Kirk-Othmer encyclopedia of chemical technology* (New York, Interscience, second edition 1963-71).

The task of discovering what guides (especially 'hidden' guides of this kind) are available for a particular subject is eased by a comprehensive bibliography which lists well over a thousand, the majority in the form of articles in journals: G Schutze *Bibliography of guides to the S-T-M literature: scientific-technical-medical* (New York, the author, 1958) and its *Supplements* (1963 and 1967). A more up-to-date listing with a high proportion of scientific and technical titles is *A literature guide to literature guides* (Manchester Polytechnic Department of Library and Information Studies, 1980). A more selective listing (itself in the form of a periodical article) is R W Burns 'Literature resources for the sciences and technologies: a bibliographical guide' *Special libraries* 53 1962 262-71. Though the ASLIB survey described above confines itself to guides produced by libraries for their own users it does include a bibliography of some 250 conventionally published guides as an appendix, the great majority in science and technology. The list also includes a number of 'hidden' guides.

The Information Department of FID is also compiling a guide to guides.

Types of guide

The student will already have appreciated that the subjects covered by these individual guides could well range from the whole of science down to the minutest corner of the field, and that in form they might vary from a volume of several hundred pages to a single-page article in a periodical. Obviously, their authors must treat their subjects with varying degrees of intensiveness. Because of the quite different uses to which they are put it is important to learn to distinguish the two main categories of guide.

a *The 'textbook' type.* Here the emphasis is on exposition, with the stress laid on types of material rather than individual titles: in the preface to B Yates *How to find out about physics* (Pergamon, 1965), for instance, we read that 'no claim is made to comprehensiveness; indeed deliberate policy has been to give indicative selections only'. The preface to H M Woodburn *Using the chemical literature: a practical guide* (New York, Dekker, 1974) makes clear that is is intended for instruction and is 'not a bibliography of sources'. B Houghton *Mechanical engineering: the sources of information* (Bingley, 1970) 'is not intended as a bibliography of mechanical engineering literature or as a listing of titles, but as a map to help the engineer find his way through the varying forms of literature'. R E Maizell *How to find chemical information* (New York, Wiley, 1979) 'presents the most important and enduring of the classical tools of chemical information; the more significant newer tools; and, most importantly, the underlying methods, principles, and keys'. J W Mackay *Sources of information for the literature of geology* (Geological Society, 1973) is an interesting example of a guide published by a learned society but written specifically in textbook form 'because researchers in universities and industry cannot normally expect any sustained professional help in their literature work'.

Designed as an integral part of a course are the Student Handbooks produced by Newcastle-upon-Tyne Polytechnic Library in the course of the Travelling Workshops Experiment, eg *Biological information: a guide to sources*. They are quite elaborately structured teaching aids, with exercises, linked with posters for display and other instructional materials in audio-visual format, such as audiocassettes and tape-slide programmes. The textbooks remain the personal copy of the students after attending the workshop.

Of course similar guides to the literature in audio-visual format are available commercially, eg *On-line searching the life science*

37

literature (Boca Raton, Fla, Science Media, 1980), comprising 96 slides, 2 audiocassettes (totalling 90 minutes), and a script.

b *The 'reference book' type.* Designed as a working tool, this kind of guide aims at comprehensiveness: the introduction to the Yescombe guide to rubber and plastics information noted above claims that 'every effort has been made to include all important sources'. Significantly, the 1976 updating volume admits 'No guide of this kind can now hope to be exhaustive'.

Of course examples can be found combining the features of both: the well-known guide to the literature of chemistry by Crane mentioned earlier states in the preface that it 'is intended to be used both as a reference book and as a textbook. Its coverage of the field of chemical literature is comprehensive ... The textbook features include discussions of basic principles and topics, emphasis on how to use each form of chemical literature, and an introduction to the art of literature searching'. And in some cases circumstances so conspire as to frustrate the author's intention: in the preface to M Schalit *Guide to the literature of the sugar industry* (Elsevier, 1970) we learn that 'What was intended to be a thorough project, an exhaustive search through and compilation of literature sources, has turned into an experiment in bibliography'. Sometimes the intention of the work is obscure: despite the clear instruction to users of C C Parker and R V Turley *Information sources in science and technology* (Butterworths, 1975) that 'This guide is not meant to be read from cover to cover — use it as a reference book', it is deliberately very selective in coverage and contains a large amount of expository text. Furthermore, much of it is in the imperative mood and it is clearly pedagogic in tone. Indeed, it is dedicated to the authors' students. It well illustrates the need to examine all such works closely in order to make one's own assessment of scope and intention: whereas the publisher's dust jacket tells us that it is written expressly for library school students, the authors' preface states 'this book is not written expressly for them'.

Designed as they are to facilitate the use of the literature within a particular field, in recent years more and more have adopted the practice of the Herner and Moody guide mentioned above and have taken to incorporating within their text facsimile pages of the works discussed: indeed of the 427 pages of F J Kase *Trademarks: a guide to the official trademark literature* (Leiden, Sijthoff, 1974) something like half are reproductions of pages from the official journals of the sixty countries covered. This is a notable feature of Pergamon's *How to find out* series.

A category of guide found exceedingly valuable by the librarian or by any user who is not a subject expert is an elaboration of the 'textbook' type which goes into considerable detail about the *subject* as well as the literature of the subject. Good examples are very few, probably because such works must be very difficult to write, combining as they do an introductory textbook on a subject and a guide to its literature, but a model worth studying is H Lemon *How to find out about the wool textile industry* (Pergamon, 1968): chapter 5 of this work, for instance, on 'Wool processing' covers 53 pages but only lists about 150 references to the literature, the bulk of the text being devoted to a summary of the actual processes, with several black and white illustrations. Where a guide takes the form of chapters by a number of contributors there is obviously more of an opportunity to achieve a useful blend of bibliographical knowledge with subject expertise: nine of the fourteen chapters in A R Dorling *Use of mathematical literature* (Butterworths, 1977) are by university mathematicians and are quite overtly state-of-the-art reviews of the different *subject* fields treated, with bibliographical references appended.

The majority of examples of both categories of guide will be found devoted to a particular discipline, or industry, usually well defined and clearly understood. The growth in recent years of new interdisciplinary areas of study has raised some challenging problems for compilers of guides to the literature of such fields, which may draw *at the same time* upon traditional subjects apparently as diverse, for example, as human physiology and engineering and law. An interesting example of an initial attempt at such a guide is J O Jones and E A Jones *Index of human ecology* (Europa, 1974), 'designed to meet the rapidly growing need for a bibliographical tool which will facilitate cross-disciplinary information search and retrieval in the complex of subjects bearing on human ecology and environmental planning'. In effect, the work is an annotated bibliography of secondary sources (mainly abstracting and indexing services) of relevance to human ecology, with a KWIC index analysing the subjects covered. In the words of the compilers, their aim is to bring some degree of 'co-ordination, at what might be called the *tertiary* level, representing the comprehensive viewpoint from which human ecology must operate'.

As with many of the information sources described in the succeeding chapters of this book, and as the last example demonstrates, the boundaries between the types are often diffuse: in particular the line that can be drawn between a guide to the literature and a subject bibliography. In theory the distinction is clear: whatever the arrangement, however detailed the annotations, a bibliography is basically a list of references to the literature; a guide must have something further, as shown in this chapter. E M Dick *Current information sources in mathematics: an annotated guide to books and periodicals, 1960-1972* (Littleton, Colo, Libraries Unlimited, 1973) is a selective bibliography of some 1,600 entries, as its preface makes clear. In practice this is not always so easy: some guides of the comprehensive 'reference book' type are nine parts bibliography, and applying the rule strictly one has to class Walford as a subject bibliography, despite the title *Guide to reference material.* Many of the so-called guides to the literature produced by libraries for their own users are better regarded as annotated bibliographies. Indeed, as the ASLIB survey found, almost a third of the five hundred examined had no annotations at all. Some of the better known are issued in series and will be discussed in chapter 8.

What is important for the student to be aware of is that this overlapping does exist. He should learn to make his own categorization of the works he examines, and not to rely entirely on how they are described. He will then not mistake the classic B D Jackson *Guide to the literature of botany* (Index Society, 1881) for a guide to the literature of botany. He will see that it is an uncompromising subject bibliography of about 9,000 unannotated short-title entries.

Further reading

A C Townsend 'Guides to scientific literature' *Journal of documentation* 11 1955 73-8.

P J Taylor *Information guides: a survey of subject guides to sources of information produced by library and information services in the United Kingdom* (British Library Research and Development Department, 1978) (Report no 5440).

Chapter 3

ENCYCLOPEDIAS

This and the three succeeding chapters are devoted to reference books, and it is worth recalling how R T Bottle, a chemist, has vividly characterized such works: 'These sources of information may be regarded as a highly fractionated distillate from the ocean of information that has appeared in the preceding primary literature'. Obviously, they fall into the category of secondary sources, and are all deliberately arranged for ease of consultation. They are the books to which the scientist or technologist turns first for his 'everyday' information needs (see chapter 1).

Of all reference books the encyclopedia is probably the best known, and the student will already be familiar with the form and function of the great general multi-volumed encyclopedias. Science and technology too have their multi-volumed encyclopedias, with a similar aim: to give in concise and easily accessible form the whole corpus of knowledge within the subject scope of the work. This is an ambitious goal and some would say an impossible task, and encyclopedias have been severely criticized as of little real use to the specialist. Not only does he find his own speciality treated superficially but he knows that by its very form the information is bound to be out of date. This judgement is probably unfair, and in any case has less validity if the argument is confined to subject encyclopedias, which can specialize more, and can usually be revised more frequently than the general encyclopedia. Before accepting the criticism it is important for the student to examine with care the various types of encyclopedia in science and technology and to relate their achievements to the demands made on them and the actual use to which they are put.

This so-called 'everyday' approach to the literature of science and technology is statistically the most frequent: the evidence is unequivocal on this point. It is also an observable fact that the need is commonly for information in a subject area peripheral to the enquirer's primary interest, a field in other words in which he is not a specialist. It is to cater for precisely such demands that many eneyclopedias have been compiled, at various levels of subject specialization. Our largest general science and technology encyclopedia, the 15-volume *McGraw-*

Hill encyclopedia of science and technology (fourth edition 1977) claims that 'Each article is designed and written so as to be understandable to the nonspecialist'. A very different work in single volume format and in a narrower field, P Gray *The encyclopedia of the biological sciences* (Van Nostrand Reinhold, second edition 1970), is equally explicit in its aim 'to provide succinct and accurate information for biologists in those fields in which they are not themselves experts'.

To attack encyclopedias for not catering for the specialist is to try to cast them for a role they are by nature unfitted to play. An encyclopedia's task is not to say the last word on a subject, but as has often been said, to give 'first and essential facts' only. They provide for neither the current nor the exhaustive approach (see chapter 1), but they do furnish a vast wealth of facts, easily found. They serve the everyday approach, with either specific information or orientation. They may offer no more than a starting point for a closer investigation of a topic, but this too is what they have been designed for, and the bibliographies to the articles can often signpost the way. It should also be borne in mind that reputable scientific and technical encyclopedias are by no means the compilations of hacks: the articles are by experts (2,700 listed in the *McGraw-Hill encyclopedia* and 500 in Gray), and are usually signed by them.

Encyclopedias for the layman
It should not be too readily assumed, however, that encyclopedias of this kind, specifically designed for the non-specialist, are therefore for the complete layman: neither of these two examples are. A number of the articles run to several pages, and into much technical detail. The preface to the *McGraw-Hill encyclopedia* is specific about level: 'Most articles and at least the introductory parts of all of them are within the comprehension of the interested high school student'. Though this represents a change from the first edition which was claimed to be 'within the comprehension of the college undergraduate in science and engineering', it still does indicate that a modicum of scientific knowledge is desirable for those enquirers wishing to read an article further than its 'introductory parts'. That is not to say that the intelligent non-scientist could not get anything out of this work, but it does remind us of one very important role of the encyclopedia in all fields, namely to explain its subject to the 'ordinary enquirer', for it is often to an encyclopedia of the subject that such a person would turn first. Moreover, in the field that concerns us here this is particularly vital, for science and technology are critical in shaping our world. This is the sphere in which the encyclopedia editor is seen, in the words of Lowell A Martin, an American librarian and a professor at Columbia University, as 'a

mediator, between the world of scholars on the one side and the individual seeking information on the other, between those who know something and those who seek to know'.

And indeed, in science and technology, encyclopedia editors have taken up this challenge, and we are seeing increasing numbers of high-quality examples of this particular category of encyclopedia, for the non-scientist. The editor's introduction to J R Newman *The international encyclopedia of science* (Nelson, revised edition 1965) says firmly that 'the needs of the common reader – the student, the teacher, the non-specialist – have been our measuring rod'. This work can also serve as an object lesson to those who may believe that a scientific encyclopedia for the non-scientist must be inferior in some way. In fact there is no intrinsic reason why such a work should not maintain the same standards of scholarship and reliability as the best of similar works for a more specialist readership. This work by Newman, indeed, demonstrates its standards by having its articles signed by a team of 450 scientists and engineers of some position. And it is likely that, if asked. they would confirm a common experience of scientific writers: it is often far more difficult to write for the layman than for fellow-scientists.

An interesting example of a single-volume encyclopedia for the broad field of science and technology, written by a scientist of repute, is J G Cook *Science for everyman encyclopedia* (Watford, Merrow, second edition, 1964): this is aimed at the home library, and in particular the student (there is a 72-page 'Student's reference section' at the end, comprising mainly tables). Another attractive example, making a feature of illustrations, mostly in colour, is the *Phoenix concise encyclopedia of science and technology* (Oxford, 1978): 'only two thirds of the space in this volume is occupied by the text articles'. Many of the layman's encyclopedias, however, are devoted to more specialized subjects, which although indisputably scientific or technical, have become traditionally the province also of the amateur enthusiast: A Hellyer *The Collingridge encyclopedia of gardening* (Hamlyn, 1976) tells us it is for 'the vast number of gardeners and plant lovers who nowadays are educating themselves towards almost professional standards of excellence'; *Summerhay's encyclopaedia for horsemen* (Warne, sixth edition 1975), although quite technical, is plainly designed for those using horses for recreation; and clearly E F Carter *The railway encyclopedia* (Starke, 1963) is for the lay enthusiast rather than the professional railwayman.

Encyclopedias for the specialist
All this is not to imply that there are no encyclopedias for the specialist. In point of fact, as the subject field narrows it obviously becomes

43

possible to cater more for the experts, and in numbers this category of specialist encyclopedia is probably the largest of the three examined in this chapter. The student who examines the massive *Kirk-Othmer encyclopedia of chemical technology* (Wiley-Interscience, third edition 1978-), to be in 25 volumes plus index, can be left in no doubt that this is a tool for professional chemists and chemical engineers. Indeed, it is clear to see in this and similar comprehensive works an extension of the usual role of the encyclopedia beyond the 'first and essential facts' only: although still alphabetically arranged, they share to some extent the nature of a treatise, that is to say an authoritative attempt to digest all the primary literature and to consolidate the whole of existing knowledge on a topic. From the same publisher as *KO*, as it is familiarly known, we have the very similar *Encyclopedia of polymer science and technology: plastics, resins, rubbers, fibers* (New York, 1964-72) in 16 volumes and *Supplements* (1976-), frequently known as 'Mark' from the name of its editor, H F Mark; and *Encyclopedia of industrial chemical analysis* (1966-74) in 20 volumes. The editors of both see their subjects as ripe for this kind of thorough-going approach: in the words of the preface to the second work, 'A comprehensive encyclopedic treatment of the present state of the art seems to be a desirable and worthwhile undertaking'. Of course the physical form they take inevitably means that they cannot furnish absolutely current information: their function is to act as the great repositories of received knowledge. In fact, no sooner was the first edition of *Kirk-Othmer* complete (1947-60 in 15 volumes and 2 supplements) than plans were being made for the second (1963-71 in 23 volumes and 2 supplements), and the third is now well under way. We have already been warned that this will be necessary also for the *Encyclopedia of polymer science and technology*.

But the student should not assume that superseded editions are thereby obsolete. The earlier editions of *KO* contain useful information omitted from the later: as R E Maizell has pointed out, 'Processes, products, and concepts once believed obsolete may again become interesting because of changing economics, raw materials strategies, or environmental reasons'.

These huge syntheses of a complete area of knowledge, however, are not typical of the usual special subject encyclopedia. The majority are handy single-volume works, representing not only a remarkable range of subjects but an astonishing variety of content. Typical examples are M G Say *Newnes' concise encyclopaedia of electrical engineering* (1962) with just under a thousand pages; *The universal encyclopedia of mathematics* (Allen and Unwin, 1964) with extensive sections of formulae and tables; N W Kay *The modern building encyclopaedia* (Odhams, 1955), a work of a very practical nature, notable for its illustrations, averaging two for each of its 768 pages. An example so far unique of a bilingual encyclopedia is *Encyclopédie des gaz/Gas encyclopaedia* (Amsterdam, Elsevier, 1976) with text in French and English. The student should remember, however, that not every specialist encyclopedia is an encyclopedia for the specialist alone: many of the three hundred articles in the *McGraw-Hill encyclopedia of energy* (1976) are taken straight from the *McGraw-Hill encyclopedia of science and technology* discussed above.

A number of publishers have seen the advantages of bringing out such encyclopedias in a series, and the most extensive range at present is the couple of dozen Van Nostrand Reinhold volumes now so familiar in our libraries in their handsome standard format. They do vary to some extent one from another, but in general they contain longish signed articles (perhaps 400 in a 1,200-page volume) by experts, with short, selective bibliographies, and are deliberately aimed at a wide spectrum of users, from school students (in the higher forms), and teachers, to specialists. Examples to examine are R J Williams and E M Lansford *The encyclopedia of biochemistry* (1967), C A Hampel *The encyclopedia of the chemical elements* (1968), or H R Clauser *The encyclopedia of engineering materials and processes* (1963). Other volumes are devoted to, for example, microscopy, X-rays, oceanography, chemical process equipment; the one on chemistry by C A Hampel and G G Hawley which has now reached its third edition (1973) states its philosophy in its preface: 'The primary function of a one-volume encyclopedia of chemistry is to introduce to this vast subject a factor of convergence instead of divergence, of focal condensation instead of scattering, of unity instead of multiplicity — in a word, *chemistry* instead of a multitude of qualifying terms'.

Indeed the non-librarian is often staggered to discover some of the specialized topics in which there exists a fully fledged encyclopedia: E Gurr *Encyclopedia of microscopic stains* (Hill, 1960), R H Durham *Encyclopedia of medical syndromes* (Harper and Row, 1966), W M Levi *Encyclopedia of pigeon breeds* (Jersey City, TFH Publications, 1965) are three examples. It is probably safe to say that no important discipline is without an encyclopedia of its own. It is significant that when the first edition of this present work appeared in 1970 I was able to point out that neither computers nor anthropology had an encyclopedia (at least in English). Both gaps have since been filled.

Content of encyclopedias
It is well worth the student's while paying some attention to the various kinds of information contained in encyclopedias, for they are by no means uniform in content. As well as the 'short article' and 'long article' approaches with which he will be familiar from his

45

study of general encyclopedias he will find for example that it is not uncommon for editors to include substantial information outside the basic alphabetical sequence of the text: A K Osborne *An encyclopaedia of the iron and steel industry* (Technical Press, second edition 1967), for instance, has 48 pages of appendices with conversion tables, signs and symbols, and information about societies in the subject field.

The number and position of bibliographical references also varies quite markedly. The special-subject single-volume works for the lay enthusiast hardly ever indicate further sources of information, and even the more substantial works covering the whole field of science and technology for the non-expert rarely list any references. The Newman *International encyclopedia* referred to earlier lacks bibliographies with the four thousand individual articles, but does have a 13-page 'graded reading list' in volume four. One famous work that has shown a change of heart since its earlier editions is the remarkable *Van Nostrand's scientific encyclopedia* (fifth edition 1976) which packs nearly two-and-a-quarter million words and 2,381 pages between the covers of a single volume. It now has brief bibliographies appended to a small number of the major articles. Most of the better encyclopedias for the specialist, on the other hand, carry bibliographies with the articles as a matter of course: with works of *Kirk-Othmer* calibre these are often as extensive as any list of references appended to a monograph or a paper in a learned periodical. Yet another approach is seen with the *McGraw-Hill encyclopedia*: even though most of the longer articles have their expected bibliographies there is also the separately published *McGraw-Hill basic bibliography of science and technology*, designed as a supplement to and uniform with the volumes of the main set. Books are listed under alphabetically arranged subject headings corresponding to entries in the *Encyclopedia*.

Further study would illustrate varying approaches among encyclopedia editors to such topics as biographical information, historical aspects of the subject, illustrations, etc.

Form of encyclopedias
Library users are constantly surprised to come across encyclopedias that are not alphabetically arranged, such as the *Cambridge encyclopedia of astronomy* (Cape, 1973), or D R Woodley *Encyclopedia of materials handling* (Pergamon, 1964) in two volumes, or O S Nock *Encyclopedia of railways* (Octopus, 1978), or E H Hart *Encyclopedia of dog breeds* (TFH Publications, 1968). And yet historically the systematically arranged encyclopedia was the first on the scene by many hundreds of years. The justification for this arrangement is succinctly given in the preface to the ambitious *Materials and tech-*

nology: a systematic encyclopaedia (Longmans, 1968-75), where the claim is made that it 'ensures that related subjects are dealt with in proximity with each other rather than separated by the random vagaries of the alphabet'. With its eight volumes it is the largest systematic encyclopedia of science and technology in English and it will be interesting to observe its use, particularly as a further claim is to furnish 'up-to-date information for the layman and technologist alike'. It has been 'written specifically with the intention that each subject should be capable of being fully understood by a person who is not an expert in that subject', although a later claim was differently worded: 'Anyone who has studied physics and chemistry to sixth form level will have little difficulty in following the text'. As a source of everyday reference the key to success for a systematically arranged work lies in its index, and *Materials and technology* has an index in each volume, as well as a general index in volume eight. It is for this reason that S S Kutateladze and V M Borishanskii *A concise encyclopedia of heat transfer* (Oxford, Pergamon, 1966) is so frustrating for the user, for not only are the contents not in alphabetical order, but there is no index to its more than five hundred pages!

What is more disturbing about this example, however, is that it is not really an encyclopedia at all, systematic or otherwise. The Pergamon catalogue described it as a 'collection of data and formulae used in calculations on all types of heat transfer problems'. A more accurate description therefore would seem to be 'handbook', and this is borne out by the wording of the original Russian title which is best translated as 'reference book'. The student should be on his guard for the occasional examples of this imprecision in titles. None of the following, for instance, are encyclopedias: T Corkhill *A concise building encyclopedia* (Pitman, 1951) is a dictionary of about 14,000 terms; the *International encyclopedia of physical chemistry and chemical physics* (Pergamon, 1960-) is the collective title of a series of monographs such as B Donovan *Elementary theory of metals* (1967) and E R Lapwood *Ordinary differential equations* (1968); the *'Modern plastics' encyclopedia* (New York, McGraw-Hill) is a yearbook and directory issued as a supplement to the American technical journal *Modern plastics*.

Conversely there are encyclopedias to be found masquerading under other titles. J F Hogerton *The atomic energy deskbook* (Reinhold, 1963) is an encyclopedia by any definition, and is, moreover, alphabetically arranged; the author's preface describes it as 'combining the features of a dictionary and an encyclopedia'. A L Howard *A manual of the timbers of the world* (Macmillan, third edition 1948) and H T Evans *The woodworker's book of facts* (Technical Press, second edition 1970) are similar cases. *The Merck index* (Rahway, NJ, ninth edition

1976) does concede its true identity in its subtitle: *an encyclopedia of chemicals and drugs*. But by far the most commonly confused ascription is 'dictionary'. Dozens of examples could be quoted of works described as dictionaries that are really encyclopedias, but two will suffice: *Thorpe's dictionary of applied chemistry* (Longmans, 1937-56) in eleven amply referenced volumes plus an index volume was the predecessor of *Kirk-Othmer* in its field; and S K Runcorn *International dictionary of geophysics* (Pergamon, 1967) comprises two large volumes of 'concise authoritative articles' (eg 38 pages on 'earthquakes') by specialists, arranged alphabetically with bibliographical references.

The student will know that in theory the distinction between an encyclopedia and a dictionary is quite clear: it is demonstrated quite neatly in the twin works by Gray. In *The encyclopedia of the biological sciences* already quoted he writes: 'This is an encyclopedia, not a dictionary. That is, it does not merely define the numerous subjects covered but describes and explains them'. In the complementary work *The dictionary of the biological sciences* (Reinhold, 1966) he explains: 'It was the infeasibility of indexing the "Encyclopedia of the biological sciences" in a manner that would permit enough individual words to be found that led me to the conviction that a separate dictionary was a necessity'.

In practice this precise line is very hazy. We have companion volumes from the same publisher, identical in format and virtually identical in arrangement, but one is *The international dictionary of physics and electronics* (Van Nostrand, second edition 1961) while the other is *The international encyclopedia of chemical science* (Van Nostrand, 1964). There are works which try to have it both ways: the Royal Horticultural Society *Dictionary of gardening* (Oxford, Clarendon Press, second edition 1956) in four volumes and its *Supplements* (1956 and 1969) is subtitled 'a practical and scientific encyclopaedia of horticulture'. And there are examples that are genuinely both: E J Labarre *Dictionary and encyclopedia of paper-making* (OUP, second edition 1952) combines a dictionary (in English with equivalents in six other languages) with an encyclopedia (the article on 'wallpaper', for instance. comprises eleven pages and nine plates).

What has also emerged is a cross-bred article known as an encyclopedic dictionary, although it only needs two examples to demonstrate the difficulty of achieving a satisfactory definition: R I Sarbacher *Encyclopedic dictionary of electronics and nuclear engineering* (Pitman, [1960]) defines 14,000 terms within one volume; J Thewlis *Encyclopaedic dictionary of physics* (Pergamon, 1961-4) is a nine-volume work with five *Supplements* (1966-), 'of graduate or near-graduate standard', aiming to 'put the whole of physical knowledge on the bookshelf'. It is not easy to see how both these 'encyclopedic dictionaries'

48

could be encompassed in one category of reference book.

Perhaps most conveniently considered with encyclopedias are compilations such as the two-volumed E P Walker *Mammals of the world* (Baltimore, Johns Hopkins Press, 1968), describing itself as 'a basic source of reference concerning all the known and present genera of mammalian life on this earth'. Like an encyclopedia it covers the whole field; like an encyclopedia it gives merely the 'first facts' — no more than a single page to each genus. B Brouk *Plants consumed by man* (Academic Press, 1975) covers over three hundred plants in its 492 pages. Probably the most useful term to describe such works is *compendium*, the word actually used by the publishers of a similar comprehensive reference book, A L Simon *Wines of the world* (McGraw-Hill, 1967).

Updating encyclopedias

The difficulty of ensuring that the information in an encyclopedia is as up-to-date as possible has been touched on more than once in this chapter: with so substantial and permanent a work as 'Thewlis' the size of the problem can be imagined. In fact the publishers have gone about this in a straightforward and workmanlike fashion, and at intervals since the completion of the main set have issued a supplementary volume of over four hundred pages in the same form as the original 'Thewlis', each complete with its own index.

The alternative to frequent supplements is of course a new edition. Naturally this is easier for both editorial and production reasons where the encyclopedia is a small or single-volume work: *Kingzett's chemical encyclopaedia* (Bailliere, ninth edition 1966) has relied on this method since its first publication in 1919. It is worth noting that both *Kirk-Othmer* and *McGraw-Hill* have tried to combine both methods: supplements (called 'year books' by *McGraw-Hill*) together with a new edition shortly after the preceding one.

In theory the most efficient method but in practice the most cumbersome is the loose-leaf encyclopedia: new information can be added in any quantity at any time at any point in the work and superseded information can be likewise discarded. Examples are rare, but one that will repay study is J F Garner and D J Harris *Control of pollution encyclopaedia* (Butterworths, 1977-).

To the ordinary citizen the encyclopedia is the reference book *par excellence*. Susan Brookes has recently reminded us that 'A good encyclopedia should be able to be used as a dictionary for the definitions of terms, as a handbook for basic data and as a bibliography for the important works in its field'. It can indeed be all these and more; but the specialist in search of information, and the librarian assisting him, often need to go further and consult an actual dictionary, or a handbook, or a bibliography. Each of these distinct types of reference work will be separately considered in the chapters following.

Further reading
    J A Clarke and R D Walker 'Single volume scientific encyclopedias' *RQ* 10 1970 27-30.

Chapter 4

DICTIONARIES

As one of our most common reference books, the dictionary is probably less in need of explanation than any other. Its concern is words: either the general words of a language, or, as in this case, the special terms of a particular subject discipline. In a field like science and technology, so dependent by its very nature on communication, there is no need to stress the crucial importance of words. What is worth recalling, however, is that increasingly over the last few generations, scientists and technologists have found the ordinary language of scholarly converse inadequate to convey what they have to say to one another. We have now arrived at a point where, to quote the preface to *Chambers's technical dictionary* (third edition 1958), 'To be safe, indeed, one must regard technical language as a language apart from ordinary speech. Technical terms are in reality symbols adopted, adapted, or invented by specialists and technicians to facilitate the precise expression and recording of their ideas'.

The extent of this 'language apart' can perhaps be gauged by the fact that there are now several thousand dictionaries in the field, and substantial volumes have been compiled devoted solely to the vocabulary of very specialized topics: R Jahn *Tobacco dictionary* (New York, Philosophical Library, 1954) and E Bruton *Dictionary of clocks and watches* (Arco, 1962) both have other two hundred pages; L L Copeland *The diamond dictionary* (Los Angeles, Gemological Institute of America, 1960) and J S Cox *An illustrated dictionary of hairdressing and wigmaking* (Hairdressers' Technical Council, 1966) have over three hundred. Until recently the only serious attempt at anything like a comprehensive dictionary of the whole field of science and technology was T C Collocott and A B Dobson *Dictionary of science and technology* (Chambers, revised edition 1974) with some 50,000 entries, the successor to *Chambers's technical dictionary* quoted above, but now we have the *McGraw-Hill dictionary of scientific and technical terms* (second edition 1978), by far the most extensive, with well over 100,000 definitions. Others that are currently available are highly selective; F S Crispin *Dictionary of technical terms* (Collier-Macmillan, eleventh edition 1970), for instance, contains only about ten thousand terms

51

mainly from the applied sciences. Indeed in this general field of science and technology the majority of titles are aimed at the non-expert, either the layman or the student. E B Uvarov and D R Chapman *The Penguin dictionary of science* (Harmondsworth, fifth edition 1979) is one of an extensive series of Penguin dictionaries for the non-specialist with about six thousand words. A Hechtinger *Chatto's modern science dictionary* (1961) is a more substantial work of over five hundred pages, compiled in the United States for school students, and defining 'in uncomplicated language the meanings of some 16,000 scientific and technical words'. W E Flood and M West *An elementary scientific and technical dictionary* (Longmans, third edition 1962) is a remarkable compilation, defining some ten thousand words using an explaining vocabulary confined to two thousand words (only sixty of which are technical). The aim is to avoid giving definitions which require a dictionary to be understood! Special attention is given to word-elements (roots, suffixes, prefixes) 'so that the reader may be enabled to break up and interpret new scientific terms for himself'. A rather special example using an even smaller vocabulary to define 25,000 terms is E C Graham *The science dictionary in Basic English* (Evans, 1965).

It is to dictionaries in his speciality, however, that the practising scientist or technologist turns more readily, and it is these that make up the bulk of titles in descriptive lists such as W R Turnbull *Scientific and technical dictionaries: an annotated bibliography* (San Bernardino, Calif, Bibliothek Press, 1966), *Bailey's technical dictionaries catalogue* (Folkestone, 1979) and C W Rechenbach and E R Garnett *A bibliography of scientific, technical, and specialised dictionaries: polyglot, bilingual, unilingual* (Washington, Catholic University of America Press, 1969). Classic works of this kind well worthy of study are *Hackh's chemical dictionary* (New York, McGraw-Hill, fourth edition 1969), with definitions for 55,000 terms; *Henderson's dictionary of biological terms* (Longmans, ninth edition 1979) first published in 1920; J G Horner *A dictionary of mechanical engineering terms* (Technical Press, ninth edition 1967).

The need for frequent new editions in order to keep pace with the coinage of new terms (and to avoid the parallel danger of reliance on out-of-date works) is demonstrated clearly in R F Graf *Modern dictionary of electronics* (Indianapolis, Sams, fifth edition 1977) which defines some 20,000 terms, nearly twice as many as the first edition only fifteen years earlier. One possible sign of the future may be glimpsed in D N Lapedes *Dictionary of physics and mathematics* (McGraw-Hill, 1978), the first of what may prove to be a family of specialized dictionaries descended from the great *McGraw-Hill dictionary*: most of the 20,000 entries are derived from the master computer file of the parent work.

It is common for dictionaries to go beyond their basic function of defining words: G and R C James *Mathematics dictionary* (Van Nostrand Reinhold, fourth edition 1976) claims to be 'by no means a mere word dictionary . . . It is a correlated condensation of mathematical concepts, designed for time-saving reference work'. The most usual way is for the compiler to extend the individual entries under each word to include further information on the subject, but a more radical method is to include extra entries, quite overtly encyclopedic in kind: L M Miall and D W A Sharp *A new dictionary of chemistry* (Longmans, fourth edition 1968) and H J Gray and A Isaacs *Dictionary of physics* (Longmans, second edition 1975) are companion volumes with articles sometimes extending to a page or more, including a number of biographical entries, and occasional literature references. And examples abound of dictionaries with substantial appendices of tables, formulae, and other non-dictionary data: K G Ponting *A dictionary of dyes and dyeing* (Mills and Boon, 1980) concludes with a 19-page bibliography on the topic; a quarter of the four hundred or so pages of L E C Hughes *Dictionary of electronics and nucleonics* (Chambers, 1969) is taken up thus. The great *McGraw-Hill dictionary* mentioned above contains three thousand illustrations.

Taking this into account with the evidence in the previous chapter of the imprecise way in which the term 'dictionary' is applied to works transparently encyclopedic, the student must be aware that in certain subjects for most practical purposes the distinction between encyclopedia and dictionary no longer has any meaning. The terms are often used interchangeably, sometimes within the same work, as in J R Stewart and F C Spicer *An encyclopedia of the chemical process industries* (New York, Chemical Publishing Co, 1956) which is the latest version of what used to be called *Stewart's scientific dictionary;* and likewise in the *International encyclopedia of chemical science* (Van Nostrand, 1964) which replaces *Van Nostrand chemist's dictionary* (1954). This is not to suggest, however, that the student should not cultivate an appreciation of the special role of the dictionary in scientific and technical communication. In any field in which he wishes to specialize he must be able to distinguish those 'dictionaries', however they may be described, which are in fact concerned with terminology from those that are mainly encyclopedias.

Scope and level
Among the many problems facing the compiler of a subject dictionary there are two in particular that it is important to have in mind when one studies an individual title. The first is the question as to what words to include. To some the answer might seem obvious: those words on the one hand that lie outside the general vocabulary and do

not appear in an ordinary dictionary, such as 'thermistor', 'bus-bar', 'thyratron', and on the other hand those words in common usage that have acquired a specialized meaning within a particular field, such as 'resistance', 'valve', 'terminal'. For example, the preface to M Abercrombie *Penguin dictionary of biology* (Allen Lane, sixth edition 1977) states that 'semi-technical terms, which can be found in any English Dictionary . . . are omitted'. The foreword to J Thewlis *Concise dictionary of physics* (Oxford, Pergamon, second edition 1979) states that 'Terms which have their dictionary meaning or which are self-evident are not defined'. Yet it can be seen that some compilers do include ordinary words in their ordinary meanings: Walford points out that E J Gentle and C E Chapel *Aviation and space dictionary* (Los Angeles, Aero Publishers, fourth edition 1961) contains definitions of words like 'electrician' and 'telephone'; and *Audel's new mechanical dictionary* (New York, 1960) defines 'sizable' and 'dry'. As for including words from related disciplines that users of a dictionary might find useful, there are two points of view: some workers prefer the subject scope of their dictionaries clear-cut; they wonder whether terms from other fields can be defined with quite the same authority; and they know that they can only be chosen on a very selective basis. But many users do appreciate this extension of coverage, particularly as boundaries between subjects grow ever more indistinct, and many dictionary compilers do follow the practice, as for example G Usher *A dictionary of botany* (Constable, 1966) which is subtitled 'including terms used in biochemistry, soil science and statistics'.

The second problem of a dictionary compiler that the librarian in particular needs to consider has to do with the level of readership at which the book is aimed, for this obviously has considerable bearing on the kind and detail of terminology used in definitions and entries. This can be illustrated by taking a well-established discipline such as mathematics and studying the range of dictionaries that are available. At the simplest level is C H McDowell *A dictionary of mathematics* (Cassell, 1961) 'written specially for young people'. For students at the secondary school level we have J Bendick and M Levin *Mathematics illustrated dictionary* (Kaye and Ward, second edition 1971). A work like W Karush *The Crescent dictionary of mathematics* (Macmillan, 1962) covers high-school and college mathematics in detail. According to its preface C C T Baker *Dictionary of mathematics* (Newnes, 1961) is 'suitable for use up to degree standard', while W F Freiburger *The international dictionary of applied mathematics* (Van Nostrand, 1960) is for the advanced scientist. T A and W Millington *Dictionary of mathematics* (Cassell, 1966) is intended for several kinds of reader, including 'the enquiring layman'. There are very many others.

Aside from questions of scope and level, it would be a mistake to assume that all dictionaries in science and technology are fairly similar. There may not be the variety of types that one finds among the general language dictionaries, such as dictionaries of slang, spelling dictionaries, rhyming dictionaries, and the like, but the range is surprisingly wide. In W E Flood *The origins of chemical names* (Oldbourne, 1963), entitled *The dictionary of chemical names* in the United States, we have an etymological dictionary, giving derivations for the words listed. J Challinor *Dictionary of geology* (Cardiff, University of Wales, fifth edition 1978) is, to some extent like the *Oxford English dictionary*, compiled 'on historical principles', with quotations showing each term in a variety of uses. Indeed it is reported that an *Oxford dictionary of scientific terms on historical principles* is in preparation. A unique example of its type is *A list of scientific terms with Scottish connections* (Edinburgh, Scottish National Dictionary Association, 1976). E C Jaeger *The biologist's handbook of pronunciations* (Springfield, Ill, Thomas, 1960) concentrates on pronunciation to the virtual exclusion of definitions. There are also a handful of what can only be described as dictionaries in reverse, which the enquirer can consult not to find the definition of a particular word, but to find the word that corresponds to a particular definition, eg W A Regal *The inverted medical dictionary: a method of finding medical terms quickly* (Westport, Conn, Technomic, 1976).

Featured in many of the titles so far quoted, and particularly useful in the field of science and technology, are illustrations, usually simple black-and-white diagrams or line-drawings, eg Petroleum Educational Institute *Illustrated petroleum dictionary* (Los Angeles, 1952). Especially valuable are those dictionaries based on more elaborate drawings, often showing 'exploded' views, with each individual part labelled or keyed to the text, eg R A Salaman *Dictionary of tools used in woodworking and allied trades* (Allen and Unwin, 1975).

Thesauri

E E J Marler *Pharmacological and chemical synonyms* (Amsterdam, Excerpta Medica, third edition 1961) is something of an exception, because dictionaries of synonyms are not at all common in science and technology, but in recent years there have appeared many thesauri, which in some ways resemble synonym dictionaries, for example, *INSPEC thesaurus, 1981* (Institution of Electrical Engineers, 1980), and *Subject headings for engineering* (New York, Engineering Index, 1972) and *Supplement* (1977). Of course these are not produced *as* dictionaries but as subject-heading lists for indexers and searchers, or, in the words of the foreword to the *Thesaurus of textile terms* (MIT Press, second edition 1969), 'to provide an effective language interface between people and manual storage systems or electronic computers capable of handling immense stores of technical information'. Indeed, the first edition of the American Society for Metals *Thesaurus of metallurgical terms* (Metals Park, Ohio, second edition 1976) was originally issued in computer-printout form only. We now have available handy listings in A Kutten *Thesauri bibliography* (Haifa, Elyachar Library, 1975), and S Rizzo *International bibliography of thesauri* (Rome, Centro per la Documentazione Automatica, 1977). A list of obviously restricted scope is *Bibliography of UN thesauri, classifications, nomenclatures* (Geneva, UNESCO, 1979) which confines itself to those used by some seventy information systems and services within the United Nations family of organizations. Lists of a collection with a policy of acquiring all English-language thesauri or those with English as one of the languages have appeared as J Walkley and B Hay 'An annotated list of thesauri held in the ASLIB library' *ASLIB proceedings* 26 1971 292-300, brought up to date by V Gilbert 'A list of thesauri and subject headings held in the ASLIB library' *ASLIB proceedings* 31 1979 264-74. Many printed thesauri, issued as aids to searching a specific data base on-line, are themselves now available on-line also. Within the UNISIST programme (see chapter 10) we now have established at Warsaw the Clearinghouse for Thesauri and Classification Schemes.

A remarkable instance of initiative from the British Standards Institution is the recent *BSI ROOT thesaurus* (Hemel Hempstead, 1981), a computer-aided attempt in two volumes at a comprehensive thesaurus for the whole of science and technology. Its storage in machine-readable form means that it can be amended and updated speedily, but even more significantly the modular nature of the data base allows it to be further manipulated to produce a range of by-products in various formats, such as a series of special subject thesauri, glossaries of terms with definitions, and versions in other languages. A French translation in computer-printout format is already available.

The value of the thesaurus arrangement has also been seen by at least one compiler of a conventional dictionary: P Gray *The dictionary of the biological sciences* (Reinhold, 1966) makes extensive use of the practice of grouping words according to their meaning rather than alphabetically. Unique in its approach is A Godman and E M F Payne *Longman dictionary of scientific usage* (1979), described as 'the first scientific reference book of its kind'. It consists of 10,000 terms arranged in 125 'sets' closely related in meaning or subject area. In arrangement, therefore, it resembles a thesaurus, but the full definitions and explanations of usage do indeed place it in a category of its own.

Standardization

A problem for users of language in all fields is that of unifying and standardizing terms and their definitions. France has her Académie Française and Italy her Accademia della Crusca to care for the purity of their national tongues, but English has never had such a body. Yet in science and technology the need has become so acute that a number of national and international organizations have taken it upon themselves to bring some order into the chaos of scientific and technical terminology. In many cases their efforts have taken the form of an official list of standardized and agreed terms, virtually an 'official' dictionary. An outstanding example is the American Geological Institute *Glossary of geology* (Washington, third edition 1972): some 33,000 terms are authoritatively and officially defined; where the 'Committee believes use should be abandoned' the entries are marked 'Not recommended'. *The IEEE standard dictionary of electrical and electronics terms approved by the Standards Committee of the Institute of Electrical and Electronics Engineers* (Wiley-Interscience, 1972) is more elaborate: it gives the preferred term first, followed by variations in parentheses in descending order of preference, with deprecated terms bringing up the rear in their own parentheses together with a 'reference mark' indicating the Institute's disapproval. The second edition (1977) abandons this practice, but without explanation. The British Standards Institution, working through its numerous committees, is very active in this sphere, and has produced over 130 glossaries: for example, *Glossary of terms used in the rubber industry* (BS 3558: 1980); *Typeface nomenclature and classification* (BS 2961: 1967); *Definitions for use in mechanical engineering* (BS 2517: 1954), and (in several parts) *Welding terms and symbols* (BS 499: 1965). Other standards bodies too are much involved, eg American Society for Testing and Materials *Compilation of ASTM standard definitions* (Philadelphia, fourth edition 1980). Two interesting examples of 'official' dictionaries are United States Air University *The United States Air Force dictionary* (Princeton, NJ, Van Nostrand, 1956) and Institute of Petroleum *A glossary of petroleum terms* (fourth edition 1967). A handy guide is E Wüster *International bibliography of standardized vocabularies* (Munich, Saur, second edition 1978).

As the student will notice, the term that has been used most commonly in the titles of these works is 'glossary'. Of course the word is far older even than 'dictionary', and means no more in fact than a list, with explanations, of specialized terms, but it is a convenience to have it applied to this particular kind of standardizing dictionary. The student should be warned, however, that this description, like so many others, is not used consistently: S Patai *Glossary of organic chemistry* (Interscience, 1962), the *Meteorological glossary* (HMSO, fifth edition

1972), A W Lewis *A glossary of woodworking terms* (Blackie, 1966), T Corkhill *A glossary of wood* (Stobart, second edition 1979), and very many others, are no more than ordinary dictionaries.

There is not the space here to go into the related but vast and thorny problem of nomenclature which plagues certain disciplines like chemistry and botany and zoology. In Britain, for example, there are over seventy names for the heron. 'Rock wrens' in the United States are quite a different species to 'rock wrens' in New Zealand. According to E S Gruson *Checklist of the birds of the world* (Collins, 1976), 'With regard to English common names a pleasant disorder reigns'. World-wide there are about 300,000 green plants and over 120,000 fungi with an average of three synonyms each to describe them. A special problem for botanists is the inconvenient fact that many wild plants acquire new and often very different names when they are brought into our gardens and cultivated. Names for new chemical compounds are a constant headache for research chemists, for with the five million known chemical substances being augmented at a rate of 350,000 each year it is obviously important that they should be named as systematically and descriptively as possible. Unfortunately, no agreement has yet been reached by the chemists on an international basis of nomenclature: one index prepared by the Chemical Abstracts Services showed that 33,000 chemical substances were described in the literature by 140,000 names. What the student should know is that efforts are being made at systematization, exemplified by publications such as the International Commission on Zoological Nomenclature *International code of zoological nomenclature* (second edition 1964) which 'indicates the criteria governing the naming of an animal or group of animals, and also regulates the use of names which have already appeared in the literature', and the British Standards Institution *Recommendations for the selection, formation and definition of technical terms* (BS 3669: 1963). Within the UNISIST programme (see chapter 10) we have had established since 1971 at Vienna the International Information Centre for Terminology (INFOTERM), charged with the task of co-ordinating terminological activity throughout the world. Studies are under way into the feasibility of the central registration of newly-coined words. UNESCO has also assisted in the preparation of M Krommer-Benz *World guide to terminological activities: organizations, terminology banks, committees* (Munich, Verlag Dokumentation, 1977).

Form of dictionaries
When asking in a library for a dictionary of a particular subject most readers expect to be given a book, usually a single bound volume,
58

devoted to their topic. In fact there are hundreds of dictionaries published as part of another work, either a reference book or a book 'in the field'; and probably almost as many not in book form at all. In A K Graham and H L Pinkerton *Electroplating engineering handbook* (New York, Reinhold, second edition 1962) there is a 9-page glossary, and there is a 24-page list of technical terms in A W Eley *Stockings, silk, rayon, cotton and nylon* (Leicester, Hosiery Trade Journal, 1953). D L Bloem 'Glossary of terms on cement and concrete technology' first appeared in instalments in the *Journal of the American Concrete Institute*, from December 1962, and T Armstrong and B Roberts 'Illustrated ice glossary' was first published as two articles in *Polar record* for January 1956 and May 1958. Directories in particular (see chapter 6) often make a feature of dictionaries of specialized terms, eg *Fishing industry index international* includes a dictionary of fish terms; the *Bottler's year book* includes 'A concise dictionary of technical terms used in the alcoholic and non-alcoholic beverage industries'.

Recently we have had dictionaries made available in the form of computerized data banks for consultation on-line: DIALOG, CHEMNAME and SDC CHEMDEX have each more than half a million compounds listed, and are far from comprehensive yet.

A unique example of a type of dictionary more often found in other fields is A L Mackay *The harvest of a quiet eye: a selection of scientific quotations* (Institute of Physics, 1977), with about 1,500 entries arranged alphabetically by author. Medicine has had a similar volume for a number of years: M B Strauss *Familiar medical quotations* (Churchill, 1968).

'Structural' dictionaries

As virtually a separate language within the mother tongue, scientific and technical English has been the object of the same scholarly attention that other variations of standard English have attracted, and we now have a number of books and papers like T H Savory *The language of science* (Deutsch, second edition 1967), B Scharf *Engineering and its language* (Muller, 1971), and P B McDonald 'Scientific terms in American speech' *American speech* 2 1926 (November) 67-70. We even have works of instruction like R W Brown *Composition of scientific words: a manual of methods and a lexicon of materials for the practice of logotechnics* ([Baltimore], the author, 1954): 'logotechnics' the author defines as 'the art of composing words' and his aim is to 'diminish the area and amount of verbicultural wrongdoing'. The bulk of the volume of close on nine hundred pages is taken up with an etymological dictionary.

A more common linguistic approach is that already described above

59

in Flood and West *An elementary scientific and technical dictionary* in which special attention is given to the elements that go to make scientific and technological words. Another dictionary by W E Flood, *Scientific words, their structure and meaning* (Oldbourne, 1960), is devoted entirely to 'about 1,150 word-elements (roots, prefixes, suffixes) which enter into the formation of scientific terms'. The entry under 'stetho-' derives this from the Greek 'stethos' meaning 'the chest'; the entry for 'scope' traces this to the Greek 'skopos', meaning 'one who watches'. A stethoscope therefore is an 'instrument for "inspecting the chest"'. There are several examples of these 'structural' dictionaries in particular subject areas, for example, F Roberts *Medical terms: their origin and construction* (Heinemann, fifth edition 1971) and E C Jaeger *A source-book of biological terms* (Springfield, Ill, Thomas, third edition 1955) which lists 15,000 of these linguistic building bricks.

Much terminology, old and new, is of course based on Greek and Latin elements, particularly in fields like medicine and botany, yet it is common to find that many workers have, like Shakespeare, 'small Latin and less Greek'. Special attention is given to this in dictionaries like R S Woods *The naturalist's lexicon: a list of classical Greek and Latin words used or suitable for use in biological nomenclature* (Pasadena, Abbey Garden Press, 1944) and *Addenda* (1947). Further assistance is available from texts like W T Stearn *Botanical Latin: history, grammar, syntax, terminology and vocabulary* (Newton Abbot, David and Charles, second edition 1973) which claims that the language is 'now so distinct from classical Latin in spirit and structure as to require independent treatment'. L Hogben *The vocabulary of science* (Heinemann, 1969) takes the form of a crash course by following which 'every student of natural science should be able to gain in a few weeks more than a nodding acquaintance with the overwhelming majority of Latin and Greek words which occur as components of internationally current technical terms'. The little book by T Savory *Latin and Greek for biologists* (Watford, Merrow, 1971) is by a zoologist who 'spent forty years leading the uneventful life of a public school master'. He was prompted to write it by his encounter many years ago with a new phenomenon: 'the arrival in my Science Sixth of a boy who had never learnt a word of Latin or Greek'.

Translating dictionaries
One tremendous problem arising directly from the demise of Latin as the language of science (and it should not be forgotten that as late as 1687 Newton used Latin for his greatest work, *Philosophiae naturalis principia mathematica*, and Linnaeus was still using Latin in 1760 for

Systemae naturae, his *magnum opus* on botany) is the 'language barrier', which will be discussed at length in chapter 17. Our immediate concern is with those thousands of interlingual dictionaries, compiled as aids to translation. We are told that up to 60% of a translator's time is spent consulting dictionaries and other reference works. There are fortunately a number of lists to help the librarian thread his way through the mass of titles: perhaps the most informative of several available is the UNESCO *Bibliography of interlingual scientific and technical dictionaries* (Paris, fifth edition 1969) with nearly 2,500 titles, although lists of an individual library's own holdings can be very useful on the spot, particularly if carefully indexed. A good example is Manchester Public Libraries *Technical translating dictionaries* (1962) which includes almost four hundred titles in alphabetical subject order with a language index. Broader in subject scope is G E Hamilton *Catalogue of the translator's library in the Department of Trade and Industry: dictionaries, glossaries, encyclopedias, books about languages* (New York, Oceana, 1976) in three volumes.

Much of what has been said earlier in this chapter about monolingual dictionaries applies with equal force to translating dictionaries, such as the difficulty of compiling a really satisfactory work covering the whole field, although there have been two substantial attempts by major publishers with A F Dorian *Dictionary of science and technology, English-German* (Elsevier, second edition 1978) and *German-English* (1970) and R Walther *Polytechnical dictionary, English-German* (Pergamon, 1968) and *German-English* (1968). Publicity for the latter does bring home just how much is required of such a dictionary, for it makes the ambitious claim that it 'should bridge the gap between the general, non-technical and the specialized subject dictionaries. Terms from all fields of science and technology are included, and the selection provides both a basic vocabulary common to many fields supplemented with vocabularies peculiar to those technological sciences which form the basis of several specialized fields. Some relevant general terminology has also been included'. Another ambitious claim is made by the massive three-volumed J L Collazo *English-Spanish, Spanish-English encyclopedic dictionary of technical terms* (McGraw-Hill, 1979), '25 years in preparation'. This sets out to be a bilingual dictionary, technical encyclopedia, and thesaurus of technical terms, all in one.

In practice, however, specialist users prefer a dictionary of narrower subject scope, preferably one confined to their speciality. Highly regarded and widely used titles are the two-volumed *Castilla's Spanish and English technical dictionary* (Routledge, 1958) which excludes the physical, chemical and biological sciences, and of which the third edition has been published in Spain as the two-volumed *Diccionario politécnico de las lenguas española e inglesa* (Madrid, Castilla, 1965); L De Vries and T M Herrmann *German-English technical and engineering dictionary* (McGraw-Hill, second edition 1965) and *English-German technical and engineering dictionary* (McGraw-Hill, second edition 1968), both of which exclude scientific terms and are complemented by L De Vries *German-English science dictionary* (McGraw-Hill, fourth edition 1978); and the even more specialized R Ernst and E I Morgenstern *Dictionary of chemistry: German-English* (Pitman, 1961) and *English-German* (1963); and United States Department of the Army *English-Russian, Russian-English electronics dictionary* (New York, McGraw-Hill, 1958).

Perhaps because the general bilingual dictionaries used by language students and scholars are usually two-way (eg from English into French and from French into English), it is commonly found (and the student will already have noticed) that many scientific and technical dictionaries are two-way also. Yet there is much evidence that use is largely one-way: the English-speaking chemist will almost invariably wish to translate papers and books *from*, say German or Russian. The professional translator likewise will spend most of his time turning material *into* his own language, whether French or Chinese or any other. Indeed, one noted authority on translation, I F Finlay, holds firmly that 'no translator, unless truly bilingual, should be asked or expected to translate out of his/her mother tongue'. Some of the finest translating dictionaries are indeed one-way only: the student should study with care I F Gullberg *A Swedish-English dictionary of technical terms* (Stockholm, Norsedt, second edition 1977); L I Callaham *Russian-English chemical and polytechnical dictionary* (Wiley, third edition 1975); A M Patterson *A German-English dictionary for chemists* (New York, Wiley, third edition 1950), which served as a model for Callaham; and the companion volume A M Patterson *A French-English dictionary for chemists* (New York, Wiley, second edition 1954).

In a category by itself is E Garfield *Transliterated dictionary of the Russian language* (Philadelphia, Institute for Scientific Information, 1980) which provides a half-way house for those with little or no Russian attempting to use a regular translating dictionary. It comprises conversion tables for turning Cyrillic to Roman, and a Russian to English translating dictionary with the Russian words already in their transliterated Roman form arranged alphabetically.

Representatives of most of the categories of dictionary described earlier can also be found among the translating dictionaries. Indeed, the illustrated dictionary using diagrams with keyed parts was pioneered by the monumental Schlomann-Oldenbourg series *Illustrated technical dictionaries in six languages* from various publishers since 1906. There are many examples of interlingual glossaries of standard terms: the International Commission on Glass *Dictionary of glass-making* (Charleroi, 1965) is in English with French and German equivalents, and indicates incorrect terms by inverted commas. There are separately published supplements for Dutch, Italian, Spanish, Swedish, Japanese, and English-Czech-Russian. English, French and Russian are the languages of the International Organization for Standardization *List of equivalent terms used in the plastics industry* ([Geneva], 1961). As with monolingual dictionaries, some translating dictionaries include supplementary matter judged useful for their users, eg S MacIntyre and E Witte *German-English mathematical vocabulary* (Edinburgh, Oliver and Boyd, 1966) has a short grammar section, two sample passages with translations, and a list of German type-faces and script. There are a large number of multilingual thesauri in science and technology, often produced by international organizations: the Commission of the European Communities has compiled several, eg *Multilingual thesaurus on veterinary science* (Munich, Saur, 1979) in four separate language volumes (English, French, German, Italian) with a quadrilingual index volume. And reminding us that today it is not only man that communicates with his fellows we have H Breuer *Dictionary for computer languages* (Academic Press, 1966): the user is told that 'using this book as one would a common translation dictionary, it is possible to translate a FORTRAN program into ALGOL, and vice versa'.

Multilingual dictionaries
The observant reader will doubtless have noted that some of the titles just quoted are multilingual (or polyglot) dictionaries. It is instructive to consider the organizational problems of a dictionary compiler once his field extends to more than two languages, and to examine some of the proffered solutions. The *International glossary of leather terms* (International Council of Tanners, 1975) has each of its five language sequences printed on different coloured paper. The method adopted by the famous *Hoyer-Kreuter technological dictionary* (Berlin, Springer, sixth edition 1932-4) is simple but expensive: a separate and self-contained edition for each of the three languages, German, English, and French, each edition of course giving equivalents in the other two languages. This approach is quite unrealistic, however, where the dictionary attempts to cover five or six or seven languages, as is frequently the case. The system used by the very successful Elsevier multilingual series is to arrange the volume by the English word, followed in tabular form by the definition, and then by the equivalent word in the other five or six languages in turn. Individual indexes for each of these languages refer (usually by number) to the English word equivalent. Well-established examples to study are W E Clason *Elsevier's dictionary of television, radar and antennas in six languages* (1955), and

G S Stekhoven *Elsevier's automobile dictionary in eight languages* (1960). Several in this series have been extended from, say, six languages to seven by a supplement, eg W E Clason *Elsevier's dictionary of nuclear science and technology in six languages* (1958) has a *Russian supplement* (1961), now incorporated in a second edition (1970). Other variations include classified rather than alphabetical order in *Elsevier's dictionary of the gas industry in seven languages* (1961) and *Supplement* (1973); the addition of an *Index Latinus* (ie, of scientific names) to G Haensch and G Haberkamp de Anton *Dictionary of agriculture: German/English/French/Spanish/Russian* (Amsterdam Elsevier, fourth edition 1975); and the use of German rather than English as the base language, as in B D Hartong *Elsevier's dictionary of barley, malting and brewing in six languages* (1961). Elsevier were not of course the inventors of this method, nor are they the only publishers to adopt it: J Nijdam *Horticultural dictionary in eight languages* (Interscience, [third edition] 1962) is compiled on a Dutch base, and was first published by the Netherlands Ministry of Agriculture in 1955. It is perhaps interesting to note that Elsevier has now taken over the latest edition of this work, which appears as *Elsevier's dictionary of horticulture in nine languages* (1970), although still under the auspices of the Netherlands Ministry of Agriculture and Fisheries. Italian is the language that has been added. From a different publisher we now have R W B Stephens *Sound* (Crosby Lockwood, 1975), an English-base dictionary with equivalents in seven European languages, and one of a series of 'International dictionaries of science and technology'.

Dictionaries of this kind have come in for two specific criticisms: they include only a comparatively small number of terms (for the tabular layout consumes a great deal of space), and the 'synoptic' coverage oversimplifies the translation problem (for the validity of a 'neat row of five foreign equivalents' is suspect). Nevertheless, they do seem to fulfil a need, and they proliferate greatly, even in quite specialized subject fields, for instance, H A Anderfelt *Technical vocabulary for the match industry: English, Swedish, German, French* (Jonkoping, Swedish Match Co, 1961), and J Brandt *Emails, enamels, émaux, smalti: a dictionary in four languages* (Leverkusen, Farben-fabriken Bayer Ag, 1960). There are even works which give equivalents in a dozen or more languages: G Carrière *Detergents: a glossary of terms* (Van Nostrand, 1960) lists only 257 terms, but covers 19 languages; A Herzka *Elsevier's lexikon of pressurized packaging (aerosols)* (1964) lists 262 terms in 21 languages! It must be said, however, that most professional translators would feel that a set of good bilingual dictionaries is a better purchase than almost any multilingual dictionary, particularly if you compare the number of words you get for each pound or dollar.

So far as arrangement of entries in a dictionary is concerned the student might think that a compiler has no alternative but to adopt alphabetical order — 'dictionary' arrangement, in fact. He should, therefore, try to examine I Paenson *Systematic glossary of the terminology of statistical methods* (Oxford, Pergamon, 1970), which covers English, French, Spanish, and Russian. He will find that 'the terms are embedded in a continuous exposition which assumes the appearance of a very short multi-lingual manual'. This systematic presentation, it is claimed, 'has the great advantage of defining the terms in the most accurate way possible in so far as their meaning becomes entirely clear from the context'.

Like monolingual dictionaries, translating dictionaries are found in a variety of forms in addition to the bound volume. Directories will frequently be found to contain brief translating dictionaries of specialized terms, eg the five-language glossary of leather terms, seventeen pages long, in *Leather guide: international directory of the industry*. In particular, much hope for the future has been pinned on the computer, which has already provided a number of examples of terminological data banks, ie virtually automatic dictionaries where the translator keys in a word at a terminal and obtains the foreign equivalent he is seeking.

It should not be too complacently assumed that English, even scientific and technical English, means the same thing throughout the world. There are considerable variations between British and American English, for instance: a car hood is two very different things on either side of the Atlantic; many drivers in Britain would be puzzled to learn that American cars use gas for fuel; even a gallon is not the same. It is perhaps not yet necessary to have a British-American translating dictionary, but many of the better scientific and technical dictionaries do take trouble to distinguish the two usages. It is standard practice in the Elsevier multilingual series, and is given particularly close attention in M Polanyi *Technical and trade dictionary of textile terms* (Pergamon, second edition 1967), German-American/English and American/English-German.

A final note of caution on translating dictionaries: only rarely do they attempt to define; far more commonly all we are offered is equivalents. The student will be aware of the pitfalls here for the user: as J E Holmstrom warns, 'technical *concepts* are one thing and the technical *terms* which serve as more or less mutable symbols for them are another'. In his own compilation *Trilingual dictionary for materials and structures* (Oxford, Pergamon, 1971) he deliberately uses lengthy definitions in some cases 'to ensure that British, American, French and German specialists in the subject-matter who were invited to decide the most appropriate corresponding terms in their

65

respective languages would all have the same concepts in mind when doing so'.

Dictionaries of names

The problems of nomenclature have already been touched on but as guides through the jungle there are several dictionaries devoted solely to names: the coverage of H L Gerth van Wijk *Dictionary of plant names* (The Hague, Nijhoff, 1911-6) in two volumes, P G Embrey and J P Fuller *A manual of new mineral names, 1892-1978* (British Museum (Natural History), 1980), and of H N Andrews *Index of generic names of fossil plants, 1820-1965* (Washington, US Geological Survey, 1970) is world-wide, while that of F A Sharr *Western Australian plant names and their meaning: a glossary* (Nedlands, Western Australian Press, 1978), and of R D MacLeod *Key to the names of British butterflies and moths* (Pitman, 1959) is obviously limited.

Some of these dictionaries endeavour to trace who it was who first coined each name, citing literature references where possible. Two classic and continuing works (both in English despite their Latin titles) are S A Neave *Nomenclator zoologicus* (Zoological Society of London, 1939-) which in the seven volumes and quarter of a million entries so far published endeavours to provide 'as complete a record as possible of the bibliographical origins of the name of every genus and sub-genus in zoology'; and *Index kewensis* (Oxford, Clarendon Press, 1893-) with some half-a-million entries in over a dozen volumes, subtitled 'an enumeration of the genera and species of flowering plants . . . together with their authors' names, the works in which they were first published, their native countries and their synonyms'. A multilingual dictionary is P Macura *Elsevier's dictionary of botany: volume 1 Plant names in English, French, German, Latin, and Russian* (Amsterdam, 1979).

Of course many plants and animals are better known by their common name than their official name. Some dictionaries (eg Gerth van Wijk above) do include such names, but there are also special dictionaries: C E Jackson *British names of birds* (Witherby, 1968), and R M Carleton *Index to common names of herbaceous plants* (Boston, Hall, 1959) are examples from each side of the Atlantic. And though the student will find that a number of the ordinary scientific and technical dictionaries include such names there are others that firmly exclude them. A W Leftwich *A dictionary of biology* (Constable, second edition 1973) explains the dilemma: his first edition 'did not include the English names of common or well-known animals, chiefly because such names are often vague and ambiguous and because there are well over a million species of animal, many of which have no English names. We have felt, however, that the complete omission of
66

such names has been a defect and we have endeavoured to rectify this'.

Names frequently encountered in everyday speech are those eponymous words like cardigan, sandwich, boycott, mackintosh that have entered the language as the names of people with whom the things or practices they stand for were associated. Scientists and technologists seem particularly prone to this habit, and the literature is dotted with terms like Ohm's law, Heaviside layer, Parkinson's disease. Medicine has so many of these (some as obviously venerable as Adam's apple and Achilles tendon) that a policy of voluntary limitation has been agreed, and in some fields such as anatomy they are now forbidden words. Once again it is true to say that the ordinary subject dictionaries do include many of these terms, but there are also dictionaries specializing in eponyms, eg J A Ruffner *Eponyms dictionary index* (Detroit, Gale, 1977) with over 20,000 entries, drawn mainly from technical fields; and S Jablonski *Illustrated dictionary of eponymic syndromes and diseases* (Saunders, 1969). Two others in contrasting subjects are D W G Ballentyne and D R Lovett *A dictionary of named effects and laws in chemistry, physics and mathematics* (Chapman and Hall, fourth edition 1980) and C P Auger *Engineering eponyms* (Library Association, second edition 1975). They present a contrast in methods too: the entries in the former merely define, for example, what the Doppler effect is, but give no information about the man, no date, and no literature references; the relevant entry in Auger, on the other hand, not only says what the Wankel engine is, for instance, but cites two references, including Wankel's own paper, with a note of an available translation.

Raising similar problems are the many thousands of trade names encountered in science and technology. The complications here are such that trade names are accorded separate treatment in chapter 15.

Dictionaries of abbreviations

Any regular reader of the literature of science and technology will have found it plentifully sprinkled with abbreviations, like VHF (very high frequency), H_2O (water), t½ (radioactive half-life). Many can be looked up in the regular dictionaries, general and subject, and the student will already know of dictionaries specializing in abbreviations, but so extensive now is the practice that there are several dozen such dictionaries solely for science and technology, such as E Pugh *Pugh's dictionary of acronyms and abbreviations: abbreviations in management, technology and information science* (Bingley, cumulative edition 1981), with a total of over 25,000 entries; P Roody *Medical abbreviations and acronyms* (New York, McGraw-Hill, 1976) with 14,000 entries; and O Brandstetter *Abbreviations and acronyms in science and technology* (McGraw-Hill, 1972). E B Ocran *Acronyms: a dictionary of abbreviations and acronyms used in scientific and technical writing* (Routledge, 1978) is of particular interest as a compilation by a librarian who culled the bulk of his entries 'from journals on open access at the Science Reference Library'. This is, however, an area of much confusion: Charlotte Schaler, for instance, has documented twelve methods of forming abbreviations.

For purposes of study it is useful to distinguish five categories of what may loosely be called 'abbreviations':

a *The contraction* is simply a shortened form of the original, such as in (inch), Chem Soc (Chemical Society), Al (aluminium), where part of the word stands for the whole. Sometimes it is taken from an alternative form of the original: lb, Ag are abbreviations of the Latin words for pound and silver. This is a common method of abbreviating the title of a scientific periodical, such as *Brit chem eng*, or *Arch biochem biophys*, and A Davidsson *Periodica technica abbreviata* (Stockholm, Petterson, 1946) was an early attempt to provide a dictionary of such abbreviations.

b *The initialism* is an abbreviation comprising only the first letter(s) of the original term(s), as UV (ultra-violet), bp (boiling point), emf (electromotive force). Contractions and initialisms of this kind make up the bulk of entries in works like E B Steen *Dictionary of abbreviations in medicine and the related sciences* (Bailliere, third edition 1971) and L W Wallis *Printing trade abbreviations* (Avis, 1960).

c *The acronym*, a comparative newcomer, is a word consisting of the initials of a group of words, that is to say, a pronounceable initialism, for instance, laser (light amplification by stimulated emission of radiation), radar (radio detection and ranging), LED (light-emitting diode), LEM (lunar excursion module), scuba (self-contained underwater breathing apparatus). Dropping the use of capital letters is usually taken as a sign that the acronym has become accepted as part of the language. So popular and useful have acronyms become that it is now common practice so to contrive the name of a new device or project or material that its initials will form a neat and appropriate acronym, eg ADMIRAL (automatic and dynamic monitor with immediate relocation, allocation, and loading), ASH (Action on Smoking and Health). Listing more than ten thousand, mainly from the fields of aviation and astronautics, is R C Moser *Space-age acronyms* (New York, Plenum Press, second edition 1969). The use of acronyms has been closely identified with government and particularly military activities so it is not at all unexpected to find a special list compiled by the (US) Defense Technical Information Center *Government acronyms and alphabetic organizational designations* (1980). The so-called 'telescope' words like Mintech (Ministry of Technology), and FORTRAN (formula translation) are not acronyms according to strict definition, since they comprise more than simply the initials of the component words, but they are a particularly common form of abbreviation in German and Russian. Perhaps the most notorious is Gestapo (Geheime Staats Polizei, secret state police).

d *The code* is the most difficult to define because it may or may not be an abbreviation and it may or may not be a pronounceable word. The point to grasp is that a code, whether a letter-code or word-code, may be assigned by a method other than abbreviation, perhaps even arbitrarily, such as TSR2 for a new military plane, or ADO 16 for a prototype car. Pluto, for instance, was the code-name during world war II for the system of cross-channel fuel-lines supplying the Allied invasion forces in Europe in 1944, but it was also the abbreviation (and indeed the initialism and acronym) for pipe line under the ocean. Apollo, on the other hand, the code-name for the US man-on-the-moon project, is not an abbreviation and not an initialism and not an acronym. Code-names are quite likely to result from arbitrary allocation: in military and security contexts the code-name may be deliberately selected as a 'cover' word, with the object of being as non-committal as possible, such as Barbarossa, for Hitler's invasion of Russia, but many others are quite fanciful choices, as the code-names given to typhoons by meteorologists, such as Edna, Lucy, Alan. A guide to some 8,500 of these is F J Ruffner and R C Thomas *Code names dictionary* (Detroit, Gale, 1963). An example of a guide to code-letters (as opposed to code-names) is L E Godfrey and H F Redman *Dictionary of report series codes* (New York, Special Libraries Association, second edition 1973), which endeavours to identify the agency originating a particular US government research report from the code letters and numbers assigned.

e *The symbol*, on the other hand, is never formed by abbreviation, and quite frequently is not in letter form at all. The plus and minus signs are probably the best known, and indeed mathematics abounds with symbols. Many of them, it is true, are in letter form, such as J (mechanical equivalent of heat), y (altitude), although in some cases the letter has been modified as in ℞ (recipe) and £ (pound sterling) and there are separate lists, eg British Standards Institution *Schedule of letter symbols for semiconductor devices* (BS 3363: 1968). Greek letter forms are commonly used: everyone is familiar with the symbol π for the ratio of the circumference of a circle to its diameter. But the most difficult of all to keep track of bibliographically are the non-literal, non-numerical symbols: some of them do have a tenuous connection with what they stand for (the biological symbol ♀ for female is said to be a stylized representation of the hand-mirror of Venus,

Goddess of Love), but many are chosen without apparent relevance. There is fortunately a wide selection of guides, usually arranged systematically because of the impossibility of alphabetical order. The most comprehensive is O T Zimmerman and I Lavine *Scientific and technical abbreviations, signs and symbols* (Dover, NH, Industrial Research Service, second edition 1949), but other examples are D D Polon *Dictionary of architectural signs and symbols* (New York, Odyssey, 1966), and R G Middleton *Electrical and electronic signs and symbols* Indianapolis, Bobbs-Merrill, 1968). The most extensive listing is W Shepherd *Shepherd's Glossary of graphic signs and symbols* (Dent, 1971), over twenty years in preparation, containing some five thousand distinct forms with over seven thousand meanings. The majority have some scientific or technological connection and the arrangement is by a most ingenious classification. Attempts at standardization have met with a substantial measure of success here: the British Standards Institution *Graphical symbols for general engineering* (BS 1553:1949-) and *Letter symbols, signs and abbreviations* BS 1991:1961-) are but two of several similar titles, both still in progress. The American Society of Mechanical Engineers *Letter symbols for units used in science and technology* (New York, 1970) has been approved as an American National Standard (see chapter 16). Another systematic compendium of symbols approved as standards in the United States is A Arnell *Standard graphical symbols: a comprehensive guide for use in industry, engineering and science* (McGraw-Hill, 1963). The compiler explains the significance of this work: 'To an engineer, the compact symbols of graphics represent the language of communication. This book is a dictionary of that language.' The Royal Society even has its own Symbols Committee, responsible for *Quantities, units and symbols* (second edition 1975).

There are also interlingual dictionaries of abbreviations, such as L Seguin *Four-language dictionary of technical abbreviations* (Ottawa, Information Canada, 1970), comprising English, French, Spanish and German in two volumes; M Azzaretti *Dictionnaire international d'abréviations scientifiques et techniques* (Paris, Maison du Dictionnaire, 1978) in the same four languages with the addition of Russian; and the rather simpler G E M Wohlauer and H D Gholston *German chemical abbreviations* (New York, Special Libraries Association, 1965) which gives the complete German form with an English equivalent. A different arrangement is used by P Wennrich *Anglo-American and German abbreviations in science and technology* (New York, Bowker, 1977) with the 150,000 German and English terms interfiled in a single sequence. They are particularly needed for Russian, for 'No other language has given birth to abbreviations of its current scientific and technological terminology to the extent which modern Russian has and

70

no other technical language is more saturated with acronyms', according to H Zalucki *Dictionary of Russian technical and scientific abbreviations with their full meaning in Russian, English and German* (Elsevier, 1968). The Library of Congress Aerospace Technology Division *Glossary of Russian abbreviations and acronyms* (1967) lists no fewer than 23,000 with English translations.

One last word of warning to the student: probably because it is so difficult for anyone (other than perhaps the compiler himself) satisfactorily to assess within a short time the worth of a dictionary, this category of work is probably more variable in quality than any other likely to be regularly consulted.

The user should note the words of W R Turnbull in the preface to his bibliography mentioned earlier: 'Like the thousands of other books being published each year, many are outstanding and the degree of backbreaking labour is evident on every page, while others are mediocre or almost totally inadequate'.

Further reading

A L Gardner 'Technical translating dictionaries' *Journal of documentation* 6 1950 25-31.

C Schaler 'Technical abbreviations and contractions in English' *Journal of chemical education* 32 1955 114-7.

M J Bailey 'The use of abbreviations and acronyms in the physics literature' *Journal of the American Society for Information Science* 27 1976 81-4.

J Alvey 'Dictionaries and reference books for the translator' *ASLIB proceedings* 31 1979 521-4.

Chapter 5

HANDBOOKS

The reference work most frequently consulted by the working scientist and technologist is the handbook. These compilations of miscellaneous information in handy form are of course found in other fields also, but they are seen at their best in science and technology, and the student should lose no time in examining for himself some of the classic titles. The *Standard handbook for mechanical engineers* (McGraw-Hill, eighth edition 1978) is still known as 'Marks' after L S Marks, editor of the first five editions from 1916 to 1951. Similarly, 'Perry' always refers to the *Chemical engineers' handbook* (McGraw-Hill, fifth edition 1973), edited from 1934 to 1950 by J H Perry. The amount of information contained in works of this kind is staggering: they both for instance have about two thousand illustrations. And *Lange's handbook of chemistry* (McGraw-Hill, twelfth edition 1979) and *Standard handbook for electrical engineers* (McGraw-Hill, eleventh edition 1978), first published 1907, each has over two thousand pages. *Machinery's handbook* (New York, Industrial Press, twentieth edition 1975) has sold well over two million copies since the first edition of 1914.

To help the librarian understand the aim of these 'one-volume reference libraries', as they have been called, it is only necessary to read what they have to say about themselves in their forewords and prefaces and introductions: O R Frisch *The nuclear handbook* (Newnes, 1958) is 'a day to day reference book'; E U Condon and H Odishaw *Handbook of physics* (New York, McGraw-Hill, second edition 1967) contains 'what every physicist should know'; D E Gray *American Institute of Physics handbook* (New York, McGraw-Hill, third edition 1972) is 'a working tool'. J H Potter *Handbook of the engineering sciences* (Van Nostrand, 1967) is more explicit about its role 'to assemble, categorize, and digest the more or less enduring fundamental considerations of the principal engineering sciences on a level approximating that of the first year graduate student in engineering'. And H H Huskey and G A Korn *Computer handbook* (New York, McGraw-Hill, 1962) maintain that 'sufficient detail is presented so that anyone competent in the field can proceed to construct a computer'.

72

Handbooks are the first port of call when a straightforward factual problem arises in a particular subject field. It has been claimed that a library with no more than a sound collection of handbooks can answer 90% of quick-reference queries. As the last title indicates, the kind of information they contain is of a very practical nature: the *Steel designers manual* (Crosby Lockwood, fourth edition 1972) aims 'to bridge the gap between the normal textbook on the theory of design and its practical application in structural engineering ... [and] provides authoritative data facilitating economic and efficient practice'. And in the words used in Cyril Long *Biochemists' handbook* (Spon, 1961), 'so far as possible, opinions and suggestions are not included'. One type of user frequently in the compiler's mind is the worker without access to a large literature collection. W G Driscoll and W Vaughan *Handbook of optics* (McGraw-Hill, 1978) was sponsored specially by the Optical Society of America to 'provide the first comprehensive one-volume coverage of technical data in the field ... Up to now engineers and specialists were forced to refer to many different and often widely scattered sources for enough optical information to make even first-order calculations'. Indeed handbooks are often described as 'bench-books'. A useful bibliography with over two thousand entries and a detailed subject index is R H Powell *Handbooks and tables in science and technology* (Phoenix, Ariz, Oryx Press, 1979).

Within one volume such works cannot hope to be comprehensive, but in any case that is not what the users want. What they seek is convenience combined with reliability: handbook editors try to provide this by the most skilful selection of data, the most time-saving arrangement and indexing, supported by regular (often annual) and painstaking revision. To become known as 'the chemists' Bible', the *Handbook of chemistry and physics* (Boca Raton, Fla, CRC Press, sixty-first edition 1980) has had to maintain the highest standards since 1913. As might be expected, designed as they are for day-to-day consultation on the job, they do not usually give literature references, though there are notable exceptions: there are several thousand in the three volumes of P L Altman and D S Dittmer *Biology data book* (Bethesda, Md, Federation of American Societies for Experimental Biology, 1972-4), still known to many as 'Spector', after the editor of the first edition, W S Spector *Handbook of biological data* (Philadelphia, Saunders, 1956). Unlike textbooks, for instance, they often assume a certain knowledge on the part of the user, reminding him of what he once knew.

Although the titles so far mentioned in this chapter make up no more than a fraction of the total, it will have been noticed that they are preponderantly American, and that many of them come from the same publisher. This indeed is a precise reflection of the total picture, which is dominated by major American houses like McGraw-Hill, Wiley-Interscience, and Van Nostrand Reinhold. One specialist publisher, indeed, CRC Press, has brought out a computer-produced index covering the fifty or so handbooks the firm issues: *Composite index for CRC handbooks* (Cleveland, Ohio, second edition 1977) which itself has 1250 pages. In Britain too, there are firms like Newnes-Butterworth with a sound reputation for their series of handbooks, eg M G Say *Electrical engineer's reference book* (Newnes, thirteenth edition 1973), L W Turner *Electronics engineer's reference book* (Newnes-Butterworth, fourth edition 1976), M J V Powell *House builders' reference book* (Newnes-Butterworth, 1979), a replacement for a well-known predecessor that went through five editions, and A J Hall *The standard handbook of textiles* (Newnes-Butterworth, eighth edition 1975). What has perhaps not been brought out so far is the extensive subject range of these handbooks. It is now common to find one or more for each major discipline or industry: representative titles are E F Kaelble *Handbook of x-rays* (McGraw-Hill, 1967), Illuminating Engineering Society *IES lighting handbook* (New York, fifth edition 1981), W I Orr *Radio handbook* (Editors, twenty-first edition 1978), F S Merritt *Building construction handbook* (McGraw-Hill third edition 1975), E E Grazda *Handbook of applied mathematics* (Van Nostrand, fourth edition 1966), H H Clay *Handbook of environmental health*, formerly *Clay's public health inspector's handbook* (Lewis, fourteenth edition 1977). The library user indeed is often surprised to find substantial handbooks devoted to quite specialized topics, eg Universities Federation for Animal Welfare *The UFAW handbook on the care and management of laboratory animals* (Edinburgh, Churchill Livingstone, fifth edition 1976), R T Liddicoat *Handbook of gemstone identification* (Los Angeles, Gemological Institute of America, tenth edition 1975), V H Laughner and A D Hargan *Handbook of fastening and joining of metal parts* (McGraw-Hill, 1956).

The reader will have noticed that these works are not always called handbooks (or manuals, which is simply the Latin-derived term for the same thing): reference book is often used as a quite satisfactory alternative. What it is important to recognize is the handbook under some less revealing title: *The agricultural notebook* (Newnes-Butterworth, sixteenth edition 1976), R M C Dawson *Data for biochemical research* (Oxford, Clarendon Press, second edition 1969), M Souders *The engineer's companion* (Wiley, 1966), E Rabald *Corrosion guide* (Elsevier, second edition 1968) and S Glasstone *Sourcebook on atomic energy* (Van Nostrand, third edition 1967) are all handbooks. On occasion the disguise is even more difficult to penetrate: despite its title, G Wyszecki and W S Stiles *Color science* (New York, Wiley, 1967) is quite clearly a handbook by any definition; the preface describes it as a 'collection of
74

concepts and methods, quantitative data and formulas' and goes on to explain that 'descriptive and qualitative material ... which would probably find a place in a textbook or treatise on color is not included'. The student should be warned, however, about the German word 'Handbuch', frequently used, as in J Durm and H Ende *Handbuch der Architektur* or S Flügge *Handbuch der Physik.* A dictionary would translate the term as handbook, thus providing a neat illustration of the limitations of dictionaries of equivalents, allready cautioned against in chapter 4. These works are treatises, not handbooks: the former came out in 62 volumes between 1892 and 1907; the latter is still incomplete but so far has reached almost 70 volumes since 1955. And as the student will now have learned to expect, there are works described in plain English as handbooks or manuals that are really something else: the American Welding Society *Welding handbook* (New York, seventh edition 1976-) in five volumes is more of a treatise than anything else; the *Manual of building maintenance* is an annual directory of manufacturers and suppliers; and A M Collier *A handbook of textiles* (Wheaton, third edition 1980) describes itself simply as a 'textbook' in the introduction, and then goes on to make plain that it is indeed for students following courses of study.

In fact, almost by definition a handbook (or manual) has to be in one volume if it is to stay 'handy'. Yet so vast and rapidly increasing is the store even of selected data that a worker 'ought to know' that a number of well-known titles have had to extend to two, three, or even more volumes. *Specification: the standard reference book for architects, surveyors and municipal engineers* (Architectural Press) has appeared annually since 1898, but in recent years has become two volumes, and then five; though the third edition was a single volume J G Cook *Handbook of textile fibres* (Watford, Merrow, fourth edition 1968) is now two. And there are even larger works: A Davidson *Handbook of precision engineering* (Macmillan, 1970-) has reached twelve volumes; The American Society for Metals *Metals handbook* (Metals Park, Ohio, ninth edition 1978-), claiming in its foreword to be the largest single source of usable information on the technology of metals ever compiled, might well exceed a dozen volumes. Walford comments that 'it has the proportions of an encyclopaedia rather than a handbook'.

There is found the occasional work which appears to have had second thoughts: C J Smithells *Metals reference book* (Butterworths, fifth edition 1976) grew from one to three volumes over eighteen years and then reverted to one again. And in fairness it has to be added that many handbooks have resisted the temptation to outgrow their description; many indeed are even pocket-sized, specifically designed to be carried about, eg *Newnes' engineers' pocket book* (sixth edition 1971), *Mechanical world year book*, and the Society of Automotive

Engineers *Automotive handbook* (Warrendale, Pa, 1978) which is an English-language version of a German original that has sold more than half-a-million copies.

The layout adopted by handbook editors is almost invariably systematic, and a feature is the widespread use of the tabular form for the presentation of appropriate data. For satisfaction in use, of course, this arrangement demands adequate indexing, and in any assessment of the worth of a handbook this is an important aspect to consider. This is not to say, however, that other arrangements are never found: alphabetical order is used in G S Brady *Materials handbook* (McGraw-Hill, eleventh edition 1977), which indeed is subtitled 'an encyclopedia', in N I Sax *Dangerous properties of industrial materials* (Van Nostrand Reinhold, fifth edition 1979), which used to be titled *Handbook of dangerous materials*, and in M Grieve *A modern herbal* (Cape, 1931).

Special types of handbook

This last example reminds us that there are a number of special categories of handbook-type reference works that are found in one discipline only. Because it is common practice in the building trade to obtain estimates before work commences, we find price-books giving up-to-date information on wage rates, prices of materials, methods of calculation, and a host of related matters. Both *Spon's architects' and builders' price book* and *Laxton's building price book* have run through over a hundred editions.

A star catalogue is another example of a special type of handbook: the four-volumed *Smithsonian Astrophysical Observatory star catalogue* (Washington, US Government Printing Office, 1966) gives the position and proper motions of over a quarter-of-a-million stars. Almost 2,500 similar works are listed in a computer-produced bibliography by M Collins *Astronomical catalogues, 1951-1975* (Hitchin, INSPEC, 1978).

A flora is a botanical handbook giving detailed information on the taxonomy (ie classification) and distribution of plants in a defined area. There are popular, amateur floras such as the best-selling W Keble Martin *The concise British flora in colour* (Ebury Press, 1965), and scholarly, professional floras like A R Clapham *Flora of the British Isles* (CUP, second edition 1962) with *Illustrations* (1957-65) in four separate volumes. Floras may be local: W G Travis *Flora of South Lancashire* (Liverpool Botanical Society, 1963); or national: H W Rickett *Wild flowers of the United States* (McGraw-Hill, 1966-75) in six volumes and index; or continental: T G Tutin *Flora Europea* (CUP, 1964-80) in five volumes. There are fortunately two helpful guides to the thousands of titles: N D Simpson *A bibliographical index of the British flora* (Bournemouth, the author, 1960) with more than 65,000 entries, and S F Blake *Geographic guide to floras of the world*

(Washington, US Department of Agriculture, 1942-) as yet incomplete. Similar compilations (which can be found in zoology also) are taxonomic indexes, species catalogues, check lists, and the like, eg the massive J L Peters *Check-list of the birds of the world* (Cambridge, Mass, Harvard University Press, 1931-68) in fifteen volumes. A number of them (including some floras) are specially designed to aid identification of biological material, often a far from simple task, eg J Hutchinson *Key to the families of flowering plants of the world* (Oxford, Clarendon Press, 1967). Many textbooks also contain these keys, eg for fungi in G Smith *Industrial mycology* (Arnold, fifth edition 1960), and 'keying out a species' in a flora or fauna key of this kind is one of the first skills the young biologist learns. A useful guide is G J Kerrich *Key works to the fauna and flora of the British Isles and northwestern Europe* (Academic Press, 1978) with some 2,500 entries for books and articles. A three-volumed bibliography of key works is R W Sims and D Hollis *Animal identification: A reference guide* (British Museum (Natural History), 1980).

Pharmacopoeias are drug handbooks, particularly those with some degree of official sanction. The *British pharmacopoeia* (HMSO, twelfth edition 1973) and *Addendum* (1978) (BP for short) appears every five years on the recommendation of the Medicines Commission under Act of Parliament. The *Pharmaceutical codex*, formerly *British pharmaceutical codex* (Pharmaceutical Press, eleventh edition 1979) is designed to supplement BP and is published by direction of the Pharmaceutical Society of Great Britain. Of less official standing (but perhaps of wider use) is *Martindale: the extra pharmacopoeia* (twenty-seventh edition 1977). Supplementing pharmacopoeias are formularies such as the *British national formulary*, which are virtually recipe books. Since drugs and their prescription are so closely controlled by law, it is usual for each country to have its own pharmacopoeia and related literature, although *Pharmacopoea internationalis* (Geneva, World Health Organization, second edition 1967) and its supplement *Specifications for the quality control of pharmaceutical preparations* (1971) is an attempt at establishing international standards. We also have a *European pharmacopoeia* (1969-75) in three volumes and *Supplements* (1973-). One interesting variation is the American Pharmaceutical Association *Handbook of non-prescription drugs* (Washington, sixth edition 1979), 'the only definitive compilation of facts on home remedies' according to its preface. Similar works can be found in specialized fields: the *British pharmacopoeia (veterinary)* (HMSO, 1977), like BP is an officially approved work issued 'on the recommendation of the Medicines Commission pursuant to the Medicines Act, 1968'. J N Le Rossignol and C B Holliday *A pharmacopoeia for chiropodists* (Faber, eighth edition 1971) is unofficial.

Formularies other than pharmaceutical make a category of reference book that could well be studied with handbooks. Distinctly practical in nature these are compilations of formulas with instructions for making particular products or producing certain reactions. The largest collection (over a hundred thousand tested formulas) is H Bennett *The chemical formulary* (New York, Chemical Publishing Co, 1933-), with twenty-three volumes so far and a *Cumulative index for v.1-15* (1972); sample recipes are for lipstick, rust-remover, kinky hair straightener, jujubes, bubble gum and powdered Worcestershire sauce. A handier one-volume work is *Henley's twentieth century book of formulas* (New York, new edition 1945). K M Swezey *Formulas, methods, tips and data for home and workshop* (Harper and Row, new edition 1979) is a more recent example. These are mainly for the amateur: for the professional there are works like R E Silverton and M J Anderson *Handbook of medical laboratory formulae* (Butterworths, 1961) or P Gray *The microtomist's formulary and guide* (Constable, 1954).

Quite different from these recipe books are the compilations of mathematical formulas: H-J Bartsch *Handbook of mathematical formulas* (Academic Press, 1974), translated from the German, is unusual for a handbook in that it is primarily an educational aid. We read that 'The book starts from secondary school mathematics and is primarily intended for students of technical schools, colleges and universities'. An example from another field is K R Lang *Astrophysical formulae* (New York, Springer, 1974).

Updating handbooks

The data about the physical world that is the stock-in-trade of the handbook is constantly being refined, updated, or even superseded. Frequent revision is essential for any handbook that wishes to remain reliable. Indexes too can rarely be completely adequate to locate every last item in such very detailed volumes; for instance, D B Dallas *Tool and manufacturing engineers handbook* (McGraw-Hill, third edition 1976) has 2950 pages packed with factual minutiae. An obvious possibility with this kind of information manipulation is aid from the computer and in 1975 the National Science Foundation funded a research study of the electronic storage and delivery of handbook information. Although distinct progress has been made in the field of data in the narrow sense, as will be described below, we still await a computerized handbook. A limited number have chosen the loose-leaf method of revision: B A Maynard *Manual of computer systems* (Gee 1964-) in four volumes; one that began its life in loose-leaf form but has now reverted is the Royal Institute of British Architects *Handbook of architectural practice and management* (fourth edition 1980). Most pin their faith in regular new editions. There are several, indeed, that are undergoing continuous modification and are issued annually: *Radio amateur's handbook* (West Hartford, Conn, American Radio Relay League) has been appearing since 1926, and *Kempe's engineer's yearbook* (Morgan) since 1894. Although described as a yearbook, the latter is without doubt a handbook through and through, and indeed one of the most widely used and respected.

Tables

As mentioned earlier, data (defined here in its strict sense to mean numeric or quantitative information relating to physicochemical and other properties) in tabular form is a feature of many handbooks, and as the proportion of tables to text increases, the handbook as a form of literature merges into the book of tables. Because science, particularly the physical sciences, and technology are largely concerned with quantification, numerical information takes up much of the literature. Tables are a convenient way to present clearly details such as melting points, atomic weights, and solubilities. Indeed in certain fields like thermodynamics, or spectroscopy, or crystallography, tables are vital to the whole study and progress of the discipline, simply because such a large amount of information has been collected in tabular form, eg J H Keenan *Gas tables* (New York, Wiley, second edition 1980), R G J Miller and H A Willis *Irscot: infrared structural correlation tables* (Heyden, 1964-   ), which is loose-leaf, International Union of Crystallography *International tables for x-ray crystallography* (Birmingham, Kynoch Press, 1952-62) in three volumes.

The purpose of tables is to save time. The information could of course be presented in a number of ways, or even left to lie where it was first reported in the primary sources. Indeed, as R T Bottle points out '. . . many details of physical properties are deeply buried in the literature, and effort, patience and time are required to retrieve them. For this reason the standard books of tables are invaluable.' Provided, he might have added, that layout and indexes are designed with this prime aim in mind.

First to be considered are the great general exhaustive compilations such as the *International critical tables* (New York, McGraw-Hill, 1926-33) in seven volumes and an index. Two characteristics are plain: this is a collection of *critically evaluated* data, ie the values given are those regarded as most reliable in the opinion of the three hundred expert consultants; secondly, full bibliographical references to the primary literature are given to enable the researcher to form his own estimate of the status and accuracy of the data. It is interesting to note that the language barrier is virtually non-existent in tabular compilations, especially if the explanatory text is multilingual, as here. In fact the great French and German compilations corresponding to *ICT* can be quite profitably consulted by a research worker with a minimum knowledge of the language.

For everyday use, however, what the worker needs by him is a selective listing, matched to his needs, and as up-to-date as possible. One of the best examples, first published in 1911, is G W C Kaye and T H Laby *Tables of physical and chemical constants* (Longmans, fourteenth edition 1973). A larger American compilation with over nine hundred tables, although still in a handy single volume is the *Smithsonian physical tables* (Washington, Smithsonian Institution, ninth edition 1954). An even more selective listing is W H J Childs *Physical constants selected for students* (Methuen, ninth edition 1972). Though there will probably always be a demand for handy, time-saving books of tables such as these, the practicability of the comprehensive compilations like *ICT* is now in question. Not only are they difficult to keep up-to-date and cumbersome to use, but the editorial problems of digesting the vast amount of new data that accumulates daily are virtually insuperable by conventional means. The trend in recent years has been towards a more fragmented approach to the compilation of tables, with separate works devoted to particular sets of values or designed for particular disciplines, eg A Seidell *Solubilities: inorganic and metal organic compounds* (Van Nostrand and American Chemical Society, fourth edition 1958-66), W Kunz and J Schintlmeister *Nuclear tables* (Pergamon, 1963-), S Eilon *Industrial engineering tables* (Van Nostrand, 1962). And of course many such data compilations appear not as books but as articles in periodicals or as research reports. Indeed, there are journals that specialize in this field, eg *Journal of chemical and engineering data, Journal of physical and chemical reference data.* The National Standard Reference Data System of the United States has estimated that 200,000 works containing properties data are published annually throughout the world. As a direct result of this piecemeal approach, however, two new difficulties have arisen: first, how to avoid wasteful duplication of effort; and second, how to keep track of the many tables being published.

As the mass of data has grown, books of tables have appeared covering more and more limited areas, but even so, some of these works represent an investment of many thousands of man-hours in calculating and manipulating and setting out. In many cases the conventional methods of typesetting have proved too slow or too expensive or both, and the tables have been printed from typescript by photo-litho-offset, eg *Organic electronic spectral data* (Interscience, 1960-), in several volumes which cover the literature from 1946. For some, mechanization has been seen as offering a partial solution to the problem of absorbing the new and improved values reported in the primary literature. Microforms have been used, for example, and also available on subscription are several services which furnish data on a continuing basis in the form of punched cards, eg *Documentation of molecular spectroscopy* (Butterworths) with some two thousand edge-punched cards each year.

Computerized data

The computer in particular has proved its worth in this struggle. It 'ias been used for a number of years now to manipulate data: the National Engineering Laboratory *Steam tables 1964* (Edinburgh, HMSO, 1964) were partly calculated by computer, and D P Jordan and M D Mintz *Air tables* (New York, McGraw-Hill, 1965) were 'photographically reproduced from the original printed computer output'. C M Lederer and V S Shirley *Table of isotopes* ((New York, Wiley, seventh edition 1978) used as source material 'over 30,000 journal articles, reports, theses, and private communications', the data from which 'was entered, edited, stored, checked for form and consistency, formatted, and finally typeset or drawn by a computer'.

Most handbooks are collaborative efforts, perhaps inevitably: D M Considine *Energy technology handbook* (McGraw-Hill, 1977) claimed as the first of its kind, and in a notably interdisciplinary field, was 'prepared by 142 specialists'. The early editions of the *Table of isotopes* was also a collaborative effort by what the preface to the current edition calls 'Sunday compilers'. But a change is taking place in the human input to such compilations: 'The data explosion has created the professional compilation project, if not the professional compiler'. This change can be seen in progress in C W Allen *Astrophysical quantities* (Athlone Press, third edition 1973), compiled since the first edition of 1955 by a professor of the University of London. By the time the third edition appeared he was 68 and retired, and his preface tells us 'approved values . . . of constants change from year to year and there can be no finality about the last digit of most of the values quoted . . . yet another revision will be justified after a lapse of about seven years and preparation for this should begin at once. The author would like to negotiate with anyone willing to cooperate'.

More recently it has been grasped that for many purposes (eg reference as opposed to current awareness) data need not be printed out and published at all, provided that it is on call when required. Computer technology now permits this facility through remote-access terminals, with the computer memory acting as a 'data bank', eg the US National Bureau of Standards Alloys data base; or the Registry of toxic effects of chemical substances, maintained by the US National Institute for

Occupational Safety and Health, and containing toxicity data on some 50,000 substances. With official encouragement 'data centres' are now proliferating, with the specific tasks not only of collecting and coordinating, but also evaluating and disseminating critical data within their subject fields, eg the Mass Spectrometry Data Centre set up in 1966 at the Atomic Weapons Research Establishment at Aldermaston and taken over in 1978 by the Royal Society of Chemistry United Kingdom Chemical Information Service at Nottingham University has made available the largest collection in the world of 21,765 mass spectra on magnetic tape, as well as its monthly *Mass spectrometry bulletin* with its five different indexes offering a range of entry points. A computer program to identify unknown spectra by matching against a library of spectra has also been developed. It has also published the four-volumed *Eight peak index of mass spectra* with a total of 31,101 spectra. Both the index and the bulletin are also available on magnetic tape.

It is important for the student to distinguish such accumulations of data, strictly defined, from the kind of information to be found in what are commonly referred to as 'data bases' (see chapter 11). Many of these are really only *bibliographic* data bases, and the data they contain is simply documentary or bibliographical, ie citations of individual scientific papers, usually with descriptors and more rarely with abstracts. As the Committee on Data for Science and Technology of the International Council of Scientific Unions (CODATA) is at pains to point out, 'the ultimate units of data services are individual pieces of data, which are units of scientific knowledge themselves ... Although, sometimes, users may wish to refer to source papers to find out how the data were measured, the main function of data dissemination centres consists in providing users with the actual data'.

Mathematical tables
The tables so far mentioned have been mainly concerned with physical and chemical data, but these are not the best known. Mathematical tables of logarithmic values, trigonometrical functions, square roots and so on are far more widely used: indeed many of us are familiar with four-figure tables from our schooldays. Even simple tables such as these are ten times as accurate as the common ten-inch slide rule used for engineering calculations, but scientists and technologists often need more extended tables: a good standard set is the two-volume *Chambers's six-figure mathematical tables* (1948-9), although it is possible to go further, eg with A J Thompson *Logarithmetica Britannica: being a standard table of logarithms to twenty decimal places* (CUP, 1952). But tables of this kind are only one part of the scene: there are hundreds
82

of others. Some publishers have extensive series that are just tables, eg 'Pergamon mathematical table series' in some forty volumes. Two useful collections are *Handbook of mathematical sciences* (West Palm Beach, Fla, Chemical Rubber Co, fifth edition 1978) and R S Burington *Handbook of mathematical tables and formulas* (McGraw-Hill, fifth edition 1973), and there are a number of bibliographies, of which the outstanding example is A Fletcher *An index of mathematical tables* (Oxford, Blackwell, second edition 1962). An interesting feature of this two-volume work is a 150-page list of known errors in specific published tables.

Mechanization has proved useful for mathematical tables also: the series of *Royal Society mathematical tables* from volume 11 (1964) are in the form of a 'photographic reproduction of sheets prepared on a card-operated typewriter'. One effect of the computer in this field has been the great increase in the number of tables; ventures such as F W Kellaway *Penguin-Honeywell book of tables* (Harmondsworth, 1969) would have been impossible without the computer, for not only have all the standard tables been recalculated, but the actual pages of the book have been typeset by computer, the first Penguin to be so produced. Indeed, according to the Chief Scientist of IBM, 'Today mathematical tables are obsolete . . . Why print such a table? By storing a simple algorithm you can order the computer to build the interesting parts of a table whenever you want them'.

The student can also observe another effect of technological advance in some books of tables: there is a marked difference between J J Tuma *Engineering mathematics handbook: definitions, theorems, formulas, tables* (New York, McGraw-Hill, second edition 1979) and the previous edition. It has been 'entirely reconstructed' on the assumption that the user 'possesses a ten-digit display, four-register stack electronic pocket calculator with the standard complement of elementary functions, known as the *electronic slide rule*'. Similarly the editors of 'the chemists' Bible' mentioned above warn the user: 'We shall continue to remove from the *Handbook* those mathematics which can more advantageously be utilized by use of the small, hand-held electronic calculators'.

Data is one field of scientific activity where the benefits of international co-operation are evident, and CODATA has done much to stimulate joint effort. In 1974 it completed a major study for UNISIST on the problems of accessibility and dissemination of data for science and technology. The main recommendation was the establishment of a world scheme or network comprising three parts: for each subject discipline, a data evaluation centre; for each country, a data dissemination centre with broad subject coverage; and a global referral centre for directing users' enquiries to the most appropriate source. Such a

World Data Referral Centre has now been established in Paris. CODATA is currently compiling a directory of data sources for science and technology.

There are many other fields of endeavour where tables are widely used, and many other kinds of data that benefit from tabular presentation. Three examples will suffice: *World weather records, 1951-60* (Washington, US Weather Bureau, 1965) is the latest in a continuing series going back to the earliest observations; J H Kenneth and G R Ritchie *Gestation periods: a table and a bibliography* (Commonwealth Agricultural Bureaux, third edition 1955); the *Admiralty tide tables* appear annually in three volumes. A very common type of book of tables is the ready-reckoner, used for currency multiplication or division and for the calculation of commissions or discounts. Each country, of course, has its own. In Britain all such works were made obsolete, at a stroke, with the inauguration of decimal currency in 1971, and new editions had to be prepared. *Warne's efficient reckoner* (1970), for instance, now contains decimal tables only, but does include a supplementary conversion table for old pennies.

Finally the student should be aware of one difficulty that is found whenever any value is expressed in numerical units, such as inches, or ounces, or degrees Fahrenheit, or gallons. It is a fact that not everyone uses the same units. Universal metrication is seen as the long-term solution (although even this is not as simple as it sounds, for there is more than one metric system used in science and technology), but in the meantime workers have to rely on aids such as conversion tables (or conversion factors if they are prepared to do the calculations themselves). There are official tables published by bodies such as the British Standards Institution *Conversion factors and tables* (BS 350: 1962-74), as well as long-established works like O T Zimmerman and I Lavine *Conversion factors and tables* (Dover, NH, Industrial Research Service, third edition 1961). An interesting multilingual work with explanations in five languages is S Naft *International conversion tables* (Cassell, revised edition 1965) which claims to have many tables found nowhere else, and is also available in a condensed edition for schools, universities, and colleges, as *Cassell's concise conversion tables* (revised edition 1965). A particularly extensive listing of conversion factors is Y Chiu *A dictionary for unit conversion* (Washington, George Washington University, 1975). Within the metric system itself there is now fairly general agreement among the scientific community on the adoption of the Système International d'Unités, known as SI units, and most works now appear in these units. J W Rose and J R Cooper *Technical data on fuel* (Edinburgh, Scottish Academic Press, seventh edition 1977), the latest revision of a handbook that has been useful for over fifty years is described as 'completely revised in SI units'. But there is still the older

84

literature, not to mention older scientists and technologists, and once more conversion tables have been provided, eg G Socrates and L J Sapper *SI and metrication conversion tables* (Newnes-Butterworths, 1969). And the ASM *Metals handbook* referred to above has decided to retain both its customary US units and the metric SI units. Finally, as a general guide to the vast array of units themselves we have works such as H G Jerrard and D B McNeill *A dictionary of scientific units* (Chapman and Hall, fourth edition 1980) and S Dresner *Units of measurement: an encyclopaedic dictionary of units both scientific and popular* (Aylesbury, Harvey Miller and Medcalf, 1971).

Further reading

N B Gove and K Way 'The data compilation as part of the information cycle' *Journal of chemical documentation* 2 1962 179-81.

B Mountstephens and others *Quantitative data in science and technology* (ASLIB, 1971).

M Slater and others *Data and the chemist* (ASLIB, 1972).

H C Pestel and S A Rubin 'Electronic handbook — fact or fiction' *Online* 1 1977 (October) 84-7.

Chapter 6

DIRECTORIES AND YEARBOOKS

Directories are basically lists of names and addresses, arranged for reference purposes in a variety of ways to match the requirements of their users and frequently updated: the annual *'Packaging review' directory* is a list of manufacturers and suppliers arranged alphabetically, for instance, but the *B & C J directory*, still known to many as the *'Contract journal' directory*, arranges its list of building and other contractors by county. The *Directory of contractors and public works annual*, on the other hand, is arranged in classified order. To this basis of name and address is often added a range of other information about, for example, the products of a company (as in the *Electrical and electronics trades directory*), the activities and staff of an international organization (as in the *Nuclear research index, incorporating the World nuclear directory*), or the academic qualifications of a scientist (as in the *Directory of British climatologists*).

The word 'directory' will not necessarily appear in the title: two well-known examples are *Ports of the world* and *Guide to the coalfields*. Very commonly, where a directory appears annually, it will be called a yearbook, even though it may not at all fit the strict definition of yearbook (see below): the *Mining international year book* annually since 1887), *Oil and gas international year book* (annually since 1910), and *Year book of technical education* (annually since 1957) are all quite clearly directories.

It is staggering to contemplate the range of subjects dealt with by directories: even the most specialized industries seem to be covered, as for example, in the *Directory of waste disposal and recovery*, the *Fur trade directory of Great Britain*, and the *Directory of shoe machinery*. In fact, directories make up the largest single category of reference books: a large library might have as many as four or five thouusand current directories. Fortunately there are bibliographical guides available: in addition to the general lists by Anderson, Ethridge, and Klein with which the student will be familiar there are guides specifically for science and technology: the most comprehensive is J Burkett *Directory of scientific directories* (Guernsey, Hodgson, third edition 1979). A much slighter 'provisional checklist' is the Library of Congress

Science and Technology Division *Directories in science and technology* (Washington, 1964), but the major US listing is now B Klein *Guide to American scientific and technical directories* (Rye, NY, Todd, 1975), though it has been criticized for including many bibliographies and other reference books inappropriate for a guide to directories. *Trade directories of the world* (Queens Village, NY, Croner) is in loose-leaf format.

Trade or industrial directories
As the directory is primarily an instrument of commerce with its main aim the bringing together of buyer and seller, it is not surprising that the majority of directories fall into this category. Because industry and trade is international, many such directories attempt world-wide coverage: the *International plastics industry directory* covers 76 countries; *Skinner's wool trade directory of the world* used to claim to be the only one of its kind in the field, and so its demise has left a gap; and the *Containerization international yearbook* is an interesting example of a guide to a huge international industry that has developed from very small beginnings in a short time. But the majority of titles are national in scope: well-established and well-used examples for the student to examine are the *Timber trades directory* (first published 1892), *'Chemist and druggist' directory* (first published 1869), *Gas directory* (first published 1898), *Chemical industry directory* (first published 1923), and *Ryland's coal, iron, steel, tinplate, metal, engineering, foundry, hardware and allied trades directory* (first published 1881 and understandably known simply as 'Ryland's').

A number of these directories have sharpened their focus by calling themselves 'buyers' guides', although of course all trade directories are by definition guides for buyers (or sellers). They are not only guides to commodities and manufactured goods but increasingly to services also. Not all of them are recent innovations: *Where to buy chemicals and chemical plant* and *'Machinery's' buyers' guide* have both been going over half a century.

A feature of this category that the student may encounter is the habit that has grown up from casual and obvious beginnings of calling a particular work the 'yellow book' (as with the *Yellow book . . . formerly Who's who in the fire services and fire brigade directory*) or the 'red book' (as with the *Rubber red book*) or the 'green book' (as with the *Green book of tractor, farm and forestry equipment*) or the 'blue book' (as with the *Blue book of the British leather clothing industry*). This can be a source of confusion in a library when an engineer asking for the 'blue book' may not realize that the *Adhesives guidebook and directory*, the *Advertiser's annual*, and of course, the *Hotel and catering blue book, Soap/cosmetics/chemical specialities blue book*, the *Blue book of optometrists*, the *Brewery age blue book*, the *Leather and shoes blue book*, the *Printing trades blue book, Davison's textile blue book*. and *MacRae's blue book . . . the buying directory for engineering* are all known to their respective trades or professions simply as the 'blue book'.

The observant student will also have noticed that many of these industrial directories are published by or have close ties with a particular periodical, as *'Flight international' directory of British aviation* or *'Chemical week' buyers' guide*, or *'Sheet metal industries' year book*. At least one example, *IEA: the directory of instruments, electronics, automation*, is issued in association with two journals: *Control and instrumentation* and *Electronic engineering*. These may be produced and priced quite separately from the periodical itself, as *'Electrical and electronic trader' yearbook*, but more commonly take the form of an annual supplement or a special issue of the periodical, included in the subscription, as *'Chemistry and industry' buyers' guide* each autumn. A handy list of such journals is C M Devers *Guide to special issues and indexes of periodicals* (New York, Special Libraries Association, second edition 1976). Sometimes the directory section is no more than part of an issue, as the 'Buyer's guide and equipment review' in the January number of *Rock products*. A useful list has been compiled by the Science Reference Library *Trade directories in journals: a list of those appearing within numbered parts of serials* (1979). Some journals even make a feature of directory-type information in *each* issue, as the so-called 'grey pages' in *Building*.

Some directories have extended their scope beyond mere listing to incorporate pages of manufacturers' own literature describing their products and services, eg *Concrete year book*. Most notably volumes 11 to 16 of *Thomas' register of American manufacturers*, known collectively as *Thomas' register catalog file* (or THOMCAT), comprises the catalogues of over five hundred US manufacturers amounting to over six thousand pages. Here trade directories begin to merge into trade catalogues, which will be discussed more fully in chapter 18.

What could perhaps be regarded as a special category within this general group of industrial or trade directories is the exhibition catalogue. An example to study is the catalogue of the great Frankfurt chemical engineering exhibition ACHEMA which extends to over a thousand pages.

Trade names will be discussed in chapter 15, but it is worth noting that many industrial directories contain lists of trade names, eg the *European glass directory* and the *Concrete year book*, and that there are examples of directories devoted entirely to them, eg the *'Watchmaker, jeweller and silversmith' directory of trade names and punch marks*.

Directories of individual scientists and technologists
It is commonly the case that directory-type information (name, address, etc) is called for about a person rather than a firm. As we have seen, information of this kind is sometimes found in the trade directories, eg the *Adhesives directory* contains a 'who's who', as does the *Printing industries annual* and the *World tobacco directory*. Indeed, some of the directories listed in this chapter have added *and who's who* to their titles, eg *Brewery manual and who's who in British brewing, Gas directory and who's who*. There are also a large number of separately published professional directories: in a class by itself (and part of the 'spin-off' from the computerized data base of *Science citation index, Index to scientific and technical proceedings, Index to scientific reviews, Current contents*, etc) is the annual *Current bibliographic directory of the arts and sciences* (Philadelphia, Institute for Scientific Information), formerly *Who is publishing in science*. This lists with their addresses and organizational affiliations the 400,000 scientists and scholars from 175 countries who have had a journal article or a conference paper or a book published the previous year. It is more usual to find examples confined to one particular discipline, although still on a world-wide or international basis, eg *International directory of marine scientists* (Rome, Food and Agriculture Organization, 1977), also computer-produced, which lists over ten thousand specialists from 90 countries; *World directory of mathematicians* (International Mathematical Union, 1979), listing 20,000 names; more frequently they list the experts within a particular country or area, eg *Fire research specialists: a directory* (Washington, National Bureau of Standards, 1977).

The most frequently encountered of such lists, however, are the membership directories of the scientific and technical societies and institutions, eg the Royal Institute of British Architects *Directory of members* and the *Combined membership list* of the 40,000 members of the American Mathematical Association, the Mathematical Association of America and the Society of Industrial and Applied Mathematics. These are particularly common in the United States and sometimes attain a vast size, eg the American Chemical Society *Directory of membership* with over 90,000 names. They are not always easy to get hold of. The joint *Register of members* of the Chemical Society and the Royal Institute of Chemistry (who have since combined to form the Royal Society of Chemistry) has over 40,000 names. It contains a note: 'This register is a valuable asset. Unrestricted use or misuse of the information it contains is financially damaging . . . and can result in great annoyance to members. It is therefore not being made available to non-members'.

Frequently such lists appear as part of a yearbook, eg in the Institution of Civil Engineers *Yearbook* with 58,000 members, or in the

Royal Institution of Great Britain *Record*. Less commonly they are found in a periodical, eg the American Physical Society *'Bulletin' membership directory issue* every two years or so. In certain professions such lists are statutory, eg the *Medical register*, issued annually under the authority of the General Medical Council and the *Register of patent agents* maintained by the Chartered Institute of Patent Agents on behalf of the Department of Trade.

The value of such tools in facilitating personal communication among individual scientists and technologists can well be imagined, but there is one type of professional directory which has a further function, ie the list of consultants. This is to some extent also a trade directory insofar as it serves to draw to the attention of possible clients the services that the consultant is selling, eg the Royal Institute of Chemistry *Directory of consulting practices in chemistry and related subjects* (1980).

These works are not really biographical sources in the full sense of that term, since the information given about each individual is usually biographically meagre, often no more than name, academic or professional qualifications, post held, and address. They are primarily location tools, and unlike the biographical dictionaries, which are almost invariably alphabetically ordered, these professional directories sometimes prefer a geographical arrangement, as in the *List of research workers in the agricultural sciences in the Commonwealth and the Republic of Ireland* published every three years by the Commonwealth Agricultural Bureaux, or even more commonly, a geographical index to the main alphabetical sequence, as in the *Fifth international directory of anthropologists* (University of Chicago, fifth edition 1976). Another arrangement encountered occasionally is by subject, with the specialists listed under their speciality, but once again this approach is more widespread in the form of the subject index to the main alphabetical sequence, as in the *Register of incorporated photographers*.

Once the entries begin to include more than mere directory information, as in the *Medical list*, or the *American architects directory* (New York, Bowker, third edition 1970), this type of reference work begins to merge into the biographical sources discussed in chapter 22.

Worth mentioning at this point are the private, unpublished 'expertise indexes' maintained by many scientific and technical organizations such as research associations, industrial firms, government departments, professional bodies, and of course libraries. Designed explicitly to identify people knowledgeable in particular subjects or having specific skills, they are found in a variety of formats ranging from 5in by 3in cards to a computer file, and are used not only for problem-solving or enquiry-answering but also for 'head-hunting' when filling vacant posts. A British Library-funded survey in 1976 found over 250 examples, a third in the UK and over half in the US. Most of them are consulted at least a hundred times a year, some of them many thousands of times – a rate which compares favourably with well-used reference books.

Directories of scientific and technical organizations
Including all kinds of organizations other than industrial firms (which are covered by trade diirectories) these make up a large and important class. Their special significance for the information worker lies in the fact that they are (with the directories of individual scientists and technologists) the guides to the non-documentary sources of information described in chapter 1, such as government departments, learned societies, universities, etc. These sources by definition are not part of the literature of science and technology, but the directories and guides to them are, and therefore receive attention here.

It is a truism to say that science knows no frontiers, and there are indeed a large number of organizations of scientists and technologists that are international in scope and which should perhaps be considered first. At the summit of world science sits the International Council of Scientific Unions, comprising bodies such as the Internaional Union of Pure and Applied Chemistry, the International Union of Biological Sciences, and many others. A useful guide to study is their *Year book* published in Rome by the ICSU Secretariat. Much wider in scope though now out of date is the UNESCO *Directory of international scientific organizations* (Paris, second edition 1953) covering 264 bodies in science and technology. More recent is the OECD *International scientific organizations: catalogue* ([Paris 1965]) and its *Supplement* (1966). A particularly valuable tool for librarians since it concentrates on some 449 of these organizations particularly as sources of information is the Library of Congress General Reference and Bibliography Division *International scientific organizations: a guide to their library, documentation, and information services* (Washington, 1962).

A common variant of this type of directory is the research guide: as most research in science and technology nowadays is institutional rather than individual, lists such as the two-volumed *European research index* (Guernsey, Hodgson, fourth edition 1977) are virtually directories of organizations of all kinds where research is conducted, or promoted, eg government research establishments, independent research institutes, university research departments, and research laboratories of industrial firms.

Of course most scientific organizations are national in scope, and are described in directories such as *Australian scientific societies and professional associations* (Melbourne, Commonwealth Scientific and Industrial

Research Organization, second edition 1979), or the National Academy of Sciences *Scientific, technical and related societies of the United States* (Washington, ninth edition 1971), or *Industrial research in Britain* (Guernsey, Hodgson, eighth edition 1976), or the *Directory of scientific research institutes in the People's Republic of China* (Washington, National Council for US-China Trade, 1977-8) in three volumes. An example of a regional guide is *Directory of European associations: part 2 National, learned, scientific and technical societies* (Beckenham, CBD research, 1979). An ambitious recent attempt at world coverage is the *Guide to world science* (Hodgson, second edition 1974-6) in 24 volumes. Aiming to 'unravel this organizational web or science', each volume takes the form of a directory of important scientific establishments of all kinds within a single country (or group of countries). The last volume is devoted to international scientific organizations, both intergovernmental and non-governmental. A valuable one-volume listing is *World guide to scientific associations and learned societies* (Bowker, second edition 1977) which covers more than ten thousand organizations in 134 countries.

All the directories mentioned so far under this heading have covered the whole of science and technology, but there are many examples concerned with particular subject fields. A guide of international scope is the *Pollution research index* (Guernsey, Hodgson, 1975); a national listing is the Royal Society *United Kingdom marine science* (1972).

There are also directories, international and national, of particular types of scientific and technical organizations. The Department of Industry *Technical services for industry* (1977) is in effect a directory of government research stations, grant-aided industrial research associations, and government departments concerned with science and technology. By way of contrast, *Industrial research laboratories of the United States* (New York, Bowker, sixteenth edition 1980) is devoted to non-governmental or non-academically supported organizations. The *Research centers directory* (Detroit, Gale seventh edition 1981) concentrates on yet another type of organization: university-related and other non-profit research bodies in the United States and Canada. The Netherlands Institute for Documentation and Filing *Research guide* (The Hague, 1963) is a key to over a hundred centres willing to undertake research on contract. The corresponding British directory is E A G Liddiard *Register of consulting scientists, contract research organizations and other scientific and technical services* (Stoke Poges, third edition 1976). Two unique directories are P Millard *Trade associations and professional bodies of the United Kingdom* (Pergamon, sixth edition 1979) and UNESCO *World directory of national science policy-making bodies* (Paris, 1967-) to be in four volumes.

Directories of scientific and technical information organizations
Of particular significance to the librarian are directories of scientific
and technological information organizations, such as libraries, information analysis centres, data centres, etc. Again lists are found of
international and national scope. In a class by itself is the International
Federation for Documentation *National technical information services:
worldwide directory* (Helsinki, third edition 1970), a guide to the major
channel in each country through which the sources of technical information in any field can be located. The UNESCO *World guide to science
information and documentation services* ([Paris, 1965]) covers 65
countries and is complemented by the *World guide to technical information and documentation services* (Paris, second edition 1975) listing
476 centres in 93 countries. An international guide to a particular
sort of information source is the *Directory of United Nations information systems* (Geneva, 1980). On a European basis we have the
Guide to European sources of technical information (Guernsey,
Hodgson, fourth edition 1976). National lists concentrating mainly on
conventional libraries are the *ASLIB directory: volume one, information
sources in science, technology and commerce* (fourth edition 1977) and
The directory of special libraries and information centers (Detroit,
Gale, fifth edition 1979), in two volumes with supplements every six
months; but the *Guide to locating US government technical infor-
mation, technology and patents* (San Mateo, Cal, TTA Information
Services, 1972), in loose-leaf format, confines itself to official sources,
the *Directory of federally supported information analysis centers*
(Springfield, Va, Committee on Scientific and Technical Information,
1970) deals solely with information analysis centres, and the Library of
Congress National Referral Center of the Science and Technology
Division *A directory of information resources in the United States:
physical sciences, engineering* ([Washington, 1971]) is an example of
a list designed to cover all possible sources. Its alphabetical arrangement makes an interesting contrast with the classified order of the
Battelle Memorial Institute *Specialized science information services in
the United States: a directory* (Washington, National Science Foundation, 1961). An alternative approach to *The directory of special
libraries and information centers* mentioned above is now available
through five volumes of the *Subject directory* . . . (Detroit, Gale 1977),
of which volumes 3 and 5 cover the health sciences, science and technology. An example of a directory within a specific field is D N Allum
*Marine transport: a guide to libraries and sources of information in
Great Britain* (Library Association and Marine Librarians' Association,
1974); a directory of international scope is the International Civil
Aviation Organization *Aeronautical information services provided by
states* (Montreal, forty-fifth edition 1977), and an example covering a

specific region is *Permanent directory of energy information sources in the European Communities* (1980) listing five hundred centres and organizations. So numerous indeed are these directories that we have for some years had a bibliography in the field: *Directories of science information sources: international bibliography* (The Hague, FID, second edition 1967) lists 360 titles from 58 countries.

The recent proliferation of computerized information services has nourished a new kind of directory, listing the machine-readable bibliographic data bases that are now available for searching by the scientific and technical community. In fact these directories, despite the way they describe themselves, are more usefully regarded as bibliographies, primarily designed as they are to list not merely names and addresses of organizations but details of what in effect are machine-readable abstracting and indexing services. They will be considered in chapter 11 in the same way that the so-called directories of printed abstracting and indexing services are treated in chapter 10.

The student has already been warned in chapter 5 not to confuse such *bibliographic* data bases with the compilations of data in the strict sense at data centres and elsewhere. The latter kinds of sources are listed in directories such as *Data activities in Britain* (Office for Scientific and Technical Information, third edition 1969) and the CODATA *International compendium of numerical data projects* (Berlin, Springer, 1969), covering 150 data centres in 26 countries. This is being updated a chapter at a time in issues of *CODATA bulletin*, eg the *CODATA directory of data sources for science and technology: chapter 4 Zoology* appeared in September 1980. When complete, all the chapters are to be combined with further updating into a single volume.

As more and more such collections or compilations become computerized the machine-readable 'data banks' thus formed increasingly become available for searching in the same way as bibliographic data bases, and lists, guides and directories begin to appear. Sometimes these bibliographies (for that once again is what they are) also include bibliographic data bases, and there are signs that the distinction may be becoming less important. This will be considered further in chapter 11.

Directories of scientific and technical collections
Because science and technology have as their sphere of action the physical world, a feature of research is the use made of collections of physical objects such as rocks, insects, anatomical specimens, trees, micro-organisms, etc. In certain fields such as biology these assemblages have especial value for purposes of identification: there are, for example, 90 million plant specimens in European collections, with a further

45 million in the United States. The guides to the existence of these collections form a small but interesting category of directories, eg UNESCO *Directory of meteorite collections and meteorite research* (Paris, 1968) which covers 49 countries, D M Henderson and H T Prentice *International directory of botanical gardens* (Utrecht, Bohn, third edition 1977), S M Martin and V B D Skerman *World directory of collections of cultures of microorganisms* (Wiley,1972), and the *International zoo yearbook*, which lists zoos and aquaria of the world, are all international in coverage. D H Kent *British herbaria* (Botanical Society of the British Isles, 1957), *Directory of collections of microorganisms maintained in the United Kingdom* (Kew, Commonwealth Mycological Institute, 1978), and M M Roberts *Public gardens and arboretums of the United States* (New York, Holt, 1962) are obviously national. Museums are listed in two periodical articles: 'Museums of science and technology', *Museum* 20 1967 150-228; and an earlier article with an identical title, *Technology and culture* 4 1963 133-47. An example of a national guide is J E Smart *Museums in Great Britain with scientific and technological collections* (Science Museum, 1977); international guides in particular subject areas are H G Rodeck *Directory of the natural science museums of the world* (Bucharest, Revista Muzcelor, 1971) and C Lubell *Textile collections of the world* (Studio Vista, 1976-), to be in three volumes. A national listing of another kind of specialized scientific institution is H T Kirby-Smith *US observatories: directory and travel guide* (New York, Van Nostrand Reinhold, 1976). A unique listing is the *Directory of British fossiliferous localities* (Palaeontographical Society, 1954), covering some two thousand sites, and planned 'to serve as a guide to the fossil-collecting opportunities in the country'.

Of course, as well as serving as non-documentary sources of information, many of the organizations listed in these directories have a major role as producers of the literature, particularly of the primary literature such as research periodicals and conference papers, and also of bibliographical sources such as abstracting services and reviews of progress. This aspect of their work will be given due attention in the appropriate later chapters.

In proportion to its numbers the directory is probably found less frequently in its pure form than any other reference book, ie as a list of names and addresses. Very commonly indeed are directories extended by the addition of, for instance, handbook-type information or encyclopedia- or dictionary-type information. *Hard's year book for the clothing industry*, for example, contains a 1,500-term glossary; the *Finishing handbook and directory* contains a lengthy alphabetically arranged 'Encyclopaedia of finishing'; the *Carpet annual* includes a dictionary of trade terms in English, French, German and Italian.

Uniquely, the *'Chemist and druggist' directory* contains a tablet and capsule identification guide. Biographical and statistical information is frequently found also, as in the *Chemical industry directory and who's who*, and the *Water services yearbook* respectively. A feature is sometimes made of special articles on particular aspects of the subject, as for instance in the *Fisheries year-book and directory*, now known as the *Fishing industry index international*. Particularly valuable for librarians are the bibliographies often included, as the list of 'Trade journals of interest to the fishing industry' in the last example, or the 21-page list of 'Books on food' in the *Food trades directory*, or the 'Comprehensive list of specialized weekly and monthly journals' in *The fruit annual*, now known as the *Fruit trades world directory*. The series of computer-produced directories published by W A Benjamin of New York (*International physics and astronomy directory, International chemistry directory*, etc) comprise not only lists of academic departments and faculties (with faculty and geographical indexes), but separate lists of laboratories, international organizations, societies, forthcoming meetings, grants and fellowships, together with a bibliography of journals, of books in print, and a directory of publishers.

Carrying as they do all this 'subject' information, such directories are clearly not merely tertiary sources (see chapter 1). It is obvious that we have here again the merging of categories noticed so frequently before. This is even more evident in such omnibus compilations as the *Trader handbook*, now known as the *Motor trader handbook*, which is subtitled 'a technical, legal and buying guide for the motor and motor cycle trades including outboard engine specifications and spares stockists'. It is probably best to regard 'directories' as a convenient term covering a heterogeneous range of very useful reference books, each to be taken on its merits, but needing particularly close examination to determine its uses. And while not forgetting the strict definition of the term, the student should not pay too much heed to the way a work is described in its title: both the *Brewery manual* and the *World radio TV handbook* are pure directories!

As the alert reader will have noticed from the two preceding paragraphs, directories are apt to change their titles, sometimes to reflect an alteration in scope, but not always. This, like the similar phenomenon with periodicals, referred to below (chapter 9) the librarian has to learn to live with.

Yearbooks

It is useful to know how to recognize a yearbook proper, for many 'yearbooks' are nothing more than directories published annually, as we have seen. The student will know from his previous studies that a

96

yearbook's concern is the events of a particular year. An obvious example is the *McGraw-Hill yearbook of science and technology*: its prime function of course is to supplement the *McGraw-Hill encyclopedia*, but it is in the form of an account of the year's developments in the field, preceded by a brief selection of feature articles on topics of 'broad interest and future significance'. *Science year: the World Book science annual* is a similar compilation, but is more directly aimed at schools. The *Britannica yearbook of science and the future*, on the other hand, devotes less than half of its 448 pages to a review of the year, the remainder comprising essays on topics of current concern such as nuclear waste disposal, plastic surgery, the Loch Ness 'monster', etc. Another common type of yearbook is the statistical: both the *Sugar year book* and the *Minerals yearbook* are mainly given over to statistics of production, consumption, imports, exports, stocks, etc. And then there are a limited number of works concerned with the forthcoming year (or years), as for example the *Yearbook of astronomy*, which predicts the phases of the moon and the planetary orbits and the like, or *Brown's nautical almanac*, which has been appearing annually for over a century, with its daily tide tables.

In the various disciplines of science and technology much of the task of digesting the developments of the year is performed by annual reviews of progress which will be considered at length in chapter 12. It should be noted, however, that there are examples of works which do try to combine the function of annual review and yearbook and directory, eg *Mining annual review, British farmer and stockbreeder yearbook and farm diary*.

And finally, there is at least one very famous 'yearbook' that is neither a yearbook nor a directory: the *Yearbook of agriculture* of the United States Department of Agriculture. In fact it is not even a reference book, strictly speaking. Each annual issue deals at length with a particular topic such as 'Climate and man' (1941), 'Insects' (1952), 'Seeds' (1961), 'That we may eat' (1975).

Chapter 7

BOOKS 'IN THE FIELD'

The five preceding chapters in this volume have been about reference books. Now of course any work, from the Bible and Shakespeare to the daily newspaper, can be used for reference, can be 'looked up' for a specific point, but the touchstone of a true reference book is that it is deliberately arranged to facilitate such consultation. The titles discussed in this chapter are not: they comprise the works on the subject, the books 'in the field'. They serve to expound or to systematize or to discuss or to reveal their subject. The forms they take most often are the treatise, the monograph, and the textbook.

Treatises
Like the encyclopedia, the treatise attempts to cover the whole of its subject field, but in most other ways it differs significantly. The encyclopedia aims to furnish a concise survey of each topic it treats, intelligible (at least initially) to the non-specialist; it tries to provide first and essential facts only and does not attempt to exhaust the subject; and it is arranged (usually alphabetically) so that its contents are easy of access. The true treatise, on the other hand, sets out to be exhaustive, aiming for a complete presentation of the subject with full documentation. It is sometimes found that a treatise synthesizes the spirit or outlook of a particular generation of scientists, often carrying the impress of a particular major thinker or school of thought. The only information likely to be excluded is elementary or introductory material. The classic work by J W Mellor *A comprehensive treatise on inorganic and theoretical chemistry* (Longmans, 1922-37) in sixteen volumes with *Supplements* (1956-) describes itself as 'A complete description of all compounds known in inorganic chemistry'. L H Hyman *The invertebrates* (McGraw-Hill, 1940-59) claims 'The intent of this treatise, then, is to furnish a reasonably complete and modern account'. J A Steers *The coastline of England and Wales* (CUP, second edition 1964) describes itself as 'a systematic approach to the whole subject . . . wider in treatment than the local monographs . . . more comprehensive than the existing works . . . This book is a physio-
98

graphical treatise'. Works of this kind are not for the beginner, although concessions are sometimes made as in G N Ramachandran *Treatise on collagen* (Academic Press, 1967-8) where 'The editors have attempted to make the treatment extensive, for those who are comparatively new to the subject, as well as intensive, so that the latest developments are discussed in particular detail for those specializing in each field'. But not all such works are so academic: describing itself as 'a comprehensive treatise' is the two-volume G Ellis *Modern practical stairbuilding and handrailing* (Batsford, 1932), 'Primarily a practical book for workshop use', and still not superseded.

The arrangement of a treatise is invariably systematic: its function is to set out its subject in all its aspects, not to serve as a source of quick reference. The five-volumed J R Partington *Advanced treatise on physical chemistry* (Longmans, 1949-54) has volumes dealing in turn with the properties of gases, of liquids, and of solids. The work by Mellor mentioned above follows the order of the periodic table. A revealing light is cast on the contrasting roles of the treatise and encyclopedia by I M Kolthoff and P J Elving *Treatise on analytical chemistry* (New York, Interscience, 1958-76; second edition 1978-), planned in three parts, each in a dozen or more volumes. Part I concerns itself with theory and practice, part II with the analytical chemistry of the elements, but when the editors came to treat the analytical chemistry of industrial materials in part III they found it was not possible to present this area with the same systematic and critical treatment attempted in parts I and II. They decided that an encyclopedic treatment would do better justice to the requirements of the practising analytical chemist: this treatment has now appeared as *Encyclopedia of industrial chemical analysis* (Interscience, 1966-74) in twenty volumes.

As has been described (chapter 1) science progresses by the accumulation and integration of facts derived from observation and experiment, and these researches are reported first in the primary literature. The main function of the treatise is to impose some sort of system on the information scattered through time and space in these primary sources. In the preface to G T Rado and H Suhl *Magnetism: a treatise on modern theory and materials* (New York, Academic Press, 1963-66) in five volumes we read that 'The need for a consolidation of almost all theoretical and experimental aspects of magnetically ordered materials is the motivation for the present work'. And the editors of M Florkin and E H Stotz *Comprehensive biochemistry* (Elsevier, 1962-), planned to be in thirty volumes but already well beyond that, write: 'Beyond the ordinary textbook the subject matter of the rapidly expanding knowledge of biochemistry is spread among innumerable journals, monographs, and series of reviews. The Editors believe that there is a real place for an advanced treatise in biochemistry which assembles the principal areas of the subject in a single set of books.'

A corollary of this function of digesting the primary literature is, of course, the full bibliographical references that are a feature of treatises. Herein lies their strength as sources of information: for the exhaustive approach to a problem (see chapter 1) they provide an excellent point of departure. Even more valuable is the treatise which is critical, that is to say which discusses the status and worth of the material presented: Mellon regards this as the highest type of treatise.

The subject scope of a treatise is broad rather than narrow: indeed it is by virtue of its wide range that it is able to perform its valuable task of synthesis and consolidation. But what constitutes 'broad' is a matter of opinion: there are treatises on what some might regard as narrow topics, eg the five-volume G C Ainsworth and A S Sussman *The fungi: an advanced treatise* (Academic Press, 1965-73), or the two-volume 'comprehensive treatise' by R Houwink and G Salomon *Adhesion and adhesives* (Elsevier, second edition 1965-7), or H Liebowitz *Fracture: an advanced treatise* (Academic Press, 1968-72), in seven volumes. On the other hand, some subjects are too extensive for such treatment: it is widely agreed that it is impossible to compile a general treatise in the field of applied chemistry that would remain reliable for long, and the practice has been to produce works devoted to a particular area within the field, eg plastics.

Indeed, it is this vast and insoluble problem of the time-lag that could spell the end for the treatise. By its very nature, it is first of all a slow and painstaking task to compile such a work, and the beginning and the end quite often span a generation or more: the 78-year-old author of the treatise on *The invertebrates* noted above announced her retirement after volume 6, and the work is to be continued by others. Secondly, by virtue of its systematic form it is very difficult to keep up-to-date. And yet scientists have always felt the need for a work that would include everything about their subject. There is, however, substantial evidence (in physics, for example) that the relative importance of the treatise is decreasing: its place is being taken by the monograph and the textbook.

Monographs
The monograph resembles the treatise in many ways, and there are a number of examples where the two categories overlap. In the General Introduction to perhaps the best known current series, the American Chemical Society monograph series, appears the explanation: '. . . in the beginning of the series, it seemed expedient to construe rather broadly the definition of a Monograph. Needs of workers had to be recognized. Consequently among the first hundred Monographs appeared

works in the form of treatise covering in some instances rather broad areas.' As an example could be instanced the four-volumed F O'Flaherty *The chemistry and technology of leather* (Chapman and Hall, 1956-65).

Traditionally the main difference between the treatise and the monograph is that in contrast to the broad subject scope of the treatise the field of the monograph (as its Greek etymological origins indicate) is a narrowly-defined single topic. This distinction is demonstrated neatly in a treatise like H F Mark *Man-made fibers: science and technology* (Interscience, 1967-8) which is in fact a *collection of monographs*, each by different authors on different topics, such as 'Acrylic fibers', 'Nylon 66', 'Fiber testing'. Indeed one way to regard a monograph is as one section of a hypothetical treatise, although this is not the complete picture. Typical monographs are A J C Andersen and P N Williams *Margarine* (Pergamon, second edition 1965), C A Thorold *Diseases of cocoa* (OUP, 1975), R P Wodehouse *Pollen grains: their structure and significance in science and medicine* (Hafner, 1959), and E M Wise *Gold: recovery, properties and applications* (Van Nostrand, 1964).

Within its limited subject field, however, the monograph strives to be comprehensive. Mellon quotes the example of Dorsey's work on *Properties of ordinary water substance* which 'devotes 673 pages to perhaps our best known chemical', although this has recently been more than matched by P V Hobbs *Ice physics* (Oxford, Clarendon Press, 1975) with 856 pages. W Francis *Coal: its formation and composition* (Arnold, second edition 1961) and G L Walls *The vertebrate eye* (Hafner, 1963) both have over eight hundred pages. And an even more striking example is J Z Young *The anatomy of the nervous system of Octopus vulgaris* (OUP, 1971), thirty years in preparation, with over seven hundred pages and over six hundred illustrations, which 'describes the nerve cells and tracts of the brain of the octopus as completely as possible with present knowledge'.

Like the treatise, the monograph attempts the same systematizing approach to its subject: the preface to Z E Jolles *Bromine and its compounds* (Benn, 1966), a volume of just under a thousand pages, states: 'One of the aims of this monograph is to bring together in accessible form and logical order, under one cover, the factual knowledge of the chemistry and technology of bromine, for the benefit of the advanced student and the research worker'. We read in D B and E F P Jelliffe *Human milk in the modern world* (Oxford University Press, 1978) that 'The purpose of this book is to review and evaluate the modern scientific information concerning a wide range of different aspects of human milk and breast-feeding, which have not been brought together before'. The preface to P Alexander and R F Hudson *Wool: its chemistry and physics* (Chapman and Hall, second edition 1963)

101

speaks of a great need for a 'comprehensive monograph ... in view of the great diversity of the sources of information which include the journals of chemistry, physics and biology, and the applied journals of the textile industry'. The preface to S Singer *The nature of ball lightning* (Plenum Press, 1971) tells us that the standard monograph on the topic is Brand's work of 1923 (in German), but goes on to assert that 'one of the goals of this work is the presentation and correlation of all the published information on ball lightning'.

Full documentation is usual with the monograph, as with the treatise: P S Lawson *Tobacco: experimental and critical studies: a comprehensive account of the world literature* (Baltimore, Williams and Wilkins, 1961), is 'compiled from more than 6,000 articles published in some 1,200 journals ... the resultant monograph is primarily directed to the research worker and specialist', and the bibliography takes up 110 of the 944 pages; J S Rinehart *Geysers and geothermal energy* (New York, Springer, 1980) 'brings together most aspects of geyser activity ... available literature has been drawn on heavily, often simply paraphrased'.

It would be a mistake to assume, however, that it is only the narrowness of the topic it treats that distinguishes the monograph from the treatise. A prime feature of the monograph is its emphasis on contemporary knowledge: although comprehensive in aim, it will not normally include the detailed historical or background material found in a treatise: W Nachtigall *Insects in flight* (McGraw-Hill, 1974), for instance, aims to bring together 'all current knowledge'; M Sivetz and N W Desrosier *Coffee technology* (Westport, Conn, Avi, 1979) is 'an attempt to summarize present knowledge of coffee in one volume'. P G Forrest *Fatigue of metals* (Pergamon, 1962), a work of over four hundred pages with a bibliography of 686 references, spells this out in his preface: '... no comprehensive British book on fatigue, as distinct from reports of conferences, has been published during the last 30 years. It therefore seemed that there was a need to provide a general account of the present knowledge.' Perhaps it was this kind of monograph that the great Dr Johnson had in mind when he declared 'there should come out such a book every thirty years, dressed in the mode of the times'. It is this emphasis on contemporaneity that partly accounts in some fields of endeavour for the increased reliance on the monograph (or rather series of monographs) at the expense of the treatise. It is obviously far easier for an editor or publisher to ensure that individual monographs are kept up-to-date than to revise a massive treatise. In rapidly developing subjects the attempt to produce even a monograph can prove abortive: in the preface to H A Liebhafsky and E J Cairns *Fuel cells and fuel batteries* (Wiley, 1968) we read that 'since we began this book five years ago, the world has spent a hundred million dollars, more or less, on research and development aimed at fuel cells and fuel batteries

'. . . Consequently we changed course: what was envisioned in 1962 as a complete account became a historical and expository guide.'

The monograph plays a further role that is not shared by the treatise: often it can be seen to stand midway between the 'repository' function of the treatise and the 'reporting' function of the periodical. It is not unknown for a monograph to contain previously unpublished material, and instances can be found where it serves to bypass the conventional primary sources of publication such as periodicals and research reports, eg much of the twenty-five years' work described in the thousand pages of E G Wever *The reptile ear: its structure and function* (Princeton University Press, 1979). This is more commonly encountered in the humanities and the social sciences where research frequently lends itself to being written up at book length, the communication of results is often less urgent and less rapidly superseded, and where monographs are usually regarded as part of the *primary* literature. More commonly, in science and technology, however, the monograph serves in a handy form to draw to the attention of practitioners the research results reported, perhaps rather obscurely or inaccessibly, in the primary literature. A good illustration of this is L Michaels and S S Chissick *Asbestos: properties, applications and hazards* (Wiley, 1979-) where we read that following the spate of public concern about the dangers of asbestos in 1976 'An attempt to find a comprehensive source book on all aspects of asbestos to provide the foundation for a literature survey met with failure. It was rapidly realized that although there was a very considerable body of literature on asbestos, this was scattered and not readily obtainable without considerable effort'. The student should not conclude from all this that a monograph is a mere compilation or digest. To begin with, the skill required to assess just what is of contemporary concern within a particular field demands an author who is an authority in the subject : the critical approach is essential in a good monograph. And, although, as has been said, the monograph deals with a single subject field, this is usually broader than the conventional specialization of the individual research worker. Furthermore, the particular role that the monograph can play in stimulating ideas, crossing disciplinary barriers, and pointing the way forward, depends upon an author who can write interesting and readable prose. J M Ziman has reminded us that 'such a task is worthy of the very best efforts of any first-class scientist, over a period of years, and if carried out satisfactorily will make him a real authority over the whole field'.

It will not have escaped attention that many monographs appear in series under a general editor or editorial board but with individual authors, and most of the major scientific and technical publishers have one or more. Examples of titles in such series are W N Christiansen and J A Hogbom *Radiotelescopes* (1969) in the 'Cambridge monographs on physics'; G V T Matthews *Bird navigation* (second edition 1968) in the 'Cambridge monographs on experimental biology'; E A Lynton *Superconductivity* (third edition 1969) in 'Methuen's monographs on physical subjects'. As with other forms of literature, the publishers are sometimes learned societies and institutions, eg the Institution of Electrical Engineers monograph series, or the 'Carus mathematical monographs' of the Mathematical Association of America. Aimed at a particular target are the three dozen or so Royal Society of Chemistry 'Monographs for teachers', 'concise and authoritative accounts of selected but well-defined topics', eg P F Leadley *An introduction to enzyme chemistry* (1980).

It could be argued that this work is not really a monograph in the full sense; the term is used loosely by many who should know better. As a description of a category of information source perhaps the word is less abused than some others, but the student will need to be on his guard against claims such as than on the jacket of R J Bray and R E Loughhead *Sunspots* (Chapman and Hall, 1964): 'the first comprehensive treatise on sunspots to be published'. The authors' preface is more modest, pointing out that the work deals with 'just one aspect of solar activity, namely sunspots', and describes itself more accurately as a 'monograph'. A more frequent occurrence is the use of the term monograph for a work which really falls short of full monograph status: the excellent little guide by F Newby *How to find out about patents* (Pergamon, 1967) refers to itself as a monograph nine times in the two-page preface, but lacks the comprehensiveness of coverage and fullness of documentation expected of a true monograph. Any British librarian soon encounters the use of the term by the British Library Lending Division apparently to mean any separately published volume larger than a pamphlet or report: it claimed for instance in April 1979 to hold 2,100,000 monographs.

The student should also be aware of the use of the term monograph in its original sense, where it was confined to biology. In this field it still is used for an account of a single species, genus, or larger group of plants, animals or minerals, eg B R Baum *Oats, wild and cultivated: a monograph of the genus Avena L. (Poaceae)* (Slough, Commonwealth Agricultural Bureaux, 1977). This work extends to nearly five hundred pages, but some monographs may be no more than a few pages in length, and many of them are never separately published: D H Kent *Index to botanical monographs* (Academic Press, 1967) lists nearly two thousand of these from nearly four hundred periodicals.

Textbooks

A textbook is a teaching instrument: its primary aim is not merely to impart information about its subject but to develop understanding of it.

If the role of the monograph is systematization, the role of the text-book is simplification. It concentrates on demonstrating principles rather than recounting details. These principles are supplemented by descriptions, explanations, examples, etc, only in sufficient number to ensure the reader's grasp of the principles. Where the monograph is used by the scientist as researcher to communicate to his fellow-scientists, the textbook is used by the scientist as teacher to communicate to his students. Since the essence of teaching is selection, a textbook may not cover the whole field of its subject, and may not even describe the latest practice, unless there is some new principle involved. P J and B Durrant *Introduction to advanced inorganic chemistry* (Longmans, second edition 1970) covers the subject only 'up to the point beyond which its study is best followed in monographs and reviews'. The material included is representative rather than comprehensive. Although each of the following three textbooks are substantial volumes, they are deliberately selective: *Imms' A general textbook of entomology* (Chapman and Hall, tenth edition 1977), despite its two volumes and extensive bibliographies, admits 'we have had to be rigorously selective ... and we apologize to those who feel our choice of references has sometimes been almost arbitrary'; F A Cotton and G Wilkinson *Advanced inorganic chemistry: a comprehensive text* (Chichester, Wiley, fourth edition 1980) is 'intended to be a *teaching* text and not a reference book', despite its 1,414 pages; W H Salmon and E N Simon *Foundry practice* (Pitman, revised edition 1957) speaks of other books on the subject as 'highly technical and cluttered up with a mass of theoretical material ... not infrequently they approach the subject less as practical guides than as academic expositions'. This selective approach is evident also in the bibliographies included in textbooks: these are most frequently in the form of suggestions for further reading rather than exhaustive lists of references. And of course many textbooks are without bibliographies – unthinkable in a monograph or treatise. A textbook may deliberately use repetition as a teaching device (as does this work), to reinforce the lesson that is being taught. Repetition in a monograph is a sign of careless writing.

A feature of textbooks is their longevity. Simply because of their concentration on principles, good textbooks will continue to sell to generation after generation of students, even in the more rapidly changing disciplines within science and technology. Of course the better examples are kept up-to-date by frequent new editions, but apart from matters of detail there is often not a great deal of difference between one edition and the next. This is understandable: the success of a textbook depends not on its worth as a source of up-to-the-minute data but on whether its method of presentation enables its users to learn about the subject. Examples still going strong that have demonstrated their worth in this way are *Holmes' principles of physical geology* (Nelson, third edition 1978) first published 1944, A W Judge *High speed diesel engines . . . an elementary textbook* (Chapman and Hall, sixth edition 1967) first published 1933, K Newton and W Steeds *The motor vehicle: a textbook* (Iliffe, ninth edition 1972) first published 1929, *Samson Wright's applied physiology* (OUP, twelfth edition 1971) first published 1926, *Fream's elements of agriculture* (Murray, fifteenth edition 1972) first published 1892, and most famous of all, *Gray's anatomy* (Churchill Livingstone, thirty-sixth edition 1980) first published 1858. As the student will have noticed they often continue to be known by the names of their original authors long after their deaths; indeed their names are commonly incorporated into the titles. A particularly interesting example is *Strasburger's textbook of botany* (Longmans, 1976), translated from the thirtieth German edition first published 1894.

On the other hand, textbooks are subject to the particular hazard of changing methods of teaching, that is to say not so much a change in content as in the selection of the content and the way it is presented. Good textbooks take account of this: H Cotton *Principles of electrical technology* (Pitman, 1967) is indeed a new book, but it is virtually the seventh edition of a similar textbook by the same author adapted in accordance with the changing approach to electrical technology, and containing more on electronics and less on machines than its predecessor. Inorganic chemistry is another subject that has seen major changes over the last few years in the way it is taught as can be seen by comparing the emphasis upon atomic structure and chemical bonds in the 1980 textbook by F A Cotton and G Wilkinson mentioned on the previous page with the more conventional approach of N V Sidgwick *The chemical elements* (OUP, 1950). Even between editions of the same text major changes are often necessary: Cotton and Wilkinson tell us 'It is remarkable how the subject of inorganic chemistry has not only grown but changed in form and emphasis since the early 1970s . . . we have made major alterations in the arrangement of our material'. Another interesting example of a book changing its character as its subject develops is O Faber and J R Kell *Heating and air-conditioning of buildings* (Architectural Press, sixth edition 1979): its preface relates that 'When this book first appeared in 1955 as a series of articles in *The architect's journal*, literature on the subject was almost non-existent . . . Now the mass of information and literature is bewildering. . . . A fresh bibliography is published every month. Thus although originally this book supplied a need otherwise unfulfilled, it must now be regarded more as in the nature of an introduction to the subject: each chapter could well form the basis of a separate book, the whole becoming an encyclopedia. But in such form it would not doubt be beyond what the average reader requires.'

A more easily perceived change in an increasing number of textbooks, particularly from United States publishers, is the emphasis on colourful and attractive physical presentation. S N Namovitz and D B Stone *Earth science: the world we live in* (Van Nostrand, second edition 1960) is a striking example of colour printing, with an illustration on almost every page. High quality illustrations are a feature of Karl von Frisch *Biology* (Harper and Row, 1964), a translation of a German school text. The most obvious indication of this bright new approach are the full colour illustrations on the front and back boards of works of this kind. Titles worthy of study include H A Guthrie *Introductory nutrition* (St Louis, Mosby, fourth edition 1979), G C Beakley and H W Leach *Engineering: an introduction to a creative profession* (Macmillan, third edition 1977), W H Wagner *Modern woodworking: tools, materials and procedures* (South Holland, Ill, Goodheart-Willcox, 1980), M D Potter and B P Corbman *Textiles: fiber to fabric* (McGraw-Hill, fourth edition 1967), R D Barnes *Invertebrate zoology* (Philadelphia, Saunders College, fourth edition 1980), H C Metcalfe *Modern chemistry* (Holt, Rinehart and Winston, 1974), H Y Carr, *Physics from the ground up* (McGraw-Hill, 1971), and J H Dubois and F W John *Plastics* (Van Nostrand Reinhold, fifth edition 1974).

In some instances this increased attention paid to book design is matched by a revolutionary approach to teaching, as for instance in the texts that have emerged from the Biological Sciences Curriculum Study in the United States and the Nuffield Science Teaching Project in Britain, eg *Biological science: molecules to man* (Boston, Houghton Mifflin, third edition 1976), and *Nuffield chemistry* (Longmans, 1966-), issued in several parts.

An even more radical change in the textbook can be studied in works like S F Dyke *Organic spectroscopy: an introduction* (Penguin, 1971). Although complete in itself, this work has been planned as an integral part of a large series of similar unit texts making up the Penguin Library of Physical Sciences: Chemistry. The editor explains: 'for some years, it has been obvious to most teachers of organic chemistry that the day is past when one author, or a small group of authors, could write a comprehensive textbook of organic chemistry suitable for the courses in this branch of the subject as taught in universities. While a number of excellent textbooks are available they must now almost of necessity be either incomplete in their coverage of the subject or massively unwieldy and correspondingly expensive. It seemed to the publishers and the editor that a useful method of coping with this difficulty was to present the material as a series of smaller texts on various aspects of the subject'.

To the unobservant the textbook and the monograph may appear similar, for both are simply books 'in the field', and neither is immediately distinguishable by alphabetical order (as is a dictionary), or by tabular arrangement (as are many handbooks) or by type of information included (as is a directory). Yet even a superficial examination of the reading habits of scientists and technologists shows that monographs and textbooks are used quite differently. It is for this simple reason that it is vital for the information worker who wishes to advise on their use that he grasp clearly the basic differences. One of these has been discussed: the selective nature of the textbook. Two others are equally significant. In the first place, it is only rarely that the author of a textbook is presenting the results of his own researches: his book is usually put together from the contributions others have made to the subject. He may not even be an authority on the topic: indeed most textbooks cover so wide an area that no man could be a real authority over the whole field. The author's role is that of teacher, scanning the whole area to select his examples, and arranging them to illustrate the underlying principles he has chosen to expound. In the words of Mellon, '. . . the author's chief function is to select, arrange, and discuss'.

The other significant difference to note about textbooks is that they are written at several levels, quite precisely calculated, although this is not always obvious without examination of the author's preface. W Morgan *Structural mechanics* (Pitman, 1980) is 'for technician and first year degree students'; A I Vogel *A textbook of quantitative inorganic analysis* (Longmans, fourth edition 1978) is 'to meet the requirements of University and College students of all grades'; and M Dixon and E C Webb *Enzymes* (Longmans, third edition 1979) is a work 'dealing with the general principles of the subject at the research level.' One unusual example, because it is actually a government publication, is the Meteorological Office *A course in elementary meteorology* (HMSO, second edition 1978), 'intended for the reader whose knowledge of physics is roughly equivalent to those of the upper science forms in school, although parts . . . are of a rather higher standard'. Some texts are even more particularly tailored for specific courses: A M Kenwood and G M Staley *Mathematics: an integrated approach* (Macmillan, 1980) is described as 'a comprehensive O level mathematics course in a single volume; A J Grove and G E Newell *Animal biology* (UTP, ninth edition 1974) is 'for the requirements of candidates for the General Certificate of Education at Advanced Level'; R J Besanko and T H Jenkins *Physical and engineering science for technicians, level I* (Oxford University Press, 1979) is for students taking the Technician Education Council certificate and diploma programme; and D C T Bennett *The complete air navigator* (Pitman, seventh edition

1967) is subtitled 'covering the syllabus for the Flight Navigator's Licence'. Such titles can sometimes be made obsolete by the frequent changes that courses undergo.

The librarian should not be too ready to dismiss the textbook as of little account compared with the monograph. J M Ziman has no doubts on this score: 'And for any physicist who regards the writing of textbooks as mere drudgery, let me remind you of the historical fact recently unearthed by Gerald Holton — that Einstein got some of the essential elements of relativity theory from reading the textbook on electromagnetism by Föppl'. Not only has the textbook a very different role in the pattern of scientific communication, but it is often more difficult to write than a monograph, which is usually produced by an expert for his fellows. It is sometimes the case that the expert is the last man to write a textbook, if all he possesses is a knowledge of the subject. Since the textbook is a work of instruction, ideally the author should be familiar with modern teaching methods, should have an understanding of the learning process, and be acquainted with the needs of students. It is little wonder that the typical textbook author is a teacher. Indeed, many textbooks have their origin in a course of lectures: the author of L M Milne-Thomson *Theoretical aerodynamics* (Macmillan, fourth edition 1966) says that the work is 'based on my lectures . . . at the Royal Naval College'; W J and N Phillips *An introduction to mineralogy for geologists* (Chichester, Wiley, 1980) was 'developed from a course for first year university students'; S Davidson *Human nutrition and dietetics* (Churchill Livingstone, seventh edition 1979) is ultimately derived from 'lectures given to medical students at Aberdeen University'. For some authors, collaboration may allow them to obtain the best of both worlds: P J Goodhew *Electron microscopy and analysis* (Wykeham, 1975) is by an expert wise enough to seek the assistance of an experienced teacher, L E Cartwright.

It is highly unlikely that teachers who thus set down the content of their courses are planning their own redundancy: such texts are obviously intended not as substitutes for attending a course but as aids to its successful pursuance. P F Kerr *Optical mineralogy* (McGraw-Hill, fourth edition 1977) expresses the hope that his text 'should be found useful by the retiring student who does not elbow his way to the front of the class but is content to sit before a microscope with a good slide on the stage and the text at his elbow'. They include detail that time does not permit the lecturer to include in class, or illustrations that can best be presented in book form, and provide the student with a permanent record of the exposition of the subject. At the very least, they can save the student's time, as is explained in the preface to R Smith and H W Heap *Sheet metal technology* (Cassell, 1964): 'As teachers of the subject the authors have found that the students have to spend valuable time taking notes. It was to avoid this and to enable the students to utilize their time more advantageously, that prompted the writing of these books'. Some are even more explicit: R R Gillies *Lecture notes on medical microbiology* (Oxford, Blackwell, second edition 1978) is one of an extensive Lecture Notes Series. Some lecture notes may not always be formally published, but are sometimes available from the universities concerned, eg 'Carleton lecture notes' from Carleton University, Ottawa. And at least one textbook has ventured to reproduce in facsimile a teacher's own handwritten lecture notes: J Holmes *Manuscript notes on weaving* ([Burnley, 1917?]). Some teachers, however, have pursued this line to its logical conclusion: M A Tribe and others *Light microscopy* (CUP, 1975) is one of a series in basic biology for first year undergraduates. It is deliberately designed to carry the main burden of teaching, replacing lectures and classes rather than supplementing them. There are also examples of textbooks (more commonly in science and technology than in the humanities) that try to provide the whole of a course within their covers: R Passmore and J S Robson *A companion to medical studies* (Oxford, Blackwell, second edition 1976-), claims that 'the student will find in it more than sufficient information to enable him to pass his examinations'.

Variations on the textbook
It so happens that medicine is one profession that it is not possible to enter by private study, but there are many disciplines where students need not attend a course, and for these the textbook has an even more vital service to give. It is probable that the majority of such students use the normal type of textbook we have been discussing, but in recent years there has been an explosion of interest in self-instruction. Not only are there many more examples of conventional texts designed for this purpose such as J E Thompson *Mathematics for self-study* (Van Nostrand, third edition 1962), in five volumes, and the scores of titles in the series 'Teach yourself books', eg L R Carter and E Huzan *Computer programming in BASIC* (Hodder and Stoughton, 1981), but we now have large numbers of programmed texts, both linear and branching. As the introduction to one particularly successful example puts it, programmed learning 'recognizes the lamentable truth that the simple desire to learn is not at all times sufficient, so that it must be fortified by attention-compelling devices, of which the challenging question is the most effective'. RCA Service Company *Fundamentals of transistors: a programmed text* (Prentice-Hall, 1966) and G L Smith and P E Davis *Medical terminology: a programmed text* (Wiley, third edition 1976) are examples of a linear programme; two branching
110

programmes in very contrasting subject areas are G Smith *Sailing: a programmed course* (Barrie and Jenkins, 1980) designed for absolute beginners and G M Berlyne *A course in renal diseases* (Oxford, Blackwell, fifth edition 1978) for postgraduate medical students.

It is possible to distinguish a number of other variants from the standard textbook. Very frequently encountered is the book of problems or questions with solutions or answers: two contrasting examples from the USSR and the USA are N M Belyayev *Problems in strength of materials* (Pergamon, 1966) and the long-established J L Hornung and A A McKenzie *Radio operating questions and answers* (McGraw-Hill, thirteenth edition 1964). I Sinclair *Q & A transistors* (Hayden, fourth edition 1980) is one of several similar Q & A, ie 'question and answer', texts from this publisher. There are many compilations of sample examination papers, eg, S S Donald *Test papers in technical drawing* (Technical Press, 1980). A textbook which presents model solutions 'as they would be required in the examination' is A N Gobby *Solutions to problems in applied mechanics* (Macdonald, 1964). Sometimes the solutions are to problems that have been set to the student in a separate textbook: P W Atkins *Solutions manual for physical chemistry* (Oxford University Press, 1979) gives detailed solutions to the 1,060 problems in his textbook *Physical chemistry* (Oxford University Press, 1978). Commonly such works are intended for the teacher's eyes only; they are but one example of a whole sub-category of works supplementary to the textbook. Another supplementary, or parallel, category is the 'revision aid', specifically designed for examination preparation. It takes the students through the same syllabus as their standard textbook, but it concentrates not so much on basic understanding of the subject (which it assumes the students already have) but those particular aspects that will need to be brought to the fore in the examination, eg G R McDuell *Revise chemistry: a complete revision for O level and CSE* (Letts, 1979).

Less common is the 'reader' which collects in one volume a number of selected extracts from other books, periodicals, etc. The preface to C S Coon *A reader in general anthropology* (Cape, 1950) explains its function: '. . . there are too many students for our library. Most of the books we would like to have the students read are out of print, and we could not buy enough copies to go around even if we had the money and the library space . . . There is only one practical solution — a reader.' The work contains twenty selections, some forty pages in length, from monographs, learned society transactions, reports of scientific expeditions, lectures, historical accounts, classic Greek authors, etc. A slightly different objective is found in works such as *Classic scientific papers: physics* (Mills and Boon, 1964) which reproduces in facsimile specifically for school and university students some twenty landmark papers reporting epoch-making discoveries, such as Rutherford's 1914 paper on 'The structure of the atom', or J J Thomson's 1897 paper on 'Cathode rays', both from the *Philosophical magazine*. As the foreword says, 'There is a certain magic in seeing an original document that has made history'. Harvard University Press has a distinguished series of a dozen such compilations, called 'Source books in the history of the sciences', eg K R Lang and O Gingerich *A source book in astronomy and astrophysics, 1900-1975* (1979); the papers are not in facsimile and some are abbreviated, but there are well over a hundred, right down to the present day. Though they are all seminal contributions they range from a popular article in *Harper's magazine* on a giant telescope to a paper on the tensor calculus of Einstein.

Perhaps the most common variation on the basic textbook is the 'how-to-do-it' book, in its various manifestations. Clearly these are works of instruction, but they aim at more than simply inculcating principles or developing understanding. With their aid the student should find himself able to *do* something, usually of a practical kind, eg *Whittaker's Dyeing with coal-tar dyestuffs* (Bailliere, sixth edition 1964). Naturally such works include less theory and more practice than conventional textbooks, but like them (and unlike monographs) they are written for students at various specified levels: E B Bennion *Breadmaking: its principles and practice* (OUP, fourth edition 1967) is for students preparing for the examinations of the City and Guilds of London Institute; the Admiralty *Naval marine engineering practice* (HMSO, 1967-9) in two volumes is a training text for engineering mechanic ratings; R Barrass *The locust: a laboratory guide* (Heinemann, 1980) is a thoroughly practical text on the insect most widely used by the young biologist in school laboratories.

Compared to the theoretical textbooks, these practical works are more frequently found directed at the actual worker at the bench or in the field rather than the student in the classroom. R W Castle *Damp walls: the causes and methods of treatment* (Technical Press, 1964), J R Mitchell *Guide to meat inspection in the tropics* (Slough, Commonwealth Agricultural Bureaux, second edition 1980), and U Langefors and B Kihlstrom *The modern technique of rock blasting* (Halsted Press, third edition 1978) are quite obviously practical texts. Much of D de Carle *Practical clock repairing* (NAG Press, 1952), regularly reprinted over thirty years, is in the imperative mood, taking the form of direct instructions to the reader. It contains four hundred drawings and the author claims he has 'described processes and operations in minute detail'; In other cases, it is made explicit in the book itself: the opening words of the preface of A Davidsohn and B M Milwidsky *Synthetic detergents* (Godwin, sixth edition 1978) are

112

'This is a work by practical men for practical men'; J D Long *Modern electric circuit design* (McGraw-Hill, 1968) is 'written for those individuals who would design circuits as an occupation'; G W Underdown *Practical fire precautions* (Gower Press, second edition 1979) is primarily 'for the manager in industry'; and in their preface J McCombe and F R Haigh *Overhead line practice* (Macdonald, third edition 1966) write of their 'hope that it will be read by foremen and linesmen, as well as by engineers and others associated with the transmission and distribution of electricity by overhead lines'. A number are in the form of compilations of tried and tested methods, 'cook-books' of science and technology in fact, eg R G Miller and B C Stace *Laboratory methods in infra-red spectroscopy* (Heyden, 1972), I T and S Harrison *Compendium of organic synthetic methods* (Wiley-Interscience, 1971), and P S Diamond and R F Denman *Laboratory techniques in chemistry and biochemistry* (Butterworths, second edition 1973). Instruction books (often published by the manufacturer) for particular models of tractors, microscopes, computers, etc, make up a similar and extensive category of texts (see chapter 18).

A number of textbooks of this kind are indeed substantial and high-grade expositions of a particular skill: for instance, R H Davis *Deep diving and submarine operations* (St Catherine Press, seventh edition 1962) is an amazing compilation of underwater lore. But the two-volumed J Stone and A J Phillips *Contact lenses* (Butterworths, second edition 1980) 'for practitioners and students' and C Rob and R Smith *Operative surgery* (Butterworths, third edition 1976-), to be in fourteen volumes, are instances where perhaps the description 'textbook' is inadequate. The word 'practice' commonly appears in titles of works in this category, eg V Serry *British sawmilling practice* (Benn, 1963), or E Davies *Traffic engineering practice* (Spon, second edition 1968), but again such works are often more than mere textbooks, as is made plain by the title of W O Skeat *Manual of British water engineering practice* (Cambridge, Heffer, fourth edition 1969). In fact these are examples of a familiar indeterminate type of work in the field of technology, part textbook (for study) part handbook (for reference), and even part monograph: C J Polson and T K Marshall *The disposal of the dead* (English Universities Press, third edition 1975), for instance, is 'intended to provide a reasonably comprehensive account of this somewhat complex process'. Two other well established titles are J H Thornton *Textbook of footwear manufacture* (Heywood, third edition 1964), which the foreword claims as 'the first book to cover the whole range of footwear manufacture', and H Thornton and J F Gracey *Textbook of meat hygiene* (Bailliere, sixth edition 1974), 'covering the whole field of meat hygiene in a simple yet comprehensive manner'.

Of course the field of 'how-to-do-it' books is a happy hunting ground

113

for the amateur enthusiast and the 'do-it-yourself' devotee, and thousands of titles are produced specifically for the non-professional car mechanic, TV repairman, carpenter, and so on, eg J Gooders *How to watch birds* (Deutsch, 1975), J A C Harrison *The DIY guide to natural stonework* (Newton Abbot, David and Charles, 1979), P H Smith *Tuning for speed and tuning for economy* (Foulis, fourth edition 1970), and G N Smith *Tobacco culture: a DIY guide* (Liss, Spur, 1977). There are at least two bibliographies devoted to works in this category: *How-to-do-it books: a selected guide* (New York, Bowker, third edition 1963) lists over 4,000 relating primarily to spare-time activities, and F S Smith *Know-how books* (Thames & Hudson, 1956) likewise concentrates on practical texts for the general reader.

For the librarian textbooks raise a number of special problems of which the student should be aware. Most imperative is the frequently expressed view that textbooks have no place in a library. As works of instruction, designed for intensive study, they should be in students' hands as their personal possessions. The normal limited loan period of most libraries (imposed of course with the aim of ensuring fair shares) is inappropriate for a book which a student may need by him throughout a course. And as a source of information, the average textbook supplies nothing that is not available elsewhere, usually in more convenient form. A further difficulty for the librarian is the vast number of similar publications, particularly in those subjects which attract large numbers of students. Very often one of these works differs from the next only insofar as it represents another teacher's view as to how the topic should be expounded. The reverse of the coin is the scarcity of textbooks for the many emerging subjects in science and technology that require to be taught to the coming generation. Perhaps even more serious are the criticisms of unreliability in matters of fact, or misleading oversimplification, or lack of awareness of modern developments, that are levelled at textbooks, particularly school textbooks, from time to time in the scientific press. And yet, as the student will also be aware, textbooks are found in large numbers in many kinds of scientific and technical library.

Introductions and outlines
Although as categories these are less precisely defined, they are often classed with textbooks and are therefore briefly considered here. An introduction is clearly a *first* book in a subject, designed to lay the groundwork for its user, and leading on to a more advanced or detailed or particular study, eg H R Broadbent *An introduction to railway braking* (Chapman and Hall, 1969). They are not necessarily for 'students': the two-volumed P W Abeles *An introduction to prestressed*
114

concrete (Concrete Publications, 1964-6) is for 'civil engineers, architects, and contractors'. A number of them are planned, in the words of the foreword to B E Waye *Introduction to technical ceramics* (Maclaren, 1967), so as to be 'suitable for students and others', eg M S Ghausi and J J Kelly *Introduction to distributed-parameter networks with application to integrated circuits* (Holt, Rinehart and Winston, 1968), which is 'for seniors and first-year graduate students as well as for self-teaching by practicing [sic] research engineers'. And of course a large number of 'introductions' are simply textbooks, eg P Grosberg *An introduction to textile mechanisms* (Benn, 1968), 'aimed in general at the ordinary and honours degree student'; K J Astrom *Introduction to stochastic control theory* (Academic Press, 1970), 'presented as a graduate course . . . at the Lund Institute of Technology during the 1968-1969 academic year'; and P E Gray *Introduction to electronics* (Wiley, 1967), 'intended to be used in introductory or first-level courses'. Not all textbooks are introductions, on the other hand: a number of them are exceedingly advanced, eg V I Smirnov *A course of higher mathematics* (Oxford, Pergamon, 1964), used in the USSR for 45 years, and filling six volumes in the English translation.

An outline covers the *whole* of its particular subject, but not in detail. Only the salient features are emphasized. Its aim is not so much to develop understanding (as the textbook), but to map out an area. Whereas a textbook (or an introduction) is designed for continuous study, and arranged on an assumption that it will be worked through in sequence, an outline can also be used quite easily for reference. The preface to E N Simons *An outline of metallurgy* (Muller, 1968) reflects this approach precisely when it speaks of the need for a book that 'without going into excessive detail would give a general view of the subject, making it unnecessary to read an entire shelf-full of textbooks in order to grasp what is involved'. It is of more than minor importance for the librarian to be able to distinguish categories such as these. It can be seen that outlines (and to a less extent introductions) are particularly useful for the enquirer seeking to orientate himself in a field relatively new to him — the background approach, in fact, as described in chapter 1.

This chapter, and the previous chapters, have been in the main concerned with books. Now it is often said, somewhat scornfully, that in science and technology books contain only second-hand information; typically, too, they are two years out of date on publication. T S Kuhn has reminded us that 'The scientist who writes one is more likely to find his professional reputation impaired than enhanced'. This, of course, is true. All the types of literature we have so far considered have been either secondary or tertiary sources: we have not yet studied the literature of science and technology at its heart,

115

at its fountainhead. But the student should beware of underestimating the role of books. Despite the world recession of the past decade the number of books published in the United Kingdom has shown a substantial and sustained increase; in particular, scientific, technical and medical books have multiplied at a considerably higher rate than the rest, and this growth is continuing into the 1980s. In the United States D W King found in his study for the National Science Foundation that the form of scientific and technical literature showing the greatest percentage rise in the fifteen years to 1974 was in fact books, with more than treble the number published each year.

The chemist is probably as fully aware of the primary literature as any man but in his recent guide to the literature R E Maizell has reminded us that 'The chemist, however, cannot afford to ignore books. Books are usually the best starting point for searching the chemical literature. They can often provide quick, straightforward answers that would otherwise take many hours to locate in widely scattered journal articles and patents on the subject'. A generation earlier in the classic guide by E J Crane (see chapter 2) we read: 'Chemical books have most important uses. They introduce the novice to the general field of the science or of some part of it, explain new theories in the light of already known facts, and help to coordinate and systematize knowledge. They furnish information, exhaustive or not, in a form adapted for quick reference, and guide the searcher back to the original source by means of citations. Historical works record the development of the science, popular books initiate the public into its mysteries and elicit interest and support, and treatises on the various fields of chemistry give the reader the benefit of the long experience or combined researches of many workers. Who shall say that the chemist can depend on journals alone? The mere fact that nearly two thousand new books of chemical interest are published annually proves the demand for them.'

Chapter 8

BIBLIOGRAPHIES

Before we leave for the time being our study of the literature of science and technology in book form, we should pause to examine those sources of information that are consulted to ascertain what books are indeed available on a subject — the bibliographies, in fact. Of course such lists of books are only one form of bibliography: as will become evident in succeeding chapters, all the various non-book materials such as periodicals, computerized data bases, patents, theses, etc, have their own bibliographies (or indexes as they are often termed). Very common also is the mixed bibliography, listing books and non-book materials together irrespective of form.

Bibliographies of course are tertiary sources (see chapter 1), and as such are particularly the librarian's province. Indeed, as has been mentioned, many of them are actually compiled by librarians, and it is from librarians that they obtain much (perhaps most) of their use. Traditionally, they are divided into current and retrospective, but for the student this can lead to confusion, for what is current today can be retrospective tomorrow: there are many examples of current bibliographical listings which do later serve their turn as retrospective bibliographical repositories, particularly when cumulated, eg the monthly *ASLIB book list* forms the basis for *British scientific and technical books, 1935-52* (1956) and its supplements.

A more illuminating line for the student to pursue is to try to match the bibliographies he is studying to the various approaches made to them as sources of information on what books exist. Whether the users are scientists, technologists, or librarians and information officers, they demonstrate in their use of bibliographies precisely those 'approaches' already observed in the general literature, ie the Current, the Everyday (and Background), and the Exhaustive (see chapter 1).

The current approach
If the available lists are any indication, keeping abreast of current books in science and technology is very much the librarian's preserve: the two longest-established are both primarily annotated book selection

117

tools, compiled by and for librarians and information officers, namely the *ASLIB book list* (1935-) and *New technical books* (1915-). The latter is a production of the New York Public Library, which reminds us that the accessions lists of appropriate libraries are widely used for the same purpose, allowing all to reap the advantage of the selection skills of others, eg *The recorder*, a 'selected list of recently acquired publications' issued twice a month by Columbia University Science and Engineering Information Center Libraries, the University of London Library *Accessions list: Section VI – Science medicine, agriculture, engineering*. A particularly useful example worth singling out is *Lewis's quarterly list*, which is a little unusual in that it lists 'new books and new editions on sale, some of which are available from the medical, scientific and technical, lending library,' which is of course a commercial library. Not all accession lists are separately published: the excellent annotated list of additions to the library of the Institution of Mechanical Engineers used to appear in the monthly *Chartered mechanical engineer*. Similarly a number of booksellers specializing in science and technology issue regular lists of recently published books in their field; some send out their announcements in the form of cards. *British science books*, quarterly from the British Council, is designed specifically as a selection tool for overseas, as is its companion monthly *British medicine*. The computer too is making its contribution: *New books in chemistry* is a twice-monthly bulletin from the Chemical Abstracts Service, one of over a hundred such alerting journals in the *CA selects* series derived from the CAS data base.

A number of periodicals too make a feature of lists of new books: perhaps the most widely used is the occasional 'Recent scientific and technical books' section in *Nature*. Many of these regular lists in periodicals are there simply to acknowledge copies of books sent for review by publishers, and surveys have shown that the book review columns in the scientific and technical press are one of the most important sources for the practising scientist of current information on new books, despite the fact that reviews commonly do not appear until several months after publication of the books. Eventually the majority of scientific and medical books do get reviewed at least once, though this is less true of technical books, particularly those of a practical or industrial character. Of course, many periodicals carry such reviews, eg *Textile month, Bioscience, Chemistry in Britain, Science,* and thousands of others, but in some journals they extend to twenty, thirty, or even more pages in each issue, eg *Physics today, School science review*. Indeed, in recent issues of the *Quarterly review of biology*, for instance, the 'New biological books' department has reached seventy pages, or over half the journal. And since 1965, in *Science books: a quarterly review*, now called *Science books and films*, published by the American

118

Association for the Advancement of Science, we have a journal consisting entirely of reviews. *Appraisal: children's science books* is another commendable American venture, which has appeared three times a year since 1967, reviewing some fifty books in each issue; specialization by subject field rather than by intended readership is seen in *Natural history book reviews*, also three times a year from 1976. Since 1980 we have had the monthly *Index to book reviews in the sciences* from the Institute for Scientific Information, citing some 35,000 reviews a year under the author or editor of the original work. Its long established predecessor in the field is the monthly *Technical book review index* (compiled by librarians), which for some 3,000 books a year gives extracts from published reviews, and, as its title states, serves as an index to the book reviews which have appeared in over 500 journals. While not exactly furnishing book reviews in the accepted sense, many abstracting services do include announcements of new books, often with indicative abstracts of their contents, eg *Chemical abstracts, Hosiery abstracts.* Others are more selective: INSPEC, for instance, generally excludes all textbooks and those treating the subject at below graduate level.

A remarkable but unfortunately short-lived project in recent years was *Sci-tech book profiles* (and its companion *Medical book profiles*), which 'profiled' some 150 books each month by reproducing from each in reduced facsimile anything up to twenty of the more significant pages, eg title page, table of contents, preface, list of contributors, index, etc.

It is also possible for the alert librarian to keep himself informed about books not yet published. Some booksellers offer a pre-publication selective dissemination of information service (see chapter 11) in subject areas specified by customers. Apart from ensuring he is on the mailing lists of the major scientific and technical publishers, he can also consult the scientific and technical headings in *Subject guide to forthcoming books* which appears six times a year.

The everyday (and background) approach
Whether by the practitioner or by the librarian, this approach takes the form of a search for a suitable book on a particular subject of interest. The emphasis here is on appropriateness to the task in hand, as the need for information usually arises in the course of daily work. If the requirement is put into words (as it has to be if the librarian is asked to help), it is usually for 'a good book on the subject', or 'the standard work', or even 'the best source'. The bibliographies of most immediate use here are the selective, annotated, evaluative lists (as opposed to the comprehensive, more purely 'bibliographical' lists used for the exhaustive

approach, described below). A classic example is the work known universally as 'Hawkins' after its first editor: *Scientific, medical, and technical books published in the United States of America: a selected list of titles in print with annotations* (Washington, National Research Council, second edition 1958). This stands in the same sort of relationship to the current listing *New technical books* as does its British counterpart *British scientific and technical books, 1935-52: a select list of recommended books* (ASLIB, 1956) and its supplements to the *ASLIB book list*. Though technically 'selective' rather than comprehensive, these are both substantial bibliographies with 8,000 and 11,000 titles respectively, and find their main use as tools for book selection and readers' advisory work in libraries (despite their need for updating). Serving a similar function is the very different *McGraw-Hill basic bibliography of science and technology* (1966), issued as a supplement to the *McGraw-Hill encyclopedia*, but subtitled 'recent titles on more than 7,000 subjects'.

A clearly discernible category within this class of bibliographies of basic books is the compilation specifically designed to suggest a representative list of titles for library purchase. Examples worth studying are E B Lunsford and T J Kopkin *A basic collection for scientific and technical libraries* (New York, Special Libraries Association, 1975) with 2,400 annotated entries arranged by subject; ASLIB *Select list of British scientific and technical books* (sixth edition 1966), about 1,300 items, without annotations, and limited to books in print; Association for Science Education *Science books for a school library* (Murray, fifth edition 1968), 650 titles; *British scientific and technical reference books: a select and annotated list* (British Council, new edition 1976), 250 entries chosen 'with the needs of overseas librarians particularly in mind'. Similar listings are available covering specific subject fields, eg American Dental Association *Basic dental reference works* (Chicago, 1975) listing 104 titles; Mathematical Association of America *A basic library list for four-year colleges* (Washington, second edition 1976) which includes about seven hundred books and journals. Lists of this kind may also appear as articles in journals, eg A N Brandon 'Selected list of books and journals for the small medical library' *Bulletin of the Medical Library Association* 63 1975 149-72. Aimed also at purchasers, though not necessarily libraries, is of course the regular British trade bibliography *Technical and scientific books in print* (Whitaker) and its American equivalent *Scientific and technical books and serials in print* (Bowker). Specifically to supplement the trade bibliographies is J M Kyed and J M Matarazzo *Scientific, engineering and medical societies publications in print, 1978-1979*, (Bowker, 1979) covering 365 US, Canadian and British organizations.

Yet another distinguishable type of selective list is the 'reader's guide'. Two established titles in this category are the National Book League *Science for all: an annotated reading list for the non-specialist* (revised edition, 1964), with its five supplements, 'prepared in consultation with the British Association for the Advancement of Science', and its precise transatlantic counterpart, the American Association for the Advancement of Science *Science book list* (Washington, third edition 1970) and *Supplement* (1978), an annotated list for the student and non-specialist reader.

At the heart of both the current and the everyday approach to bibliographies lies the problem of choice, of selecting from the mass of books listed those titles not only appropriate in subject matter but also in level and treatment. Perhaps most bibliographies of this kind that librarians use do give sufficient indication of subject matter to assist in making sensible choices; very few suggest level or treatment — the *ASLIB booklist* is a rare exception. This is why the recent proposals of the International Federation of Library Associations for an International Target Audience Code (ITAC) have aroused such interest. Prompted by the ineffectiveness of machine-readable bibliographic records as indicators of the audience at which a work is directed, this code for distinguishing intellectual level and function has now been developed under a research contract within the UNISIST programme.

The exhaustive approach

As pointed out in chapter 1, 'exhaustive' in this context is a relative term, for even where total coverage of the literature is the aim, a search is usually curtailed at the point where the law of diminishing returns begins to operate. Nevertheless, this is the area of activity where bibliographies really come into their own: to locate a bibliography of the subject is often the vital first step in the pursuance of a research project. Indeed, George Sarton, probably the best-known bibliographer of the history of science, and himself by training a mathematician, has stated that 'Each investigation must begin with a bibliography and end with a better bibliography'. Although the number and range of bibliographies already published frequently astound even the librarian, it is often necessary, however, for the investigator to compile his own if he finds himself the first in the field.

Such a worker, systematically seeking to trace all the books published in his subject would probably look first for comprehensive retrospective bibliographies covering the whole field of science and technology. He would soon discover that such lists do not exist (for English titles at any rate) as separate publications. *American scientific books* (New York, Bowker) used to be 'a complete record of scientific and technical books published in the United States of America', and in four volumes covering 1960 to 1965 cumulated some 4,000 titles per year from the monthly *American book publishing record*. Since then, however, these scientific and technical titles are incorporated into the general annual, five-yearly and larger listings *BPR cumulative* in a fashion similar to the *British national bibliography*. As the student will be aware, subject access is simplified to the relevant titles in both these general works (and their cumulations) by their arrangement in Dewey Decimal Classification order. For the last half-a-dozen years, however, we have had a comprehensive annual listing of technology books, utilizing computerized data retrieval and typesetting techniques: *Bibliographic guide to technology* (Boston, Hall) includes all the publications catalogued by the New York Public Library with additional entries from the Library of Congress MARC tapes.

Library catalogues
The fact that this last title serves as a supplement to the *Classed subject catalog of the Engineering Societies Library* (see below) reminds us in point of fact that the most comprehensive broad-based retrospective bibliographies of science and technology are the published catalogues of the great scientific and technical libraries, eg British Museum (Natural History) Library *Catalogue of the books . . .* (1903-40) in eight volumes; Wellcome Historical Medical Library *A catalogue of printed books* (1962-), described as 'perhaps the greatest collection of medico-historical works in the world'; and Harvard University *Catalogue of the Library of the Museum of Comparative Zoology* (Boston, Hall, 1967) with 153,000 entries in eight volumes and *Supplement* (1976). Particularly worthy of study on account of its form is this last example: unlike the others, which are conventionally printed, it is made up of facsimiles of the actual cards from the catalogue in the library, reproduced by photo-litho-offset, several to the page.

Printed library catalogues such as these have a particular value as bibliographies, for not only do they provide information about the existence of a book and its date, publisher, format, etc, but by way of a bonus they locate an actual copy. A grave disadvantage with many works of this kind, however, is the lack of subject access: the three examples quoted, for instance, are all arranged alphabetically by author, with no subject index. All the other general bibliographies mentioned in this chapter, both current and retrospective, are arranged by subject (or have a subject index): indeed the various 'approaches' so far described demand such access. But of course to produce a subject catalogue calls for much more intellectual effort than an author list, and until recently good examples in the science and technology field were few: the twin Science Museum Library short-title lists *Books on engineering: a subject catalogue* (HMSO, 1957) and *Books on the chemical and allied industries: a subject catalogue* (HMSO, 1961) contain together about 10,000 books arranged by the Universal Decimal Classification; a particularly handy list is the *Catalogue of Lewis's medical scientific and technical lending library* (1975) and its *Supplement* (1976), of almost 50,000 titles, with an author/title sequence in one volume and the subject index in the other.

What has revolutionzed the situation in the last few years has been the extensive use of the photo-litho-offset method described above to reproduce in book form the card catalogues of several dozen of the world's major libraries, many of which are subject lists of one form or another. The largest in the field of science and technology, with well over a million books and pamphlets, is the John Crerar Library (Chicago) *Catalog* (Boston, Hall, 1967) in 77 volumes, 35 of which make up the *Author-title catalogue,* and 42 the *Classified subject catalog* (with subject index). A contrasting arrangement is seen in the *Dictionary catalog of the National Agricultural Library, 1962-1965* (Boston, Hall, 1968-70) in 73 volumes, with a 12-volume supplement covering 1966-70), though it is interesting to note that a classified arrangement has now been adopted here also. In the *Classed subject catalog of the Engineering Societies Library* (Boston, Hall, 1963-), in 13 volumes and annual supplements, it is the author approach that is denied. Recently an American university reference librarian, Bonnie R Nelson, complained that works of this kind 'have too often been ignored in reference work. Examination of dozens of printed library catalogs reveals that many of them contain the depth of cataloging, the inventive and generous use of subject headings, and the analytic entries that reference librarians often ask for but seldom get in their own libraries' card catalogs. They can thus be used to find material in one's own library that cannot be identified through the card catalog'.

The continuing advance of technology has now enabled library catalogues to take a further step towards wider availability with the advent of computer-output-microfilm (see chapter 21). The Science Reference Library offers its author/title and its classified catalogues from 1975 for sale on COM-fiche, reissued every quarter completely updated.

The catalogues of smaller specialized libraries are also sources it is unwise to ignore in an exhaustive search, particularly if they cover subject fields which lack a conventional bibliography, eg J Bromley *The clockmakers' library: the catalogue of the books and manuscripts in the library of the Worshipful Company of Clockmakers, London* (Sotheby, 1977), which lists 1151 items.

As has already been pointed out, this chapter concerns itself with bibliographies of *books*; yet in an exhaustive search, by definition,

123

books make up only one part. And it is an observable fact that many published bibliographies are 'mixed', inasmuch as they combine the features of a bibliography (ie of books) and an index (ie of periodical articles, research reports, patents, etc). They are themselves, as often as not, the published results of an exhaustive search of the literature. Each of the following, for instance, covers pamphlets, articles, and other printed material as well as books: G S Duncan *Bibliography of glass (from the earliest records to 1940)* (Dawsons, 1960) has over 15,000 references; H S Brown *A bibliography of meteorites* (Chicago UP, 1953) goes back as far as 1491 for its first entry; and A M C Thompson *A bibliography of nursing literature, 1859-1960* (Library Association, 1968) and *1961-70* (1974) claims to be a contribution to history rather than a working tool. On occasion, however, special circumstances may dictate the exclusion of periodical articles for instance, even in a comprehensive bibliography: E A Baker and D J Foskett *Bibliography of food* (Butterworths, 1958) does so because 'They are easily traced by searching the abstract journals'; H C Bolton *Select bibliography of chemistry, 1492-[1902]* (Washington, Smithsonian Institution), with 18,000 titles included, would obviously be too vast an undertaking if not confined to separately published works.

Bibliographies of bibliographies
It is easy to understand why this kind of specific subject bibliography is so highly regarded by the research worker obliged to carry out an exhaustive literature search during the course of a project. His particular problem is to find out whether such a search in his particular field has been carried out before; or, in practical terms, whether there is a bibliography. Here the librarian is particularly well placed to help, with his acquaintance with sources of information, and in particular with those intriguing tools, the bibliographies of bibliographies. He will of course be thoroughly familiar with the general tools such as Besterman, but he should also know the various subject editions that have been derived from it by collecting the appropriate subject entries from the main work, eg *Physical sciences: a bibliography of bibliographies* (Totowa, NJ, Rowman and Littlefield, 1971) in two volumes, and the similar works covering *Agriculture*, and *Technology, including patents*. Apart from these there are a limited number of such titles in science and technology also, eg C J West and D D Berolzheimer *Bibliography of bibliographies on chemistry and allied subjects, 1900-1924* (Washington, National Research Council, 1925) and *Supplements, 1925-31* (1929-32); L H Swift *Botanical bibliographies: a guide to bibliographic materials applicable to botany* (Minneapolis, Burgess, 1970); R Lauche *World bibliography of agricultural bibliographies* (Munich, BLV, 1964). He
124

will also be aware of the difficulties of locating the so-called 'hidden' bibliographies, ie those that appear as articles in journals, or as part of a monograph, or are otherwise not separately published. He will probably know that the Science Reference Library maintains a subject index on cards of bibliographies, including hidden bibliographies, and Walford in his science and technology volume makes a particular point of listing hidden bibliographies. An even greater problem is bibliographies that are not published at all: a large number of these exist in universities, research institutes, libraries, etc, typically in the form of a file of references on cards or slips; an example is the card handlist of British manuscripts in medieval science compiled by D W Singer which is now in the manuscripts department of the British Museum and is available on microfilm at the Library of Congress, Cornell University, and elsewhere. As often as not they are freely available for consultation, and would indeed be published, if the time and effort could be spared. Instances of files of this kind that have been published are the Boyce Thompson Institute for Plant Research, Yonkers, NY, collection of 30,000 references built up over 20 years which appeared as L V Barton *Bibliography of seeds* (Columbia UP, 1961); and the Plutonium Kartei at the Karlsruhe Nuclear Research Centre, a comprehensive bibliography on 11,000 cards on the element plutonium, which from January 1967 was converted into a monthly abstract journal *Plutonium-Dokumentation*. And like library catalogues, such files can now be comparatively effortlessly reproduced by photo-litho-offset, eg the Gray Herbarium Index to new plants since 1886 on 265,000 cards at Harvard University published in ten volumes (Boston, Hall, 1968).

It will be evident to the student reading this that he has stumbled into an area where bibliographical control is deficient. It may surprise him to learn that there is no fool-proof way of tracing a bibliography on a particular subject without actually carrying out an exhaustive search. Even the most knowledgeable librarian cannot guarantee that the relevant bibliographies will be encountered at the beginning rather than during the course of a literature search. There is some hope here that can be glimpsed in the greatly increased manipulative power that computers can offer: it is comparatively simple at the input stage to ensure that citations consisting of or containing bibliographies are 'tagged' as such. Then at the retrieval stage an appropriate search strategy will see to it that such references emerge first. On the other hand, mechanized systems (like MEDLARS and *Chemical abstracts*) are adding to the problem of control with the facilities they offer for on-demand, custom-produced bibliographies. An indication of how the computer can help is provided by the *Bibliography of agricultural bibliographies: a categorized listing of bibliographies indexed in AGRICOLA* (Beltsville, Md, US Department of Agriculture, 1977-).

125

That there is a need for coordination of bibliographical effort is evidenced by flagrant instances of duplication, eg J H Batchelor *Operations research: an annotated bibliography* (St Louis UP, second edition 1959), and Case Institute of Technology *A comprehensive bibliography on operations research* (New York, Wiley, 1958), both supplemented by further volumes; and V E Coslett *Bibliography of electron microscopy* (Longmans, 1951) covering 1927-48, and C Marton *Bibliography of electron microscopy* (Washington, USGPO, 1950) covering 1926-49. The compilation of bibliographies is an area eminently suited to co-operative endeavour: by its nature it lends itself to precise divisions of labour and of responsibility and even the smallest contributions can play their significant part in a total scheme. The most obvious manifestation of such cooperation is the union list, eg American Chemical Society Rubber Division Library *Union list of books relating to the fields of rubber, resins, plastics and textiles held by* [eight] *technical libraries . . .* (Akron, Ohio, 1962), and Cement and Concrete Association *Bibliography of cement and concrete: list of books and papers in London libraries* (third edition 1952). Other instances are: K M Clayton *A bibliography of geomorphology* (Philip, 1964) which covers books and articles 1945-62, and is deliberately designed to dovetail with *Geomorphological abstracts*, which commenced in 1963; W H Mullens *A geographical bibliography of British ornithology, from the earliest times to the end of 1918* (Wheldon and Wesley, 1919-20), the arrangement of which under counties neatly complements W H Mullens and H K Swann *A bibliography of British ornithology from the earliest times to the end of 1912* (Macmillan, 1916-17) and *Supplement* (1923), which is arranged alphabetically by author; and the US National Advisory Committee for Aeronautics *Bibliography of aeronautics, 1909-1932* (Washington, USGPO, 1921-36), which in fourteen volumes and following the same basic plan serves as a continuation to the remarkable P Brockett *Bibliography of aeronautics* (Washington, Smithsonian Institution, 1910), which revealed that by July 1909 there were no fewer than 13,500 titles extant on this comparatively novel subject.

As has been mentioned, the range of subjects that do have their own bibliographies is quite extraordinary. Even more impressive in many cases is the extent of the published literature on quite specialized topics that these bibliographies unfold: L M Cross *A bibliography of electronic music* (Toronto UP, 1967) lists 1,562 writings in five languages; the modestly titled *A partial bibliography of oysters, with annotations* (St Petersburg, Fla, Florida Department of Natural Resources Marine Research Laboratory, 1972) nevertheless extends to 846 pages; D T Hawkins *Physical and chemical properties of water: a bibliography, 1959-1974* (New York, IFI/Plenum, 1976) with its 556 pages suggests

126

that 'our best known chemical' has other attributes beyond being colourless, odourless and tasteless; W F Clapp and R Kenk *Marine borers: an annotated bibliography* (Washington, Office of Naval Research 1963) takes 1,148 pages to list the world's literature on shipworms; the National Institute of Neurological Diseases and Stroke *Parkinson's disease and related disorders: cumulative bibliography, 1800-1970* (Bethesda, Md, 1971) has over two thousand pages in its three volumes. Recently, within a few years of each other, we have seen both C Necker *Four centuries of cat books: a bibliography* (Metuchen, NJ, Scarecrow, 1972) with 511 pages and E Berman and C G Liddle *Bibliography of the cat* (Research Triangle Park, NC, US Environmental Protection Agency, revised edition 1976) with 639 pages.

As the student will be aware, the art of compiling and arranging bibliographies, and a consideration of the variety of form, content and coverage of published examples is a study in itself, beyond the scope of this chapter. Nevertheless, there still remain a number of distinct categories of bibliography commonly encountered in science and technology that are deserving of particular attention. One of the most interesting is the 'narrative' bibliography, where the citations are dispersed through a continuous explanatory text. The best of these can provide a useful introduction to the subject as well as a list of books, but the added comment sometimes brings them close to the guides to the literature discussed in chapter 2, eg J L Thornton and R I J Tully *Scientific books, libraries and collectors* (Library Association, third edition 1971) and *Supplement* (1976). Even more than the unadorned enumerative bibliographies, such works benefit greatly where the compiler is an authority on his subject as well as a bibliographer: the result can often turn out to be a 'tour de force', as in the case of E T Bryant *Railways: a readers guide* (Bingley, 1968), and D M Knight *Natural science books in English, 1600-1900* (Batsford, 1972).

A possibly unique example of another type of bibliography is G W Black *American science and technology: a bicentennial bibliography* (Southern Illinois University Press, 1979), a compilation of over a thousand references to the journal literature of the bicentennial year of 1976, culled from five H W Wilson indexes and ranging over the whole scientific and technical field.

Library bibliographies
A fruitful source of information are those bibliographies (usually of highly topical and very specific subjects) which are issued by libraries as items in a series. Perhaps the best known is the Science Museum Library Bibliographical Series which has been running since 1931 and now has

some eight hundred titles to its credit: examples of topics covered are semiconductors, manufacture of glass mirrors, papermaking. Since most of such lists are compiled originally in response to a request, it is obvious that they hold considerable current interest. They are often available from the issuing library merely for the asking, they are almost invariably based on extensive subject collections, and frequently they are compiled by subject experts. Well established series are the (US) National Agricultural Library Lists (no 95, published in 1969, is *Sunflower: a literature survey, January 1960-June 1967*, with over 1,800 references); the Iron and Steel Institute (now Metals Society) Bibliographical Series (no 24 (1965) comprises 600 references on *Vacuum metallurgy of steel*). Newer series are the Science Reference Library Guidelines (*Bloodstock, equine breeds and types* was published in 1979; *Electric vehicles* in 1980), and the Science Tracer Bullets of the Library of Congress Science and Technology Division, each of which, (eg *Acupuncture* (1972), *Lead poisoning* (1977), *Microcomputers* (1977)), contains from two to a dozen pages. The first 51 from 1972 to 1975 have been reissued in a 320-page volume by Science Associates/ International of New York. An interesting example of initiative is the British Library series of short reading lists on topical subjects known as BLAISE ALERT. These are made available not in printed format but in videotext *via* PRESTEL, the British Telecom system providing access to computerized information over a telephone line to the domestic television set.

Perhaps the most sophisticated examples of such bibliographies are the 'Library Pathfinders' produced by the Massachusetts Institute of Technology, defined as 'a checklist of references to those basic sources representing the variety of forms in which information on a specific topic can be found ... A Pathfinder enables the user to follow an organized search path ... Pathfinders are not exhaustive subject guides or bibliographies ... Pathfinders are aids for the first three to five hours of literature searching'. Not the least interesting feature of this scheme as it developed was its cooperative nature: compilers were eventually drawn from a range of other libraries (and from library schools) and publication and distribution of what finally amounted to over four hundred titles was undertaken by Addision-Wesley, a commercial publishing house. Examples of topics covered are tunnelling, offshore structures, nylon, Boolean algebra.

Bibliographical series produced merely by searching back through particular abstracts or indexes are also found: *Noise control in the pulp and paper industry* (1975) is number 264 in the Institute of Paper Chemistry Series and is simply the product of a fifteen-year back search of the Institute's *Abstract bulletin*. In many ways similar are the bibliographies resulting from a computer search of a machine-readable
128

data base. These may also be issued in series, eg the American Society of Metals bibliographies derived from the METADEX data base, such as *Metal scrap, Electroplating, Hot rolling*, and fifty others.

Exhibition catalogues
Similar in some ways to the library catalogue insofar as it is not primarily compiled as a bibliography is the exhibition catalogue. From time to time special displays of books are gathered together for a limited time, sometimes to mark a particular occasion such as a centenary, and then they are dispersed, perhaps for ever. If the exhibition is really important, its catalogue may have permanent value as a subject bibliography, eg Science Museum *A hundred alchemical books* (HMSO, 1952); H D Horblitt *One hundred books famous in science: based on an exhibition held at the Grolier Club* (New York, 1964); National Book League *Do-it-yourself: a touring exhibition* (1962). Catalogues of private collections too can serve as valuable additions to the bibliography in their field, where the collection is sufficiently worthy, eg *Catalogue of botanical books in the collection of Rachel McMasters Miller Hunt* (Pittsburgh, 1958-61), two volumes in three; *Bibliotheca alchemica and chemica: an annotated catalogue of printed books . . . in the library of Denis I Duveen* (Weil, 1949). Regrettably, such catalogues are sometimes compiled as a preliminary to the sale or dispersal of the collections, but J Ferguson *Bibliotheca chemica: a catalogue of the alchemical, chemical and pharmaceutical books in the collection of the late James Young* (Glasgow, Maclehose, 1906), in two volumes, and *Bibliotheca Osleriana: a catalogue of books illustrating the history of medicine and science, collected . . . by Sir William Osler* (Oxford, Clarendon Press, 1929), both describe collections that were bequeathed as a whole (to Strathclyde and McGill universities respectively). Of similar bibliographical interest are some of the sale catalogues of antiquarian booksellers: an interesting bibliography compiled from just such sale catalogues of a London bookseller is H Zeitlinger and H C Sotheran *Bibliotheca chemico-mathematica: catalogue of works in many tongues on exact and applied science* (Sotheran, 1921) in two volumes with *Supplements* (1932, 1937, 1952), with over 50,000 titles, and still sufficiently valuable to warrant a 1980 reprint.

Author bibliographies
Obviously, catalogues of exhibitions and of private collections tend to be of mainly historical interest, and the same is probably true of most author bibliographies in the field of science and technology. Of course they are not *subject* bibliographies at all, strictly speaking, but they can

serve as a further source to consult on occasion. They fall into two main categories: those which confine themselves to listing the writings by a particular scientist or technologist, and those which in addition list writings *about* him. Examples of the first kind are N Boni and others *A bibliographical checklist and index to the published writings of Albert Einstein* (Paterson, NJ, Pageant Books, 1960) and A E Jeffreys *Michael Faraday: a list of his lectures and public writings* (Chapman and Hall, 1960). Examples of bibliographies with works by and about an author are J F Fulton *A bibliography of the Honourable Robert Boyle* (Oxford, Clarendon Press, 1961), A L Smyth *John Dalton, 1766-1844: a bibliography of works by and about him* (Manchester UP, 1966), P and R Wallis *Newton and Newtoniana: a bibliography* (Dawson, 1977), and R E Crook *A bibliography of Joseph Priestley, 1733-1804* (Library Association, 1966).

A feature of these last two titles is the location of copies of the works listed. Their different approach provides an instructive contrast: the Priestley bibliography locates copies of the books *by* him (other than periodical articles) in over two hundred libraries in a dozen or so countries; the Newton bibliography gives locations for works both *by* and *about* (but again excluding periodical articles) in seven hundred libraries (including eight private collections) in over twenty countries. This reminds us that for practical purposes the tracing of a wanted item in a bibliography is only half the battle, and compilers who take the trouble to indicate where a copy can be consulted (or borrowed, or photocopied, or microfilmed) perform a valued service, eg R M Strong *A bibliography of birds* (Chicago, Natural History Museum, 1939-46), with 25,000 entries in three volumes.

Further reading

M P Canfield 'Library pathfinders' *Drexel Library quarterly* 8 1972 287-300.

C-C Chen *Biomedical, scientific and technical book reviewing* (Metuchen, NJ, Scarecrow, 1976).

R Sweeney *International Target Audience Code (ITAC): a proposal and report on its development and testing* (IFLA International Office for UBC, 1977).

Chapter 9

PERIODICALS

A well-nigh universal finding of the many surveys of the literature habits of scientists and technologists is that the most frequently used of all sources of information are periodicals: typically they account for well over half of all their reading. Known variously as journals, serials, magazines, transactions, proceedings, bulletins, etc, and commonly appearing weekly, monthly or quarterly, their value cannot be over-estimated. They form the heart of most specialist collections and in many scientific and technical libraries more is spent on periodicals than on books. On a broader front D W King's survey for the National Science Foundation found that almost two-thirds of US 'scientific and technical communication resource expenditure' (including manpower costs) was accounted for by periodicals, compared to a little over a quarter attributable to books. Furthermore, the content of most books in science and technology is based on the periodical literature of previous years.

It is generally agreed that modern science had its beginnings in the seventeenth century: it is not without significance that the invention of the periodical dates from that time also. Although the printed book had been serving as an increasingly important medium of communication for two hundred years, having rescued scientific learning from the decline into which it had fallen during the early fifteenth century, the developing community of scientists found it necessary to supplement its inadequacies in 1665 by founding in London the *Philosophical transactions of the Royal Society*, the earliest learned journal in any subject still in existence. The book, it was felt, was more suited to the formal and proper publication of mature reflections or a completed *opus*, and was not for reporting current observations or experiments. This new medium, on the other hand, was specifically designed to print without delay short research reports of this kind, which scientists until that time had been disclosing in the more fugitive medium of personal correspondence or at meetings with their colleagues. It was indeed as much a social as a scientific invention, a means of organizing in a formal way the previously informal correspondence among members of a society. This intimate connection between scientific discovery and the scientific periodical has continued until the present day. A hundred years ago T H Huxley was still able to claim that 'If all the books in the world except the *Philosophical transactions* were destroyed, it is safe to say that the foundations of physical science would remain unshaken, and that the vast intellectual progress of the last two centuries would be largely, though incompletely, recorded'. Of course this was not to remain true for long, and we have seen the exponential growth of science paralleled by a similar increase in the number of scientific periodicals. By 1800 there were about 100 titles, by 1850 about 1,000, and by 1900 it is estimated that this figure had reached 5,000. The most reliable recent estimates indicate that in science and technology the number of current periodicals is 30,000, of which the United States accounts for one fifth and the United Kingdom for one eighth. If defunct periodicals are also counted the grand total exceeds 75,000, of which 25,000 might be regarded as 'serious', ie significant for scientific and technical research.

The continued fragmentation of science is another noteworthy feature that can be matched in the periodicals. The early journals, like early science, were undifferentiated, embracing the whole of 'natural philosophy', as it was then called, and much else. Indeed at that time even the boundary between science and technology was often ignored: the charters of the Royal Society (1662 and 1663) specifically refer to 'further promoting by the authority of experiments the sciences of natural things *and of useful arts*'. But as we know, 'general science' was soon supplemented in the eighteenth century by the individual disciplines of chemistry, biology, physics, which in turn developed divisions such as pure and applied chemistry, and then further aspects like organic and inorganic chemistry. Sub-division still continues apace, as the price we pay for increasing knowledge, and now to the layman seemingly minute specializations like chromatography or carbohydrate chemistry are established disciplines in their own right. Studies have shown that it takes no more than a hundred scientists working in a particular field to generate enough research papers to keep a journal going. It is an illuminating exercise for the student to trace these developments as reflected in the establishment of new journals in a discipline over a period of years. A study carried out for the Royal Society on new UK primary journals in science founded in the period 1969 to 1979 found that the major reason given was the lack of an existing journal for the subject field and the consequent shortage of adequate outlets for publication of research. As the investigator commented, 'this lends support to the hypothesis that the major motivation behind learned publication is the author's desire to publish rather than the reader's desire to know'. There is no doubt that this is a very important aspect of the role of the scientific journal and it will be considered again later.

132

This specialization of science is demonstrated even more dramatically by the increasing number of established journals showing fissiparous tendencies: in 1966 the *Journal of the Chemical Society* split into three parts after more than a century as the major periodical in its field, and by 1972 these had become six; more recently the *Proceedings of the Physical Society* have similarly divided into the tripartite *Journal of physics*, later becoming seven, A to G; the *American Society of Mechanical Engineers transactions, Physical review, Industrial and engineering chemistry*, and the *Institute of Electrical and Electronics Engineers transactions* all took a similar course some years ago. In 1971 what is probably the most prestigious scientific weekly in the world, *Nature*, proliferated into a Monday issue concentrating on physical science, a Wednesday issue devoted to biology, and a general Friday issue, though it later reverted. In one recent period of ten years the American Chemical Society started twelve new journals: all except two of them resulting from the splitting off of separate fields of chemistry from existing journals. There are now a host of similar examples.

All this is not to imply that periodicals covering the whole field of science are now superfluous. On the contrary, they are needed more than ever to play two specific roles: firstly, to act as a counter to the centrifugal tendencies of modern science by providing a forum for different kinds of specialists and attempting co-ordination across disciplinary boundaries, and secondly, to give scientists an opportunity to explain, particularly to non-scientists but also to specialists in other fields, the implications of their work, eg *New scientist, Scientific American* (with its circulation of well over half-a-million), each of which, to quote from the policy statement of what for many years was one of the best examples of this category of journal, *The advancement of science*, are 'of interest to laymen wishing to understand something of the significance of science and its impact on society'. Its transatlantic counterpart, *Science*, the weekly journal of the American Association for the Advancement of Science, makes a point of 'including the presentation of minority or conflicting points of view'. Recently as a popular complement to *Science* the AAAS has begun to issue *Science 80* to appeal to an even wider public. *SciQuest*, a monthly from the American Chemical Society, is also specifically written at the layman's level; it 'interrelates all sciences with current events for high school and college students as well as all others interested in understanding more about the world they live in'.

It is through the pages of such 'general interest' journals that the alert scientist or technologist finds it productive to browse, as described above (chapter 1). For technologists in particular, what have been described as 'broad' journals have an important role in suggesting analogies in other fields. It is such cross-pollenation which will often

133

furnish the stimulus for new ideas. In a study of the contribution of scientific and technical information to a sample of a hundred inventions in the chemical, electronics, and mechanical processing industries, it was found that '*Scientific American* or another general science publication was read by a majority of those interviewed and this seems the most specific linkage in information use within the group . . . these innovators use widely-read science publications to stay aware of general developments in science and technology'. Examples can also be found of journals that are even more specifically interdisciplinary: the stated purpose of the *Journal of chemical physics* is to 'bridge a gap between journals of physics and journals of chemistry'.

The importance of periodicals

Before proceeding further it is essential for the student to understand why periodicals hold so pre-eminent a place in the literature of science and technology, and a good way to start is to examine the ways they differ from books. R L Collison has written: 'The extent of knowledge in any field consists . . . of the information given in the books on the subject *plus* the periodical articles that have been published since the latest book was written'. In science and technology at least, no worker would contemplate writing up his *latest* research in the form of a book. In the first place, the simple speed of printing and distribution of the periodical compared with the book, deriving from the comparative brevity of the papers, the guaranteed (usually pre-paid) circulation, and the unbound format, means that it is usually months and often years more up-to-date than its rival; secondly, the regular and frequent appearance of the periodical ensures that it can *remain* close to the frontiers of knowledge by continuous addition, correction, or even retraction. In the words of E J Crane *A guide to the literature of chemistry*, 'A book is soon out of date, but a live journal can and does keep up with the onward march of scientific discovery'. It should not be too readily assumed, however, that the periodical article is the *most* up-to-date form in which scientific progress is reported. It is certainly the earliest *formal* presentation in most instances, and it is usually the first genuinely public unveiling, but as studies at the Center for Research in Scientific Communication at Johns Hopkins University have clearly shown, four-fifths of all authors make some kind of pre-publication report of the main content of their articles, most commonly to small audiences orally, but also at larger conferences, or in written form as 'preprints', theses, or research reports. Each of these less formal media of dissemination will be discussed later.

It would appear that the scientific information in most journal articles is up to a year behind the *real* research front at the time they

134

are published. Nevertheless, it is these various distinctive features of the periodical which have also enabled it over the years to perform the social rather than truly scientific function of establishing priority for the work of the individual scientist. In the same way as an inventor takes out a patent to protect his legal property rights in the fruit of his ingenuity, so the scientist publishes a paper in a journal to protect his intellectual rights to have his achievement preserved forever in the archive of science. We have already seen some evidence that 'journals are as important in meeting the scientist's need for publication outlets as they are in meeting the scientist's need for information'. It is on the basis of his published papers that he makes his reputation, and the primary journal plays a very important role in science by 'maintaining the discipline of a field through peer group judgements'. R K Merton has drawn a distinction between the mere *printing* of scientific work and its *publication*, which involves a much more rigorous scrutiny of content, principally beforehand by editors and referees (see below), but afterwards by reviewers. And there is a more personal element to this also, as Eugene Garfield has more recently reminded us, 'Peer judgment and praise are critical ingredients in the satisfaction derived by scientists'. It cannot be too strongly emphasized that 'journals serve two very different constituencies, the subscribers and the contributors'.

The average citizen is often astonished to learn that there are considerable numbers of topics on which no information has been published in book form. It is true that many are new subjects which will appear in the books in due course, but that still leaves thousands which are too brief, or insignificant, or local, or ephemeral, or recondite, ever to warrant full-scale treatment in books. For topics such as this (as for new subjects) the only source of printed information may well be an article in a periodical. Furthermore, even where books are available on a particular subject, periodical articles can often furnish more detailed (not to say more accurate) information on background, history, methods, apparatus, etc. And it is often only in periodicals that one may read of methods tried unsuccessfully or experiments that failed. After all, it is the periodical that is the primary source : the book is secondary, based largely, if not entirely, on the periodical literature.

By their nature, periodicals perform a further range of very useful tasks impossible for books. They contain correspondence columns often of great importance, permitting exchange of views about papers published (or any relevant topic). Many carry book reviews, obituaries, editorial comments, abstracts. Current news, professional announcements, advertisements, are also commonly found. With their varied content they provide rich pasture for the browser: according to a recent authoritative report on scholarly communication, 'There is simply no substitute in the scholarly process for browsing through current journals as a stimulus to thought and further research'. In many cases they serve as 'organs', often of a society or group, but in some cases merely of that band of like-minded individuals making up the subscribers, much in the same way as does the *Daily express* newspaper, which once described its readers as members of one large, happy family.

Paradoxically perhaps, in view of some of the difficulties which will be outlined later, the contents of the major periodicals in science and technology are often easier of access than books; that is to say they are better indexed, abstracted, and located. Because of their prime importance in science and technology the bibliographical care normally exercised on books has been devoted to the production of a remarkable range of bibliographies, indexing and abstracting services, and location tools. And it is mainly to the aid of the periodical rather than the book, that the powerful aid of the computer has been summoned, as will be seen in chapter 11.

Categories of periodicals
Faced with the tens of thousands of current periodical titles in science and technology, the student could well be excused his sigh of despair. And it must be confessed that any attempt at classification by type can only be arbitrary and at times conflicting, although for purposes of study it must be attempted.

There is no dispute, however, about the basic division into primary and secondary journals. The primary journals, of course, devote themselves to reporting original research. Known also as archival or 'recording' journals, they form the bedrock of scientific and technical literature, eg *Biochemical journal, Journal of physiology, Journal of mechanical engineering science, Institution of Electrical Engineers proceedings* (now in nine parts, A to I), *Computer journal, Tetrahedron, Philosophical magazine, Annals of botany, Molecular physics*. The task of the secondary journals, on the other hand, is to digest, comment on, and interpret the research reported in the primary literature. They have been called 'newspaper' journals, but they make up a far more heterogeneous collection than the research journals, eg *Chemistry in Britain, Glass, Colliery guardian, American machinist, Textile industries, Chemical and engineering news*. And of course there are many hybrids, containing both research papers and secondary material, eg *Spaceflight, Production engineer, Journal of the Electrochemical Society, Chemistry and industry*.

There is, however, a particular manifestation of the secondary journal that has been the object of such close attention in recent years
136

and is obviously destined to play so crucial a role in scientific and technical communication in the future that it is worth elevating into a special third category. This is the 'review' journal, comprising articles that briefly survey developments in a particular field of endeavour over a period, eg *Biological reviews, Advances in physics, Science progress.* The review journal will be considered at length with other kinds of reviews of progress in chapter 12.

One commonly used *ad hoc* classification of periodicals for purposes of study is by their origin. It should be explained that in most cases this is the same as classification by publisher, but not always: the *Journal of the Geological Society* is actually published by a commercial house, Blackwell Scientific Publications; the *Journal of zoology*, the organ of the Zoological Society of London, is published by Academic Press. Many learned societies have such an arrangement.

a *Learned societies*: as we have seen, the periodical as a form of literature owes its origin to the learned society, and since the earliest days such societies have been responsible for a significant proportion of the total, eg *Journal of the Chemical Society, Journal of heredity* (American Genetic Association), *Journal of physical chemistry* (American Chemical Society), *Proceedings of the National Academy of Sciences* (USA), *Mathematical proceedings of the Cambridge Philosophical Society* have all been established for over fifty years. Learned societies are responsible for some two-thirds of all natural science journals in the UK. The main purpose of such periodicals is to furnish an opportunity for authors (usually members of the learned bodies concerned) to publish the results of their investigations, and perhaps the majority of titles in this group are research journals, but there are also a number of secondary journals issued by the societies, frequently alongside a primary journal; eg in addition to its research quarterly *Computer journal* the British Computer Society also brings out *Computer bulletin* as its 'organ', with reports of meetings, data on new equipment, additions to the library, etc. As well as its century-old *Proceedings* (now *Mathematical proceedings*) the Cambridge Philosophical Society produces the quarterly *Biological reviews*, surveying four topics of current interest in each issue. A large society like the American Mathematical Society finds it needs *two* research journals (*Proceedings* and *Transactions*) and two others (*Bulletin* and *Notices*). And the American Chemical Society issues no fewer than 22 journals of various kinds. The usual explanation is that the main journal, devoted to publishing papers in full, has become too ponderous; briefer accounts can more speedily be got into print in a supplementary journal. Such was the reasoning of the Geological Society of London, for instance, in establishing the *Proceedings* to supplement its *Transactions*. Nevertheless, the main contribution of these societies to the literature is

137

their still very substantial role in disseminating the reports of original research through their splendid scholarly journals such as *Applied optics* (Optical Society of America), *Journal of physics* (Institute of Physics and Physical Society), *Journal of applied chemistry and biotechnology* (Society of Chemical Industry), *Journal of organic chemistry* (American Chemical Society). Publication, indeed, is frequently the major activity of a learned society. Many of them look on the publication of a journal as one of the main reasons for their existence; some even have it written into their by-laws.

b *Academic institutions*: it is a commonplace of science today that research has become institutionalized, and that the lone scientist is now a very rare bird. Since much of this new research is undertaken in academic institutions it is perhaps not surprising to find it increasingly reported in university and college research journals, many of which have been established for a number of years. As yet they have not the prestige or circulation of the great learned society titles, but they are indicative of a trend, eg *Journal of geology* (University of Chicago), *Annals of tropical medicine and parasitology* (University of Liverpool).

c *Governmental bodies:* as the role played by government, both national and international, in our lives increases, so does the volume of official publication, particularly in science and technology (see chapter 15) where vast sums of public money are currently being spent on research and development. Some of these publications are periodicals, eg *Meteorological magazine, Marine observer* (both Meteorological Office journals), *Post office telecommunications journal, Plant pathology* (Ministry of Agriculture), *Australian journal of biological sciences* (Commonwealth Scientific and Industrial Research Organization), *Canadian journal of chemistry* (National Research Council of Canada), *Unasylva: an international journal of forestry and forest products* (Food and Agriculture Organization).

d *Independent research institutes*: a small but interesting group of periodicals emanates from research institutes that are basically of independent foundation (even though they perhaps have links with universities, or possibly undertake government work under contract). They may have been established with a particular subject orientation or a particular role to play, or they may be 'think-tanks' on the American pattern, willing to undertake research in a variety of disciplines on a bespoke basis. Examples of periodicals so produced are *Battelle today* (Battelle Memorial Institute, Columbus, Ohio), *Textile research journal* (Textile Research Institute, Princeton, NJ), and *Polar record* (Scott Polar Research Institute, Cambridge).

e *Professional bodies*: As a category, bodies like the Institution of Mechanical Engineers, the Royal Institute of British Architects, overlap with the learned societies, and much of their work (and the periodicals they produce) is indistinguishable. Indeed in some cases, like the Royal Institute of Chemistry in 1980, they have actually merged with the corresponding learned society, in this case the Chemical Society, to become the Royal Society of Chemistry. Nevertheless, it is possible (and useful) to consider them separately as disseminators of scientific and technical literature. In broad terms (and at the risk of oversimplifying) it can be said that the concern of the learned society is the promotion of its subject, whether biology or physics or mathematics; in the professional body consideration of the education, welfare, status, etc, of the practitioners is added to a concern for the subject.

Periodicals in this category can range from primary research journals of a calibre and prestige fit to match any learned society publication to what are little more than news bulletins. And the fact that the scientific and technological professions vary so much in their size, cohesiveness, sense of professional responsibility, and status leads to an even wider range of publications. Examples to study and compare are *Mathematical gazette* (Mathematical Association: 'an association of teachers and students in elementary mathematics'), *Rubber international* (Plastics and Rubber Institute), *Structural engineer* (Institution of Structural Engineers), *Quarry management and products* (Institute of Quarrying), *Post Office electrical engineers' journal*, *Concrete* (Concrete Society), *Journal of basic engineering* (American Society of Mechanical Engineers), *Chartered mechanical engineer* (Institution of Mechanical Engineers). And like the learned societies, some of these professional institutions find they need more than one journal to serve their needs. An instructive instance is the long-established and prestigious Textile Institute, with its three regular publications neatly illustrating the three categories of periodicals described at the beginning of this section: *Journal of the Textile Institute*, for the publication of original research work and detailed accounts of practical investigations; *Textile Institute and industry*, with briefer papers of a more practical bias, 'especially acceptable to the busy executive'; and *Textile progress*, each issue of which is devoted to a full-length critical review of a major topic of technological interest, eg the structures and properties of fibres.

f *Commercial publishers*: the majority of periodicals in science and technology are produced as a business venture by commercial publishing houses. Since they must make a profit to survive, the titles are understandably heavily concentrated at the applied, industrial and commercial, technical and trade end of the spectrum, eg *Pigment and resin technology*, *Traffic engineering and control*, *Commercial motor*, *Water services*, *Welding and metal fabrication*, *European rubber journal*, *Electrical and radio trading*. But this group contains all types and levels of periodical publication and even of the most significant scientific

139

journals the commercial publishers produce a substantial share: of the top 165 British titles (ranked in 1968 according to the number of citations they receive) over half were commercial publications and a further 26 were society or professional journals published by a commercial publisher.

So wide is the variety of such periodicals that it is expedient to subdivide them further into:

(i) *Learned and research periodicals*: examples have been commercially produced for a hundred years or more, particularly in Germany, but until recently they have always been overshadowed by the famous titles issued by the learned and professional societies. Over the last two or three decades it seems that these societies have lacked the financial resources (and possibly the commercial enterprise) to initiate further titles, despite the vast increase in the amount of research work in need of a literary outlet. Into the breach have stepped scientific and technical publishing houses such as Elsevier, Pergamon, each with some three hundred titles, Academic Press, Springer, each with over a hundred, and we now have the phenomenon of several hundred primary research journals, a significant sector of the pattern of scientific communication, appearing apparently successfully as a commercial undertaking. Very considerable efforts have been made to ensure that the highest standards are maintained, and it is usual for each such journal to have an international editorial board of eminent scientists, insulated so far as is possible from commercial pressures. While some commercial journals contain too large a proportion of inferior papers, there is no doubt that others have reported work of the highest quality. Representative titles are *Journal of molecular biology, Talanta, Polymer, Journal of chromatography, Journal of the mechanics and physics of solids, Annals of physics, Microchemical journal*. As the Physics Information Review Committee pointed out, commercial publishers and learned societies 'are not always as different as they might appear to be. Greater differences exist between the smallest and largest learned societies than between some of the large societies and commercial publishers'.

This development coincided with a huge absolute and relative increase in institutional (mainly library) buying, and unfortunately there was evidence that this captive market was being exploited by exorbitant prices, typically in terms of cost per word from five to fifteen times that of comparable society journals. It is interesting to note where some would place the responsibility for this. R S Kahn of the Chemical Society wrote in 1965 'It is idle to blame the publisher for making as large a profit as he can: it is his business to do so; but it is equally the business of the scientist to prevent excesses. The captive librarian is helpless in this situation, for he must buy what his readers require and is not free to renounce any journal of sound content; it is the scientist

140

alone who carries the responsibility, and it is his duty to prevent extortion and exploitation; if high-ranking editors will refuse to collect papers, or senior authors to submit them, then the price of the journal will inevitably be reduced to the "fair profit" level'. Far from this plea being heeded, there were signs that the scientists themselves in some of the learned societies were jostling to board the bandwaggon. After years of charging one cent a page to libraries while its competitors were charging anything up to seven cents, the American Geophysical Union asked its President to proclaim at a well-attended public luncheon in 1973: 'In the future the Union intends to obtain a larger share of the library's dollars'. Many scientific societies now have a dual pricing system for their journals, with institutions (mainly libraries) being charged much higher rates than individuals.

Other scientists have taken the alternative approach in trying to stiffen the librarians' resistance: a 1973 memorandum signed by eleven very distinguished chemists from six countries proposed that 'libraries should be urged to show more reluctance in buying new commercial journals'. And the worm did indeed begin to turn: librarians started looking much more critically at their journal subscriptions, particularly where use by readers was low. Then the economic recession combined with soaring inflation forced them to do this anyway, with the result that many libraries have been shedding subscriptions by the dozen. A survey in the mid-1970s by the Library Association of libraries of all kinds in Britain showed that the thirty libraries responding cancelled a total of 1,221 journal subscriptions in one year. In its first nine months of operation the four members of the US Research Libraries Group (New York Public Library, Yale, Columbia, Harvard) cancelled 781 subscriptions and avoided 855 that they would otherwise have started. Perhaps the scientific information community has something to learn here from its counterpart in the humanities: what emerged very clearly from the 1979 US National Enquiry into Scholarly Communication was that 'the various constituencies involved in scholarly communication – the scholars themselves, the publishers of books and learned journals, the research librarians, the learned societies – are all components of a single system and are thus fundamentally dependent on each other'. There are signs that this lesson is being taken in. Among the recommendations made by the highly prestigious Scientific Information Committee of the Royal Society in 1981 was that all concerned should take note that 'the scientific information system in the United Kingdom is complex and delicately balanced so that changes in one part of the system have repercussions in other parts'. And the Committee went on to warn: 'the scientific information system, so long taken for granted in the United Kingdom, can no longer be regarded as stable'.

(ii) *Technical journals*: these are very closely linked with the needs of industry, and although as secondary sources are of limited interest to the research investigator, they are invaluable to manufacturing, sales, and commercial personnel. They are aimed much more at the technologist than the scientist. Some have vast circulations, eg *Aviation week and space technology* prints nearly 100,000 copies every week. Their role is to digest, interpret and comment as well as to inform, and their level is that of the practising chemist, or working engineer, or technically-minded manager: commonly they contain generously illustrated papers on new processes, equipment, products and materials. Many of these, especially if they are regular features, are written by full-time technical journalists on the staffs of the periodicals themselves. The fact that they are less academic than the research journals, and do not perhaps function at such a high level of scientific activity, does not mean that the papers they print are not often very technical, advanced and scholarly. Indeed, it is possible to draw too firm a line between some titles in this group and the primary research journals. Let the student examine, for instance, *Ultrasonics* or *Chemical engineering*.

Much of their current value lies in their other features, such as news columns, letters to the editor, announcements, obituaries and other personalia, book reviews, lists and abstracts of other literature, especially patents and trade literature, etc. Particularly useful are the advertisements, of which there are usually a great number. Most of these are deliberately designed to inform, making them quite different from consumer advertisements. It is quite common to find in such advertisements for new machinery, or chemicals, or processes, for instance, details unavailable elsewhere in the literature. The index to advertisers frequently provided in each issue of a journal is but one indication of their value. Indeed, advertisement revenue is the lifeblood of such journals, as it is of the mass media, and there is in some quarters the same concern that it should not divert the technical journal from its main function. There are signs that some 'technological glossies' have allowed this to happen. Of course, advertisements also assist the finances of some of the non-profit journals issued by learned societies and professional bodies, particularly where the scientific discipline covered coincides with an advertising market, and especially in a major discipline which relies on expensive equipment, eg *Analytical chemistry* (American Chemical Society), though some readers find this intrusion of the marketplace offensive in a scholarly publication. Such journals could perhaps be regarded as hybrids. The buyers' guides and directories sometimes issued as supplements (see chapter 6) are a further well-appreciated service, eg *'Laboratory animals' buyers' guide*.

Their retrospective importance in the literature of science and technology is obviously less than their current value, and depends on the proportion of material with some claim to permanence that appears within their covers. Examples of titles to study are *Electronic engineering, Foundry trade journal, Food engineering, American dyestuff reporter, Vacuum, Wire industry, Surveyor, Shipping world and shipbuilder, Design engineering, Chemical week*.

(iii) *Trade journals*: between these and the previous group the dividing line is vague, with a great deal of overlap and many cross-bred examples, but in general they are more commercial than technical, and more news-orientated than subject-oriented. Otherwise they are very similar to the technical journals, with an equal reliance on advertisements. There are vast numbers of them: currently the Science Reference Library takes over five thousand from all over the world. Many cover a small well-defined sector, and a substantial number were founded in the nineteenth century, eg *Poultry world* (1874), *Printing world* (1878), *Drapers' record* (1887), *Contract journal* (1879). For some specialized topics, the file of the appropriate trade journal may form the greater part of the total published literature on the subject.

Such journals are particularly useful sources for market news (commodity and share prices), company news (forecasts, dividends, mergers, expansions), and general trade announcements. A large number appear weekly, eg *Shoe and leather news, Machinery market* and *Mining journal*, as well as the four titles mentioned in the previous paragraph, and are often officially registered as newspapers. Some have even adopted newspaper (tabloid) format, eg *Construction news, Electronics weekly, Farmers guardian*. They are surprisingly little known outside their respective trades, and deserve far more attention from libraries and a much wider readership among the scientific and technological community than they get. Excellent examples for the student to examine are *Chemical age, Timber trades journal, Sewing machine times, Horticulture industry, Bakers' review, Motor trade executive, British farmer and stockbreeder, Wool record and textile world*.

A comparatively little-known service that they provide is the answering of enquiries from readers. A 1972 survey of such services concluded that 'Trade journal editors are probably one of the best sources for trade information, since it is part of their job to be up to date and well informed'. Many editors, indeed, look on the answering of readers' queries as part and parcel of their work, and are willing to go to considerable lengths to find information, although there are signs that this attitude may be changing. One major publisher with a range of trade journals in his stable is contemplating setting up a central readers' enquiry bureau.

A feature of a number of technical and trade journals in recent years has been the physical changes they have undergone as a direct result of major technological advances in printing production. As well as the switch to newspaper format just noted, it will also be observed that many have changed their page size and layout, commonly to the A4 international standard. A growing number are now printed by offset litho and a few have begun to experiment with computer typesetting.

In his exploration of technical and trade journals, the student is likely to come across a number of 'controlled-circulation' periodicals, such as *Electrical equipment selector, Industrial equipment news, Food engineer, Process engineering, Transport manager's journal.* These are largely advertisement media, and are usually sent free of charge to those considered to be qualified readers, ie actual prospective customers for the goods advertised, or at least those in a position to influence purchasing decisions. Some are available on request to libraries as well as to individuals, and they do have value as sources of otherwise unavailable data. An increasing trend in recent years is for subscription-based journals to become controlled-circulation: probably the most striking instances are *Engineer* and *Engineering*, after more than a century of independent existence as two of the most widely read technical journals in their field.

(iv) *Popular subject journals*: these are familiar to everyone, and include all the titles for the amateur, the hobbyist, and the enthusiast that are to be found on railway bookstalls, as well as a large number of esoteric (not to say crank) publications catering for the most unusual preoccupations. Typical titles are *Popular mechanics, Yachting world, Flying saucer review, Railway magazine, Practical motorist, Amateur gardening, Radio and electronics, Racing pigeon, Model engineer, Speleologist, Inventor.* Again it should be pointed out that this is not a watertight group. Serving the needs of the amateur is a trade for some, and for certain areas of interest it can be an industry. Periodicals in the field often double as trade or technical journals and popular journals, eg *Tape recorder, Motor boat and yachting, Bee craft, Personal computer world, Entomologists' record, Microscope.* There are instances of subjects like woodworking or horsebreeding or horticulture or geology where many amateurs are as knowledgeable as the professionals, and their journals reflect this. *Wireless world* (founded 1911) is often quoted as an example of an originally popular magazine that is now a highly technical journal. And those inclined to scoff at the lunatic fringe should study what has happened to the status of the *Journal of the British Interplanetary Society* since its establishment in 1934.

g *Industrial and commercial firms*: there are in English perhaps twelve thousand 'house journals' or 'house organs', issued primarily for advertising purposes by manufacturers and dealers and public corporations, eg *Educational focus* (Bausch & Lomb Optical Co), *Ball bearing journal* (Skefco Ball Bearing Co), *Phillips technical review, Ford times, Sikorsky news, Hexagon La Roche* (Hoffman-La Roche and Co). Like other forms of trade literature, however, they often contain information of value which may be unobtainable elsewhere. They are described in more detail in chapter 18.

h *Individuals*: common in the eighteenth century, these are now quite rare, although a number of today's most important journals commenced as the personal venture of a single man. They were particularly prevalent in Germany, and perhaps the best known instance (which still retains the founder's name in the title) is *Justus Liebigs Annalen der Chemie.*

The problems of periodicals

Of the importance of periodicals to the scientist and technologist no one has any doubts, but as R L Collison has pointed out, '. . . .the full exploitation of periodicals is one of the most important problems which faces the librarian in any type of library'. As we have seen, the periodical owes its origin to the seventeenth century scientists' dissatisfaction with the printed book as a medium of communication, but it never provided the complete answer, and scientists are still far from content with their literature. The periodical in particular has come in for severe criticism in recent years.

a *Proliferation.* In the first place there are too many: writing in 1965 about medical journals Sir Theodore Fox complained that 'Despite the invention of photography and microfilms and television and computers, we cling to the publishing habits of our forefathers. Though the amount of material to be communicated is fabulously greater, a very large proportion is transmitted through journals which . . . are essentially the same as they were a century or more ago. And where one journal does not suffice, we seldom think of anything cleverer than to create another like it; and another'. The Chemical Information Review Committee found little change in 1978: 'new journals tend still to be traditional in their methods, style and appearance'. A generation earlier Sir Edward Appleton, Nobel prize-winning physicist and discoverer of the ionosphere, wrote: 'If anyone set himself the task of merely reading – let alone trying to understand – all the journals of fundamental science published, and worked solidly at his task every day for a year, he would discover at the end of the year that he was already more than 10 years behind! If the same constant reader had included the technical literature as well, he would find himself about 100 years behind in his work after 12 months' effort.' Since then, of course, the number of journals has continued to increase, according to the estimates doubling in total every ten or fifteen years.

A 1968 estimate, since confirmed by other studies, was that some 850,000 papers are published in scientific and technical journals every year. Another measure to contemplate is that 'Every 24 hours, enough technical papers are written throughout the world to fill seven sets of the *Encyclopedia Britannica*'. B C Brookes has concluded 'My personal view is that the so-called "information explosion" in science and technology is largely a "periodical-publishing explosion" — and little more — which artificially increases the complexities and therefore unnecessarily inflates the costs of all information work. The increasing specialization of science discussed at the beginning of this chapter is only one of a number of reasons. We are told that this explosion will be further fuelled over the next generation by the output of the presses of the burgeoning nations of the Third World, most of which are committed to massive programmes of industrialization. They are already responsible for 10% of the world's scientific journals, compared with 1% sixty years ago. The incalculable intellectual resources of China (population 970 millions) and India (population 663 millions), for example, have as yet scarcely been tapped.

This inflation owes much of its impetus to the 'publish or perish' syndrome exhibited by many scientists, who are only too aware that their status (and their promotion prospects) so often depends on the number of their published papers. It is acknowledged of course that every scientist worth his salt nourishes the ambition to make a creative contribution to scientific knowledge, and one accepts that to ensure that each new advance passes scrutiny it has to be published in the literature. That so many succumb to the temptation to publish as many papers as they can, regardless of the real worth of their contents, explains what the SATCOM report calls 'the widespread contempt of scientists for the bulk of their colleagues' writings'. It was reported in 1970 that of a sample of 540 published articles scrutinized for possible inclusion in the *Annual review of psychology*, two-thirds were 'utterly inconsequential', a quarter were 'run-of-the-mill', that is to say 'technically competent variations on well-known themes', and only 10% could be regarded as worthwhile. The view of the two expert investigators was that 'some writers, faced with the decision of whether to publish or perish, should have seriously considered the latter alternative'. In an oft-quoted passage J M Ziman has pronounced against the 'proliferation of semi-literate, semi-scientific, half-baked and trivial material which threatens to swamp the whole system'.

It is the social rather than the strictly scientific pressures that are also to blame for the piecemeal publication of research results, with what could have been one complete and rounded paper spun out in three, four, or even more brief and inconclusive interim reports, gleaning

146

what one critic of the system has called 'pellets of prestige'. It was widely agreed at the first International Conference of Scientific Editors, held at Jerusalem in 1977, that editors should refuse to publish mere reports of work in progress and try to prevent authors from expanding one legitimate paper into three or more articles. Even worse, as C M Weiske of Chemie Information Dokumentation has reminded us, 'some authors publish their results of research twice or even more times. Mostly, the first article appears in a well-known journal and after this it is published in journals which pay for the article'. Sometimes of course the authors hope that by publishing in different journals they will reach different categories of readers. But few perhaps can match the achievement of the joint authors of an article on antifertility compounds which was published twice by the same journal, the *Indian journal of science.*

Evidence from citation studies tends to support this view of widespread 'information pollution'. Unfashionable, elitist, and antidemocratic though it might seem, only a select few workers contribute to scientific progress, and nearly all the other papers represent so much 'noise' in the system. As Luis Alvarez, winner of the Nobel prize for physics, so pungently puts it, 'There is no democracy in physics. We can't say that some second-rate guy has as much right to opinion as Fermi'. This of course is not new; Galileo said much the same thing 350 years ago: 'In questions of science the authority of a thousand is not worth the humble reasoning of a single individual'. But even then this was by no means original: a hundred years before Christ the Chinese historian Sze-ma Ch'ien wrote that 'The assent of a thousand people is not worth the blunt criticism of one scholar'. Paradoxically, as the number of journals grows, their contribution to the advancement of science is becoming relatively more constricted. J M Ziman's view is that 'The leading journals are as important to the progress of science as the leading scientists'. A study in the United States suggested that the reporting of important original research is probably confined to no more than four hundred journals. By using the number of times an article is cited by another author as a rough-and-ready measure of its influence, studies invariably show a large majority of the key articles concentrated in a minority of journals, eg in computer literature 75% of references are to be found in 17.6% of the journals; 84.3% of the references in 26 volumes of *Physical review* are to 15 journals, less than 2% of the total number of journals referred to. A major ASLIB survey investigated references to British journals in a sample year's issues of *Science citation index*, and discovered that 95% of the 68,764 citations relate to a mere 9% (165 titles) out of a possible 1,842 British scientific journals. With D J de Solla Price, the student is 'tempted to conclude that a very large fraction of the alleged 35,000 journals now

147

current must be reckoned as merely a distant background noise, and as very far from central or strategic in any of the knitted strips from which the fabric of science is woven'. This of course is the whole philosophy behind *Science citation index*, with coverage attempted of no more than 2,400 titles. Indeed a useful tool that has recently become available from the same source is *Journal citation reports*, giving an assessment of the status of the world's thousand most cited scientific periodicals. Research results of this kind dating from at least the 1920s have suggested the concept, of great interest to librarians, of a collection of 'core' journals, limited in number, but sufficient to meet perhaps 90% of the demand. The ASLIB investigators conclude that 'a rough estimate of the world's number of core science journals lies between 2,300 and 3,200'. There are now available a number of published lists of such core journals in particular subject fields, eg the International Council of Scientific Unions Abstracting Board *Core list of journals in engineering* (Paris, UNESCO, 1979), ranking the 732 journals most cited by *Engineering index*, 1974-8.

There are dangers of course in equating frequent citation of a journal or of an individual paper with 'value' or 'importance' just as there are similar hazards in drawing conclusions from frequent use in a library. Nevertheless, studies of library use do show a similar pattern of distribution. A pioneering survey at the time and one of the largest ever undertaken, covering 87,255 issues from the Science Museum Library in 1956, showed that 'Even in a library which is designed to deal with the residual demand from libraries, about 1,250 serials (or less than 10% of those available if the non-current serials are included) are sufficient to meet 80% of the demand for serial literature'. A more recent survey in the US of 80,000 interlibrary loan requests for journals in science and technology found that half were for articles in only 850 journals (out of an estimated total of 30,000), and three quarters were for articles from 2,600 journals (about 9% of those currently published). A later survey, of 61,333 requests for serials to the British Library Lending Division in 1975 showed over half the demand concentrated on less than 10% of the titles requested. A 1981 Polish study investigated references retrieved from one of the *Chemical abstracts* data bases in the course of routine processing of two thousand SDI profiles (see chapter 11). It was found that half the relevant papers came from a very small number of journals: 212 out of 6,209 and 220 out of 7,309 titles respectively for each of the two years searched.

A number of other surveys of scientists' and technologists' literature habits indicate that the core journals in any specialist discipline may not amount to more than half-a-dozen: the Library Association study of mechanical engineers noted below (chapter 12) found that only 12% see more than ten journals regularly and 37% see less than five; and an earlier survey in the electrical and electronics industries discovered that the average was 4.7. Harold Wooster of the US Air Force Office of Scientific Research has come up with the following formula, based on investigation of literature in his field: ten is the maximum number of journals an average scientist can be expected to keep up with; a hundred is the number of journals that will meet 90% of the needs of any reasonably specialized information centre; a thousand is probably the number of first class scientific journals in the world today.

What is really wanted of course is a ranking of journals according to their quality. Some members of the Chemical Information Review Committee did suggest that this could be achieved by consulting 'a group of prominent authorities in the field', though the Committee report did recognize this as a controversial proposal. The eleven university chemists from six countries who signed the 1973 memorandum referred to earlier were not so hesitant: they prescribed strong medicine in the shape of 'an impartial mechanism for evaluating the need for a new journal' with criteria set for level of quality, refereeing practice, rejection rates, and for the termination of a journal should that be necessary. Some objective measures have already been tried by the American Institute of Physics: in addition to library use these included rejection rate analysis and variations of citation analysis such as the impact factor (the number of citations per paper) and the immediacy index (showing the rapidity with which articles are used by the scientific community).

Yet, as we have seen, new journals continue to appear, mainly to provide adequate outlets for new research. Part of the pressure for this growth may well derive from the sense of grievance felt by many authors over the delay between their submission of a paper and its publication, which is rarely less than six months and not uncommonly over a year. Delay is particularly frustrating in science and technology where the pace of advance is so great: the very pressures towards rapid publication which created the periodical in the first place are now being brought to bear on the periodical itself. To some extent the cure is in the authors' own hands: if they insist on concentrating their attentions on the better-known core journals a queue is inevitable. A 1968 study by Brian Vickery concluded that 90% of scientific and technical articles appeared in about eight thousand periodicals or about 30% of the total. Nevertheless, as will be seen later, an important objective in many of the innovations in the form of the periodical is the reduction of publishing delays.

b *Refereeing.* Part of the delay is caused at the editorial check-point, for it has been common practice with learned journals particularly in English-speaking countries since the very earliest days of the scientific journal for editors to send submitted papers before publication

149

to one or more independent 'referees' for an authoritative opinion on such matters as originality, technical competence, or significance. Referees are normally selected as the leaders in their field, and usually remain anonymous. This certainly helps to preserve standards by eliminating the frequent obscure, trivial or worthless papers and the occasional fraudulent paper, but authors naturally find it burdensome, and even the most august referees are not infallible, particularly with innovatory or speculative papers. It is known that in the 1930s at least three major theoretical contributions to nuclear physics were initially rejected by referees. Studies at the Center for Research in Scientific Communication revealed that one in eight of the articles published had been rejected by other journals previously, some as many as four times. Sir Theodore Fox tells us that one of the discoverers of blood-groups revenged himself by keeping the *Lancet*'s letter of rejection framed on the wall of his consulting-room.

Critics have dismissed the whole system as a 'sweepstake', with so much depending on personal judgement − firstly of the editor in selecting a referee, and secondly of the referee in making his recommendations whether or not to publish. Often, and to some extent inevitably, the referee is a professional rival of the author; there have been dark suggestions (by the young T H Huxley, for example) of referees deliberately delaying competitors' papers by sitting on them, and more sinister accusations of suppression, pilfering, and even outright theft. But there is little doubt that the practice does lead to the beneficial amendment of many papers sent for comment, and like the jury system it does broaden the base of opinion by which an important decision is made. Referees, like the anonymous jurors, also have the far from negligible role of shielding the very visible and vulnerable editor (or judge) from the wrath of rejected authors. Their task, in the words of R K Merton, is 'sorting out good science from bad'; for D Crane they are 'the gatekeepers of science'; J M Ziman describes them as 'the lynchpin about which the whole business of science is pivoted'. And it should not of course be forgotten that referees customarily give their services free, as a disinterested contribution to scholarship. Full-time paid editors often express their wonder at such altruism.

After years of obscurity, this whole process of quality control has recently been brought into the light: a study by the editors of *Applied mechanics reviews* into varying refereeing practices was funded by the National Science Foundation in the United States, and in the United Kingdom the British Library sponsored a two-year study at the University of Leicester Primary Communications Research Centre to ascertain whether refereeing procedures exert the influence on the formal transmission of research results that is often claimed. Paradoxically enough, one finding that has emerged is that the fundamental role in the evalu-

150

ation of manuscripts submitted is played by the editor not the referee. Many papers that are obvious non-starters are rejected out of hand without being sent to a referee, but editors take much more care over other rejections than they do over acceptances. Commonly they require the approval of a publications board before they can turn down a paper, even when referees report unfavourably. Indeed, they would in general prefer to risk accepting a paper that was subsequently shown to be below standard than mistakenly to reject a good paper: this is their recurring 'editorial nightmare'. If in doubt they prefer to publish. Interestingly, editors of medical journals hold the opposite view: perhaps their profession is constitutionally more wary of errors.

As might be expected there is considerable variation among individual referees as to promptness in replying, amount of detail in reports, and willingness to make a firm recommendation one way or the other. No evidence of gross bias was found, but perhaps not surprisingly, total objectivity was also lacking.

There are signs of change: some journals, characteristically from commercial publishers, have substituted a board of editors for the referees. Such a board can of course operate as rigorously as any referees, but may be able to do it more speedily. Some journals have to some extent lifted the customary veil of anonymity from their referees: the *Journal of applied mechanics* invariably uses two referees for each paper but now lists them all at the beginning of the journal, with thanks for their assistance. But most journals still guard the anonymity of their referees, claiming that it prevents personal and subjective influences from inhibiting forthright comment. And by no means all scientists object. A survey of working scientists for the Royal Society published in 1981 found them insistent on continuing the refereeing system.

c *Wastefulness.* One of the most telling arguments against the periodical is its wastefulness. A UNESCO study has shown that a single article in a highly specialized periodical may interest no more than 10% of the workers in the subject area of that periodical, while a single article in a general periodical will interest only about 2% of its readers. A survey of the readership of the *Journal of organic chemistry*, the major English-language title in its field, showed that 'the average subscriber glanced at or began to read about 17% of the papers of a typical issue, and read half or more of only 4% of the papers in the issue. The average time spent in reading a 413-page journal was 2.2 hours'. Research by the Institution of Mechanical Engineers in 1976 found that wastage in the shape of journals routinely received by members but not read or even wanted can amount to as much as a quarter of the total circulation. A 1978 US study showed that of the articles in scientific and technical journals distributed to individual subscribers only one out of eight will be read; of those distributed to libraries only one in twenty

151

will be read. The American Psychological Association discovered that the average article in one of its journals is read by only 17 people. J D Bernal has said that in writing papers for journals instead of personal letters, scientists have replaced communication by dissemination, ie by broadcast scattering. One is bound to suspect that many articles are not read by anybody, but are published just for the record, often merely to establish a worker's priority in the field. What can be asserted is that the great majority of papers published are never cited by anyone else. As H V Wyatt has written, 'For the individual, the modern scientific paper is a social device for proclaiming intellectual property, not primarily a technique for advancing science'. The Director of Publishing for the Institution of Electrical Engineers draws an interesting distinction between the 'shelf periodical' (the periodical of record) and the 'desk periodical' (meant to be read). A *Nature* editorial is more pithy: '. . . the habit of writing for posterity is often an impediment to communication with those still alive'.

d *Library photocopying*. The inability of the periodicals themselves to provide the kind of communication format required by users is demonstrated by the vast subsidiary distribution system, unapproved and of suspect legality in many cases, made possible within the last three decades by the spread of photocopying machines which can provide high quality copies of articles in periodicals rapidly and cheaply. As C Bradley of the Publishers' Association admits, 'a major form of publishing of learned journals has now become publishing in the form of a photocopy'. In the Federal Republic of Germany it was estimated that five billion pages of copyright material were photocopied in 1979. In other European countries it must be almost as high; in the United States it amounts to tens of billions. Many of these copies are made by libraries for their readers, either on request or as interlibrary loans. A survey by King Research found that US library staffs copied 16 billion pages of library materials in 1976. In the words of Bernard Williams, Director of the National Reprographic Centre for Documentation, 'What it amounts to is that we have all become our own printers'.

Some have put this forward as a major reason for the decline in subscriptions endured by many scientific and technical periodicals in recent years. It is not without significance that the proposal for a National Periodicals Center in the US to be modelled to some extent on the British Library Lending Division was firmly opposed by the publishers, with the Information Industry Association urging 'an open, competitive, pay-as-you-go, market-place' alternative. The bitterness with which some publishers regard this particular example of the new technology was well expressed by Gordon Graham in a 1981 leading article in the *Bookseller*: 'The photocopier has stumbled, through the brilliance of the inventor, into easeful performance of one part of the
152

publishing process – the reproducttion of the finished message – and, having so stumbled, arrogates to himself the total privilege of publishing; shrinking the creative arts of authorship and publishing to mere mechanical dissemination. He reaps where others have sown and pays not a penny to the husbandman'.

This is not the view of D J Urquhart, former head of the library with the largest periodical subscription list in the world, who believes uncompromisingly that 'the expansion of periodical publishing is not an essential requirement for the development of intellectual activity. What is essential is that the potential user should have easy access to the records of progress. Library copying facilitates this. If a conflict arises between library copying and periodical publishing we should remember that, as the development of report literature has illustrated, periodical publishing on the present scale is not an essential requirement'. His successor at the British Library Lending Division, Maurice Line, argues strenuously that any restriction of photocopying facilities would simply cause his library to 'cancel journals that are rarely wanted, and transfer the money to buying duplicates of heavily used journals', which he claims would 'do more harm to journal publishers than the present situation'. He points out that a 1975 survey of 61,333 requests for serials showed that over half were for articles from volumes over three years old, and therefore unlikely to be available from publishers anyway. He has also shown that the two million photocopies per year supplied by BLLD are distributed over 52,000 current journals and 80,000 that have ceased. As he told a Commission of the European Communities Workshop on document delivery in 1979, 'Interlibrary photocopying in aggregate may appear to be large, but if it is broken down among the very large number of journals in existence it is evident that the great majority are copied very little, and the minority that are copied more extensively are mainly large journals with big circulations'. And it is a fact that the reason BLLD is currently satisfying two thirds of its requests with photocopies is simple cost-effectiveness: the cost of photocopying plus postage is less than the average cost of posting the original to the user, at least in the UK. Even more confident is the Head of Photoduplication Services at the Centre National de la Recherche Scientifique, France's major provider of photocopies from sceintific journals: he insists that he is benefiting the periodicals concerned by advertising their existence.

Similar fierce debates between publishers and libraries can be found raging in many advanced industrial countries, for instance in the Federal Republic of Germany, where each side accuses the other of posing a fundamental threat to democracy and constitutional rights. It is useful perhaps for the student to consider the opinion of a detached observer in the person of A J Meadows of the Primary Communications Research

Centre: 'If frequent requests are made for photocopies of a title that is not taken by a library, consideration is usually given to the purchase of the journal (at the expense of an existing title). If photocopies are only requested infrequently, in present circumstances the library will not, in any case, purchase the title. Hence, photocopying can only have a marginal influence on institutional sales'. Indeed, as even the Senior Vice-President of McGraw-Hill admitted, 'No publisher has gone broke over photocopying'.

One of the most important objective studies of the question was carried out in 1977 by ASLIB at the request, interestingly enough, of a joint working party between the British Library and the Scientific, Technical and Medical Group of Publishers. The aim was to 'demonstrate whether a direct causal relationship exists whereby increased inter-library lending leads to an overall decrease in periodical subscriptions by virtue of associated photocopying'. A detailed survey was undertaken covering 250 academic, research and industrial libraries, but no evidence could be found of such a causal relationship.

Of course photocopying is controlled by copyright legislation. In general, and certainly in the UK and the US, the law permits 'fair use', the making of a single copy of any particular article by, or for, an individual for his private study or research. This provision also extends to non-profit libraries making a copy at the direct request of the user. This is not the place for legal argument but it would appear, firstly, the widespread practice of libraries substituting photocopies for inter-library loans was not envisaged when the original legislation was drafted, and secondly, the well-nigh open access by the public to photocopying machines has rendered the law unenforceable anyway. We are told that 'The photocopying machines that crowd our libraries, airports and hotels are open invitations to break the law'. As the Master of Emmanuel College, Cambridge, wrote to the *Times*, 'Every scholar knows that for every properly safeguarded Xerox machine in a British University library there are at least half-a-dozen, perhaps many more, elsewhere, totally unsupervised. I remember a university library abroad with at least four such unsupervised machines outside its very doors'.

In a number of countries legal moves have been made in the direction of firmer control. In 1977 the Whitford committee set up by the Department of Trade to consider UK law on copyright admitted that 'The widespread availability of simple-to-operate and relatively cheap-to-use photocopiers has made the law unworkable' and recommended the abolition of 'fair use' and the introduction of a 'blanket' licensing scheme for libraries, educational establishments and others. In 1980 the House of Commons Select Committee on Education, Science and the Arts urged the government as a matter of urgency to tackle the copyright problems posed by the new technology. A very cautious

154

government consultative document was published in 1981.

In the United States the new Copyright Law, effective from 1978, has by no means solved the problem, although it has provided the framework for the Association of American Publishers' Copyright Clearance Center. 'Fair use' is still permitted under section 107, though it still remains vague and undefined; photocopying by libraries for inter-library loan is permitted by section 108, provided it amounts to no more for each library than five articles copied in any calendar year from issues of a single journal title published during the preceding five years. Royalties are payable to CCC for copies outside these provisions. Increasingly learned journals are appearing with a warning to this effect and sometimes a code statement on each page or at the beginning of each paper showing the charge payable, eg $1.00 for the *Journal of the American Chemical Society*, $2.00 for the *Journal of the Geological Society*. But a nationwide survey predicted in advance that most library copying would be within the guidelines laid down; in other words, for all practical purposes libraries were in compliance already. Publishers had to modify their expectations severely: indeed the Center lost money in its first year; even in its second year royalties were paid on only 200,000 copies. By the third year annoyance was being expressed by the Special Libraries Association that most of the fees paid by their members had not gone to the copyright owners, as expected, but had been retained by CCC to sustain itself.

Dissension is still rife. The view of Richard de Gennaro, University of Pennsylvania Director of Libraries, is that the publishers 'had – and continue to have – exaggerated notions of the amount of "illegal" copying that was going on in libraries'. At their mid-1980 conference the Association of American Publishers Copyright Committee was still insistent that they were the 'victims of illegal photocopying', and in 1981 the Association promised to commence litigation against certain special libraries described as 'going their merry way' without regard to copyright. The issue is by no means closed: the US Register of Copyrights is obliged by the legislation to conduct a five-year review of the impact of the new law and report to Congress by 1983. Widespread hearings are in progress.

Often forgotten in this struggle are the originators of the material over which the parties contend, the authors of the scientific and technical articles in the periodicals. Only rarely do they receive payment for their work: they seek little more than the widest possible dissemination of their research, by any means, including photocopying. Indeed if page charges are insisted on (see below), they have to pay the publishers! Following revision of the law many US journal publishers, eg the American Institute of Physics, require all their authors formally to transfer their copyrights before their articles are accepted for publication. In learned society publishing generally it has been the practice for many years to require this; indeed for society members such a stipulation is often embodied in by-laws and regulations. Some protagonists in the dispute appear to ignore the *droit moral*, the inalienable right, enacted or not, that legislation cannot effect, that every author must have in the creation of his mind and in the intellectual product of his labour. Though primarily a matter of ethics, it was recognized at common law centuries before the first copyright legislation.

Whether we shall see a resolution of the issue is debatable. One simple view that many take is that all that photocopies do is to replace not purchases but handwritten notes. What is certain is that rising costs and dwindling library budgets in the 1980s with the subsequent emphasis on 'resource sharing' and 'networks' will add a further turn to the screw.

e *Economics.* A far less debatable reason for the decline in journal subscriptions is simple economic stringency. During the decade of the 1970s both the UK and US have witnessed a five-fold increase in the price of scientific and technical journals of the kind purchased by specialized and learned libraries; this has exceeded and continues to exceed the rate of inflation. The average price of such journals in 1981 was almost £80 a year; for chemistry the average was just below £200. There are several essential titles now costing £300, and some exceed £500. It is true that this is not a simple equation because in many cases the size of the journal has increased also but a limited survey covering 1972-9 carried out for the Royal Society found that the increase in price per page was still ahead of the retail price index. One noticeable effect has been a steady decline in circulations; less obvious is the fact that subscriptions from individuals have dropped away even more sharply. And of course the trend described at the beginning of this chapter towards more specialized journals has aggravated the decline. Many workers are indeed beginning to agree with the opinion that '. . . the entire apparatus of commercial publication is too expensive a mechanism for scholar-to-scholar communication'.

Economic necessity has obliged more and more US learned and professional societies to make controversial 'page charges' of up to perhaps $85 per page payable by the contributors to their journals in an effort to keep subscription prices down. Some societies, eg the American Institute of Physics, also levy a submission fee or article charge, of perhaps $20. The philosophy behind such charges is that the publication of the results of scientific research is a final step without which the whole effort would be ineffectual; the comparatively trivial expense of publishing should therefore be treated as one of the costs of research. Some journals, for example, a number in physics, derive as much as 40% of their income from such charges. Societies do not expect individual authors to pay; they look rather to their parent institutions. Even for these, such charges are voluntary, although in practice most of them are honoured. Recently, however, the proportion refusing to pay has increased (from 30% to over 40% in the case of the American Chemical Society, for example), and at least one learned society has made them compulsory. A more common reaction to such authors is simply to push their papers to the end of the publication queue; the American Institute of Physics openly states 'Non-payment of the publication charge may lead to delays in publication'. Some US authors have responded by looking abroad for journals that do not impose page charges, particularly British journals.

Page charges have been considered in the UK, but there are major differences in the pattern of publishing compared to the US: in physics, for instance, 60% of the authors in UK-published journals are from overseas and collection of charges would cause considerable problems. Furthermore, over 80% of sales are abroad and UK authors would feel they were subsidizing overseas subscribers. But the matter is by no means closed: after a three-year study of the pressures on the scientific publication system the Royal Society in 1981 suggested page charges.

Bibliometrics
As the student knows, science and technology by its nature is progressive and cumulative, and inevitably much of its literature is ephemeral. But the time when any given article is going to be superseded cannot be predicted, and even the average rate of obsolescence differs greatly according to subject, with perhaps papers in electronics enjoying the briefest span. As a general rule of thumb, every year perhaps 10% of scientific and technological papers 'die', but this rule does not apply to a number of important disciplines where classic papers may retain their importance for generations, eg botany, geology, mathematics. Many chemists bank on the significant journal literature becoming incorporated in the book literature within twenty years, but this would certainly not hold good for biologists. Although the student should learn to distinguish this decline in value from 'obsolescence' indicating merely a decline in use, eg readership or borrowing frequency, there is certainly a connection. The massive survey of Science Museum Library issues in 1956 mentioned earlier showed that journals published within the previous seven years accounted for 57% of the use, while journals over 27 years old made up less than 9% of the issues. Measures based on citations rather than library statistics reveal a similar pattern of decline, though once again the student should understand that citation and use are not the same: the citation practices of authors are by no means consistent, and may range from citing papers that have not been

157

used to not citing others that have been used, perhaps because they are 'obvious'. Another insidious problem calling for constant vigilance is the periodical which deteriorates in quality: fortunately, as they get older most periodicals seem to improve. Considerations of this kind are of vital concern to the librarian, who has to decide, often in the face of conflicting demands for space, how long to retain back runs of journals.

Bibliographical control of periodicals and their contents will be discussed later in this chapter and the next, but it is appropriate here to look at one of the major obstacles to adequate control, and therefore to satisfactory use. This is the interesting phenomenon of 'scattering', ie the extent to which articles on a given subject actually occur in periodicals devoted to quite other subjects. This has been observed for very many years, but was first studied in detail by S C Bradford of the Science Museum Library, who codified his findings in what has become known as Bradford's Law of Scattering. His work has been refined and added to by a number of investigators since, but in very crude and over-simplified terms the Law states that of all the articles on a subject, one-third are to be found in a small core of journals on that subject, one-third in a larger group of journals on neighbouring subjects and one-third in all the remaining journals. The implications for a scientist or technologist making an exhaustive approach (see chapter 1) to a subject are obvious: he cannot hope to locate all the required information without ranging widely outside the journals in his own field. And inevitably as the number of journals increases the scattering of papers becomes wider. One might add here that a problem of a similar nature is sometimes encountered with regard to the quality of research papers: the well-known journals can ensure that a poor paper is never published in their pages, but what no one can prevent is a paper of the highest class appearing in a lowly and obscure journal.

At this point the student should take a note that these brief mentions of citation patterns, obsolescence, and scattering, no more than sketch in the outline of the matter. The quantitative investigation of these bibliographical phenomena, together with that of the growth of the literature of science and technology, has intensified greatly over the last twenty years, and in 1969 Alan Pritchard suggested the name 'bibliometrics' for this emerging discipline. It has acquired not only considerable refinement, largely drawn from mathematics, but also an interesting tendency to scholarly dissension. In 1974, for instance, the editor-in-chief of the French journal *Documentaliste* denounced Bradford's Law as a 'pseudo-scientific imposture'; shortly afterwards the former Director General of the BLLD referred to recent attempts in this field as 'a good example of the worst aspect of the academic approach'. With similar frankness, his successor as Director General expressed the

158

view that much of the writing about 'obsolescence' is characterized by imprecise definition, inadequate data, unsatisfactory analysis and invalid conclusions'; more recently a sour little correspondence on the topic in the *Journal of documentation* closed with the taunt 'none so blind as those who will not see'. Even more fundamentally, the very idea of a bibliometric approach is rejected by some philosophers of science.

One feature of periodicals little commented on is their tremendous variation in size: it has been calculated that the largest publish about five hundred times as much in bulk per year as the smallest. The spread in price is also wide: in 1968 some journals were charging their subscribers over ninety times as much per word as those at the other end of the scale. Circulations too vary widely, from titles like *Popular mechanics* with 1,670,000 or *Science* with 160,000 to many with circulations in the low hundreds.

A word too should be said here on the bibliographical idiosyncrasies of periodicals: the student should school himself never to show surprise at any aberration of titling, numbering, dating, frequency, format, pagination, or publisher. The first issue of *Traffic engineering and control*, for instance, was number 2; the first volume of *The mathematical intelligencer* was numbered 0; the first volume of the *Journal of mechanical design* was numbered 100; volume 40 (1966) of *US government research and development reports* was followed immediately by volume 67 (1967); in 1974 two successive issues of *Technical journal* were numbered 6; the index of at least one volume of *Automotive design and engineering* is two inches taller than the journal itself; in 1957 the periodical known for nearly twenty years as the *Muck shifter* changed its title to *Public works*, but in 1962 it changed back again only to switch once more in 1972 to *Construction plant and equipment*, despite a plea in a letter to the *Times* that its original title was one of its chief assets; there is one periodical which changed its name 41 times in 14 years. In 1973 frustrated librarians founded a quarterly newsletter, *Title varies*, for those of their colleagues who want to fight back, in other words, to prevent 'costly, silly, unnecessary serial title changes'. Even journals which resist the urge to change their titles will sometimes succumb to the temptation to start a new series, for reasons which are usually shrouded in mystery. Between 1843 and 1878 the youthful *Journal of the Chemical Society* started a new series no less than five times. An added complication sets in when the later series, in addition to numbering its volumes from one onwards, also retains the original numbering of the earlier series: in the case of the venerable *Annalen der Physik* from 1819 to 1977 each volume of the latest series carried no less than *three* numbers — that of the first, that of the second and that of the third series.

Another common cause for complaint from library users is the variation between the date on the journal and the actual date of publication; some journals go so far as to print two dates on the cover — the nominal date and the actual date. Equally misleading are those journals which express the date as a season of the year: summer in Australia is winter in the UK, and in many tropical countries there are no seasons as such. And even more impenetrable are the many Japanese journals using a calendar based on the Emperor's birthday. It is very common for periodicals to cease publication without any notification whatsoever. Others peter out over a period, sometimes measured in years, during which it can be extraordinarily difficult to discover whether a particular title is dead or alive. Failure to receive issues is by no means an infallible sign: the *Entomologist* lay dormant thus from 1842 to 1864. Not every journal shows consideration enough to print its own obituary notice in the final issue. One quite classic case is the *American journal of venereal disease* which closed down for lack of anything to publish in the 1950s when venereal disease was thought to be under control. This of course proved not to be so and the journal was revived under a new title, *Sexually transmitted diseases.* And despite the three-hundred-year history of the *Philosophical transactions*, one should not forget that the *average* life span of a journal is less than ten years.

Periodicals are subject to a bibliographical hazard that most books escape — the very common practice of abbreviating their titles in citations, eg J Chem Soc, Engr, Paper Tr J, Mfg Chem, Sci Prog, Phys Rev. Not all are as self-evident as these, eg Text Rec, JACS, ASME Proc, and one still encounters unofficial, home-made, inconsistent abbreviations, eg *Chemical technology*, published by the American Chemical Society, is universally known as Chem Tech (indeed this is the title printed on the front cover), yet the official abbreviation used by *Chemical abstracts*, also published by the American Chemical Society, is Chem Technology! The abbreviation recommended by the International Standards Organization is different again: Chem technol. Unfortunately, there are at least half-a-dozen official systems, all different but still widely used. The kind of problem that this can cause is exemplified by IRE Bull, which could stand for either the Institute of Radium Extraction *Bulletin* or the International Radio Engineers *Bulletin.* Sterling efforts are being made towards standardization at the national and international level. Britain and the United States have consulted closely over the British Standards Institution *Abbreviation of titles of periodicals* (BS 4148:1970-75) — which completely agrees with the International Standards Organization *International list of periodical title word abbreviations* (ISO 833 — 1971-6) and the *American national standard for the abbreviation of titles of periodicals* (ANSI Z39.5 — 1969), and there is substantial agreement on the International Council

160

of Scientific Unions Abstracting Board *International list of periodical title word abbreviations* (Paris, 1970), containing over seven thousand words and word roots with their recommended abbreviations. But even when standardization is universal it cannot alter the millions of confusing abbreviations in the literature to date.

Alternatives to the periodical

So widespread have been the criticisms of the periodical that the student will not be taken aback to hear that serious suggestions have been made to do away with it in favour of some other form of communication. The fact that most of these alternatives are open to objections even graver than those made against the periodical, and the knowledge that where any suggested replacements have been tried they have invariably failed, lead most observers to believe that the periodical will be with us for some time yet. Any proposal that ignores the social role of the journal in scientific communication will disturb the authors; any alternative that eliminates the publisher/editor/referee checkpoint will produce chaos; any solution that disregards economics will fail anyway. It is also to be observed, as J H Kuney of the American Chemical Society has pointed out, that 'attachment to the traditional system is still very strong and any change in the system will be accepted with reluctance'. Nevertheless, the suggestions proffered are worthy of study, for individually some of them do hold out remedies for individual ills. Furthermore, while in no sense supplanting the periodical, a number of the more successful innovations of recent years have very usefully supplemented it.

a *Distribution of papers as 'separates'.* This method preserves the individual paper unchanged, but makes it the normal unit of distribution instead of the usual arbitrary collection of such papers in the form of a journal issue. In 1950 the American Society of Civil Engineers started to issue its papers as separates *instead* of in the usual bound form; some years later the Physical Society, the Chemical Society, and the Institution of Electrical Engineers began issuing their papers as separates *in addition* to the normal bound form; all the schemes failed through lack of support. As J R Smith of *Biological abstracts* explains, 'This highly desirable solution is generally economic disaster for the publisher. It all but eliminates advertisement revenue and requires the printing of a lot of papers for which the demand will be so small that the economic return will in no way recover the costs of production'.

More successful have been the various schemes for providing preprints (not to be confused with the limited-circulation 'preprints' described later) of articles about to appear in regular journals, but of

161

course these do not replace the journals in any way; they merely supplement them by getting important papers into circulation with the minimum of delay. The method adopted by the Society of Automotive Engineers displays one significant variation from most other preprint schemes: only about a third of the nine hundred printed separates end up in the permanent bound volumes of the *SAE transactions*. The remainder are eliminated by means of a rigorous two-level screening by Readers' Committees and Validation Panels. All are abstracted, however, in the *Index-abstracts* volume, and are available on microfiche as a complete set for those who wish.

Preprints should be distinguished from reprints, which are extra copies of an article (usually for the author, and therefore sometimes known as author's extras), run off at the same time as the main printing (and therefore sometimes known as offprints). It is common for scientific journals to provide perhaps fifty or a hundred such reprints to the author as a matter of course, and further copies can be ordered and paid for. Authors use these for sending to their friends, or to other specialists in their field, or to those who request them (for to request such a reprint from the author is an accepted though increasingly burdensome practice in certain disciplines). Some libraries, particularly in biology, collect such reprints on a very large scale, eg the Royal Botanic Gardens, Kew, has well over 100,000. They are of course only reprints of what has already appeared in the literature, but they are usually supplied by the author without charge and in a well-organized collection can be more convenient of access. It has been claimed, however, by W O Hagstrom, that 'Extensive use of reprints reflects a kind of failure of the journal system of disseminating information; the individual researcher does the dissemination normally done by the journal'. There is no doubt that prior to the greatly improved access to journal literature as a result of the more extensive library provision of the last thirty years, such world-wide exchanges of reprints played a major role in scientific communication for the individual worker. There are signs in recent years that the availability of cheap photocopies either from the libraries of their own institutions or from BLLD has been seen by some as offering a speedy alternative to requesting a reprint, although very long papers, or those with photographs, might still be preferable in reprint form. But surveys show that many scientists still maintain private reprint (or photocopy) collections and rely upon them for their work. A survey by D W King for the National Science Foundation found that a quarter of the articles cited by authors of scientific papers had actually been obtained as reprints, rather than consulted in the author's own subscription copy or in an institutional copy. Scientists also appreciate the opportunity for personal contact with fellow-scientists that the reprint request system

162

requires. Change, however, is in the air. The surge in scientific man-power in certain fields has meant that a single paper may stimulate as many as five hundred requests for reprints, and the hapless authors have proposed a Central Reprint Clearing House, using standard request cards, prepaid coupons, and self-addressed labels. Though we still await such an all-embracing service the Institute for Scientific Information has taken one step along the way by offering personalized 'Request-a-Print' cards which can be ordered by the thousand. They are simply mailed to the author whose reprint is wanted and they are ready printed with the requestor's own return address on a peel-off label for the author to use. Costs press upon the publishers too: the American Institute of Physics ties its entitlement to a hundred 'free' reprints to the 'page charge (if honored)'; for some years the American Psychological Association has charged for *all* reprints. The demand in certain major fields can perhaps be gauged from the American Chemical Society which sells about a million and a half reprints from its stable of journals each year.

An interesting compromise between the traditional journal and the distribution of separates is the repackaging of individual papers in a grouping more attuned to particular readers' needs. 'User journals' of this kind have been tried experimentally by the American Institute of Physics, for example: papers selected from its large range of primary archival journals have been reassembled into smaller journals designed for specific user groups. Each subscriber within a sub-disciplinary group receives a journal comprising research papers pertinent to his interests, regardless of their source of original publication. Yet another alternative would be the proposed 'superjournal': the criterion of selection here would be quality, with the new package containing the best papers from a dozen or more primary journals. In one projected scheme this would be a cross-disciplinary journal of Really Important Papers chosen by citation counting and validated by an authoritative body.

A more elaborate system of distributing separates is the scheme tried by the Institute of Electrical and Electronics Engineers: each month a subscriber receives a list of recently published articles from 35 IEEE journals which match his Reader Interest Profile. He can then request reprints of any of the complete articles that he wishes to read. An example of an even more sophisticated approach was the now defunct provision by the American Mathematical Society of 'customized pack-ages or articles produced through the use of computerized profile-matching techniques': the Mathematical Offprint Service offered to the individual subscriber in this way papers selected from sixty journals. One drawback to all such schemes of selective dissemination is that it diminishes the opportunity to browse, for which the conventional journals provide the favourite ground.

b *Synopses journals*: this method is far more radical in its approach inasmuch as it not only replaces the conventional journal but it also requires a fundamental change in the traditional form of the scientific paper. This alternative to the journal contains only synopses of papers: these are lengthier than abstracts, extending perhaps to two pages with illustrations if required, prepared by the authors themselves and able to stand on their own as scientific reports. The original papers, which can be far fuller than at present, are available on request in a variety of forms, perhaps from some central repository but sometimes from the publisher of the synopses journal or even from the author. The idea has been around for a long time: when a plan of this kind was put as a hypothetical suggestion in the 1963 survey by the Advisory Council on Scientific Policy quoted in chapter 12, a majority of scientists were in favour. Since then a number of schemes have been tried, and of course many variations are possible. What is interesting so far is that it is the normally conservative learned societies that have been most prominent in the experiments.

Simplest have been the schemes using synopses within existing journals. In 1970 the American Institute of Aeronautics began including 'Synoptics' in its three major journals. Taking the form of 1,200-word summaries of the key ideas and results of larger papers or reports, this method of publication was offered as an option to authors anxious to see their work in print with a minimum of delay. The back-up papers were of course available for the asking. These 'Synoptics' were found valuable by 8% of the readers of the journals, though 56% said they would not like to see them replacing all papers. At about the same time the Canadian Aeronautics and Space Institute experimented with what were called 'condensations'; in this instance back-up took the form of full text in 'miniprint' in the back pages of the journal. It was assumed that readers had either very good eyesight or a hand lens. Another significant variation was tried in 1974 by *Chemie-Ingenieur-Technik* which began to publish some of its longer papers in the form of one-page synopses in the journal, with the subscriber getting not only the full text on microfiche but the right to reproduce it as often as he required. It was claimed that this change cut publication delays from twelve months to three.

The American Chemical Society has conducted a number of experiments with its journals, of which the most interesting is the 'dual journal'. This comprises a summary journal intended for individual subscribers, and an archival journal for libraries. In 1976 three sample issues of the *Journal of the American Chemical Society* were issued in this double format: the response gave no mandate for permanent conversion but did show strong interest in a summary journal alongside

JACS. Yet another minor variation of the synopses approach is to let later publication of a paper in full depend on the demand produced, *via* tear-out pre-paid postcards, by the appearance of the synopsis.

The first full free-standing synopses journal is the monthly *Journal of chemical research*, started in 1977 by the Chemical Society (now the Royal Society of Chemistry) and jointly sponsored with the French and German chemical societies. Described as 'an attempt to separate the archival and current-awareness roles of the primary journal', it comprises two versions available separately: a printed synopses journal (Part S); and an archival version (Part M) with full papers and background experimental data reproduced directly from author's typescripts, either on microfiche (at a reduction of 1:24) or in miniprint (1:3, ie nine pages of typescript on one journal page), backed up where necessary by full-size hard copy available from the library of the Royal Society of Chemistry. The synopses are normally in English, but the full texts may be in English, French or German. Sometimes the full text can be accommodated within the two pages allotted in the synopses journal, in which case there is no microfiche/miniprint version at all.

Initial reaction was very favourable, with a circulation for the synopses journal of three thousand within six months. Part of its enduring success is due to the speed of publication with a delay of no more than six to nine months from the receipt of an author's manuscript. It has been discovered, however, that those who do not write for it believe that their full texts would not be read; they also think it has lower standards of acceptance than other chemical journals. Contributors too have their worries: they find burdensome the double chore of writing synopses and preparing copy that is 'camera-ready'.

While the Chemical Society were so successfully launching their venture the British Library were funding a two-year experiment by the Institution of Mechanical Engineers which showed some distinctively different features. Entitled *Engineering synopses*, this comprised summaries of papers submitted to the Institution's main journal, the *Proceedings*, that had already been refereed and accepted but not published. Full texts in the form of microfiche or full-size photocopies were available on demand from the library of the University of Bath, which was also given the task of evaluating the reactions of authors, readers, publishers and libraries. What they discovered, not surprisingly, was that 'there is little doubt that authors would opt for full-text publication unless they could be convinced that synopsis publication indeed meant wider dissemination'. Again not surprisingly, there was virtually no demand for microfiche from readers, and even in the hypothetical absence of the full text in any other form there was a distinct lack of enthusiasm. The publishers found costs reduced by half, with significant reductions in publishing delays. Demand was low from librarians who seemed to prefer full text. However, some three-quarters of those surveyed did favour synopses publishing as an alternative provided that the full text was also available. Overall conclusions were that the synopses journal is 'likely to fulfil its best role in serving the individual wishing to update himself rapidly in the broad areas of interest within a single discipline, and that it is less likely to lend itself successfully to highly structured and theoretical subjects, or to the needs of small communities of specialists'.

The view taken by the Biological Information Review Committee appears broadly in agreement: 'The attraction of the synopsis journal is that many scientists read periodicals, particularly those of broad subject coverage, primarily for general rather than detailed awareness'. It seems likely that synopses journals could play an important role in cross-fertilization: scientists are genuinely interested in hearing about work in other specialized fields but, as S W Terrant of the American Chemical Society puts it, 'only from the viewpoint of the main thrust of the work reported and the conclusions reached'.

As was only to be expected, not all experiments have turned out successfully. *Production engineering synopses* was issued by the Institution of Production Engineers in 1978 as 'an expedient rather than an experiment', in the words of R J Millson. Started as an attempt to clear a backlog of unpublished papers it was forced to close through lack of support from authors. At just about the same time the long-established *Geological Society of America bulletin* changed to a two-part format: part 1 contained synopses, typically three pages in length, amounting to some hundred pages a month; part 2, on microfiche, contained complete articles each between 30 and 75 pages. Despite the reduction of the publication delay from two years to four months the change was not popular; authors in particular showed their displeasure by withholding manuscripts to such an extent that the publisher was forced to compromise and accept medium-length papers in part 1.

These cautionary examples remind us that we must not disregard human nature: scientific authors do prefer to publish in highly regarded, refereed, nicely presented and widely circulated journals, and in full rather than in synopses. A recent survey for the Royal Society found that an author's main concern in deciding where to submit his work was the scientific standard of the journal; speed of publication was of much lower significance.

There seems no doubt too that what may suit chemistry, for instance, may not fit other subjects. A A Manten has pointed out that in mathematics and physics, for example, 'reasoning is built up step by step, leaving little possibility for rigid abridgement'. It has also been asserted
166

that 'some types of engineering paper, for example those describing large projects, cannot be summarized severely without losing the main interest for the reader'. One science journal editor is even more positive: 'Basic science contributions must be lengthy, persuasively argued, and based on unequivocal facts. Brief reports, abstracts . . . will not do it'. Possibly the true domain of the synopses journal lies in the descriptive sciences where research is reported in narrative form rather than in a structured argument.

Bibliographical control would be a doubly difficult task: as the editors of the Chemical Abstracts Service recently explained, 'the summaries or whatever constitutes that part of the original disclosure that is published, must be connected by a clear reference to the deposited part. Different names for the two parts will result in different biblio-graphical citations and will double the entry records for the accessing service'. There are many who argue that the full text, *however published*, should be the definitive text and should furnish the prime reference. Librarians too do not relish the doubling of their task. The Chemical Information Review Committee was also very questioning: 'If primary journals mostly become synopsis journals, is there any need for abstracts journals, or should the secondary services concentrate purely upon indexing? Is it wasteful for primary journals to have subject indexes? Can the expense of preparing two abbreviated versions of each paper, the synopsis and the abstract, be justified? Should the whole operation of producing primary and secondary literature be merged over a period of time?'

Should a full synopsis-based scheme ever be introduced on a large scale, complete with central depository, a major difficulty will arise if it is also proposed to permit authors to deposit their papers without the full scrutiny that regularly published papers currently receive. We have been warned by the Head of the Office of Science Information Service of the National Science Foundation, an ardent promoter and stimulator of alternatives to the periodical, that 'Without even more stringent quality-control procedures, the general quality of reporting will decline, and original, quality reporting will be lost in the increasing mass of trivial and redundant material'.

c *Depositories for supplementary material*: we do have some experience of how depositories can assist the periodical in its key role of reporting research through the Supplementary Publications Scheme operated by the British Library Lending Division (and the similar earlier National Auxiliary Publications Service in the United States). Under this scheme authors submit with their papers any supplementary data (background, experimental, statistical, etc) in support of their work but which would not normally be published as part of the paper in question. The journal editors forward this supplementary data to

167

the BLLD where it is stored, normally, on microfiche. With the parent article when published a note is added to the effect that supplementary data is available from the BLLD. This scheme has been adopted by a number of the major research journals, eg *Biochemical journal, Journal of crystal and molecular structure*, as well as by the Royal Society, the National Research Council of Canada, the American Psychological Association, the Institute of Physics and the Physical Society, etc. The American Institute of Physics has a scheme of its own, the Physics Auxiliary Publications Service, described as 'a low-cost depository'. Similar large schemes are well established in the USSR. An experimental scheme in some ways similar is operated by the American Chemical Society which makes space available to authors in the microfilm editions of the nine ACS journals which they can use for the publication of material supplementing that published in the printed journals. This currently amounts to perhaps two thousand pages annually, mostly charts and tables, and is photographed directly from the authors' text. The student should be careful to distinguish this from the 'dual journal' discussed above.

There are signs that the possibility of such 'archiving', as it has been termed, is encouraging change in the nature of the primary journal. J H Kuney of ACS has speculated that 'editors, under heavy economic pressure, may adopt a more stringent attitude on what they will publish in a printed journal since there is now an outlet for whatever additional material might be required in the microfilm edition'; or, one might add, in the depository. Of course, this development does not please everyone. In the course of a 1975 study of the scientific periodical, sponsored by the National Science Foundation, the investigators wrote 'As working scientists we consider having to keep up with paper journals bad enough, but the prospect of also having to be aware of and read this supplementary material is absolutely horrifying'. In fact, if the experience of the BLLD is anything to go by, the deposited material is scarcely ever used.

But any use of depositories inevitably increases the user's dependence on the librarian. Though this may not be disagreeable in principle to librarians, an innovation that merely transfers costs from the producer of the literature to the library is not going to be welcome and may intensify the pressure for user charges. In any event many observers believe that user resistance would be even greater than the reluctance of librarians. This is not mere obstructionism: a contributor to a 1976 workshop on Trends in scholarly publishing told us 'it is well known that even slight inconveniences in present libraries can have drastic effects on the use made of them'. In this of course he was just reminding us of the Principle of Least Effort. But then he went on: 'Resistance to change on the part of users is often regarded as a nuisance by

168

managers, but in many instances it is based on good reasoning. Loss of browsing and circulation facilities would make life more difficult for the reader, especially in those areas of human activity such as creative thinking, the benefits of which are virtually impossible to measure.'

An even more solid obstacle here could be the author. Naturally, he seeks maximum visibility for his work. According to the Publishing Manager of Macmillan Journals, 'The resistance of authors to microform is infinitely greater than the resistance of users and, although it is true that certain authors may accept publication in microform if that is the only possible medium, the general attitude is to withdraw the material and submit it elsewhere whenever microform is mentioned'. Byron felt just the same in the early nineteenth century when he wrote ''Tis pleasant, sure, to see one's name in print'. Perhaps the young scientist struggling for recognition is a softer touch; according to Harold Wooster, who has observed many of them, they soon learn 'it is better to publish something in anything, even if only a government report, than not to publish at all'. Maurice Line says much the same thing: 'if they have to choose between non-conventional publication and non-publication, they will probably opt for the former'.

Certainly depositories seem to be on the increase: Chemical Abstracts Service has identified over thirty such depositories in its field. One journal, they report, 'instructs its readers to write to the author for the full text; thus, every author who publishes in that journal becomes a depository'. And scientific communication returns to the system of personal correspondence of three hundred years ago.

d *Research reports*: as has been mentioned, in certain ways the research report functions as an alternative to the periodical article and is now of course a major source of information in many fields. It will be dealt with at length in chapter 14.

Additional types of periodical
Where some positive progress has been made in the last few years is in the establishment of additional types of periodicals, specifically designed to overcome one or more of the problems of conventional journals as related above. Most successful have been those attempts to reduce the normal delay in communicating new results.

a *Previews journals.* For many years a number of journals have given previews of papers that are to appear later in full in their pages, eg *Analytical chemistry, Biochemical journal*: there are now journals entirely devoted to such previews, eg *Previews of heat and mass transfer*. An example which previews papers that are to appear in a range of journals is *Current physics index*, which summarizes the contents of almost forty American Institute of Physics journals (and a number of European

169

physical society journals) as much as two months before the appearance of the complete text.

b *Communications (or letters) journals.* In a similar way a number of journals have encouraged the rapid publication of urgent research results in brief form as 'letters to the editor' or 'short communications', eg *Journal of the American Chemical Society, Nature, Science*: It is worth a thought that in 1939 it was a series of just such letters to the editor of *Physical review* that proclaimed the birth of nuclear science. This practice has been taken a stage further by a number of titles (mostly offshoots of conventional journals) solely given over to such rapid preliminary communications, eg *Electronics letters, Chemical communications, Life sciences, Biochemical and biophysical research communications.* Such periodicals certainly allow speedy publication in unedited form (to the initiated they are jocularly known as *Acta retracta*): *Tetrahedron letters* claims to be publishing within four weeks, and *Chemical physics letters* within fourteen days of acceptance. Some journals manage to make the best of both worlds: *Physical review* still publishes 'rapid communications' which are themselves announced in advance in its sister communications journal *Physical review letters*. Evidence would suggest that these communications periodicals have a higher actual readership than the conventional types of journals, yet despite their popularity some scientists regard them as 'pernicious and undesirable because of the absence of full experimental data'. A large number of the letters never appear in any other form; in other words, rather than genuine *preliminary* communications of real interest and importance, many of them represent 'dead-end research'. Criticism has been voiced that this facility is 'abused by publication-hungry scientists whose future promotion and grants depend on the quanitity of their publications, since this is more easily assessed than quality'. The editors of *Physical review letters* announced in 1960 'We do not take kindly to attempts to pressure us into accepting letters by misrepresentation, gamesmanship, and jungle tactics'. In 1978 the Physics Information Review Committee warned that when used in this way 'a letters journal becomes a medium for quick and easy, but slack, publication'. They urged 'those who publish letters to make every effort to distinguish which items submitted are preliminary accounts and which are short, self-contained accounts requiring rapid publication to stimulate interest in the field. Separate sections within the journal could be devoted to each type, and refereeing procedures adjusted accordingly'.

'Preprints'
One of the most controversial alternatives to the periodical has been the attempt in various forms to revive individual scientist-to-scientist

communication. Not that this has ever died out completely: so long as the function of science and technology is to push back the frontiers of our knowledge of the intricacies of nature and the application of its secrets to our environment, there will always be a small group of pioneers at these frontiers, in close personal touch with one another. But according to D J de Solla Price, 'There now exist dozens of what we call invisible colleges, each consisting of the few hundred persons who make up the international body of real leaders in their subjects. They are power groups, albeit often unwittingly, and the more power they have the more they gain.' They get their name from the original 'Invisible College' formed by the seventeenth century pioneers who later founded the Royal Society, eg John Wilkins, Robert Boyle, Sir Christopher Wren. Part of this power derives from their custom of circulating within their closed group what are misleadingly called 'preprints' reporting their recent work. Their aim, to spread the good word with the minimum of delay, is entirely laudable, and of course as the most prominent workers in their various fields they do generate much of the significant new information. How widespread is this practice was shown by a study at the Center for Research in Scientific Communication at Johns Hopkins University: it was found that more than half of a sample of authors in scientific journals had distributed at least one 'preprint' before publication. One-sixth of the authors had distributed over 24 preprints each. Unfortunately, these so-called 'preprints' have roused the ire of the editors and publishers of the journals, of librarians, and of course of those scientists outside the magic circle. Not to be confused with real preprints, which as described above are actual printed copies of articles, taken off during the main run, these circulated papers may never in fact appear in print at all (although some of them obviously will). They are often preliminary, conjectural, tentative reports, sometimes handwritten in the early days, distributed as much for comment and criticism as for information, and the feedback from their recipients in many cases induces the author to modify his work. Obviously they do not pass through the normal refereeing process, and may therefore, as one critic claimed, 'propagate erroneous conceptualizations based on shoddy work'. By their nature they cannot be subjected to any form of bibliographical control, and unlike reprints, they cannot be asked for by an outsider, even if (which is unlikely) he gets to hear of their existence. Unfortunately, they are collected, used, and sometimes cited (often in the form 'personal communication'). Herbert Coblans of ASLIB refers to this kind of 'preprint' as 'that bibliographical freak misbegotten out of war restrictions by editorial slowness, and nurtured by human vanity'; in the opinion of J M Ziman they are the 'bastard progeny of priority neurosis and reproductive technology'; and the editor of *Physical review letters*

171

considers that 'plans to add unrefereed preprints to the mainstream of scientific communication will further increase storage of worthless information'. Many would go further, seeing in this reversion to 'the privacy of the seventeenth century' the beginning of the breakdown of the basic science communication system: in his trenchant way D J de Solla Price says: 'I think it means that the old-style function of the scientific paper is a dead duck'.

An interesting attempt was made in 1961 to formalize arrangements of this kind in the field of the biomedical sciences by establishing a number of Information Exchange Groups, with a secretariat at the US National Institutes of Health taking the responsibility for duplicating and distributing the papers to each member of the group. Known as *IEG memoranda*, these communications took a number of forms, eg letters, requests for information, protests, papers already submitted to the journals and awaiting publication: a typical group (IEG no 1: Electron transfer and oxidative phosphorylation) had 725 members from 32 countries, and circulated in six years some 800 *Memoranda*, including an estimated 90% of all the important papers in the field. They were not available to libraries, or to scientists outside the group. Not surprisingly, this experiment soon ran foul of the established journals, understandably aggrieved at being expected to publish articles that several hundred of the top people in the field had already seen, perhaps many months before. In 1966 their editors issued an ultimatum that they would no longer accept papers previously circulated through the Information Exchange Groups. Bitter words were exchanged for several months, the groups being accused of 'an offence against scholarship' and the editors being charged with their 'unwillingness . . . to face up to the obvious facts that the journals can satisfy only a diminishing part of the needs of the scientific disciplines in respect to communication of information'. It was decided to terminate all the groups by February 1967.

A further attempt to solve the preprint problem derived from an investigation of the American Institute of Physics Division of Particles and Fluids which revealed that something like three thousand preprints in their special field of interest were being distributed every year in the usual informal way. As a possible first step towards a Physics Information Exchange in 1969 the US Atomic Energy Commission sponsored *Preprints in particles and fields*, an experimental weekly listing. In alternate issues appeared a particularly convoluted bibliographical development in the shape of a list of 'anti-preprints', described as 'erstwhile preprints which have been published in current journals'. A bibliographical citation to the published version was provided, and the reader was instructed 'The preprints may now be discarded, and reference made to the journal publication'. In the event the proposed Information Exchange and central preprint office came under heavy fire from critics claiming that the effect on the literature would be the same as the establishment of an unrefereed journal without any editor. The plan was abandoned. Nevertheless, the *informal* distribution of preprints remains widespread in many fields of scientific and techno-logical research, and a number of physics laboratories and research institutes still issue lists of preprints received, despite the views of scientists like J M Ziman: 'For my part, I don't count preprints as belonging to the literature of physics, and value 90% of those I receive chiefly as good scribbling paper'. The opinion of Elliot Leader, like Ziman a Professor of Theoretical Physics, is that 'This method of communication is, by now, so firmly established and so widespread throughout the world that not only can it not be ignored as a "private" activity, but it must even be accorded a leading position as one of the principal modes of present-day communication'.

What is interesting to note is that the practice has spread to the social sciences: since 1970 there has been a flurry of activity in the field of economics, for example, with the circulation of thousands of what are called 'working papers', unpublished and distributed privately to fellow workers.

Similar in some ways to *IEG memoranda* insofar as they are privately circulated are research newsletters. Like the *Memoranda* they are found mainly in the biological sciences, eg *Human chromosome news-letter, Laboratory primate newsletter, Microbial genetics bulletin, Mouse newsletter*, but they differ in their general availability: hardly any of them have a restricted circulation, the BLLD at Boston Spa has an extensive collection, and though not published, a number are included in the *World list of scientific periodicals*. Commonly they contain social news, addresses, technical notes, brief research reports, and parti-cularly valuable bibliographies. As basically *news* media they do not compete with the primary journals and have not encountered the same criticisms as the IEG.

Technology and periodical production
Technology made little or no impact on the physical appearance of the scientific journal for the first three hundred years of its life, apart from the introduction of photographic illustrations: the 1966 *Philosophical transactions of the Royal Society* looked remarkably similar to its ancestor of 1666. In 1976, however, that same Royal Society together with the British Academy, its counterpart in the humanities, examined 228 UK learned societies and their activities: continually referred to in their report was the need for greater cost-effectiveness in the methods of printing and distributing society publications. In fact, over the last

fifteen years or so many journal publishers have begun to respond to the charges made by Sir Theodore Fox and others, quoted above, about their neglect of modern technology. Though advancing technology in the shape of the photocopier (and, as we shall see, the computer and the communications satellite) is seen by some as a threat to the journal, other manifestations of technology are coming to its rescue. They have had two kinds of impact: firstly, they have improved the means of providing for conventional needs in conventional form; and secondly, they have provided novel ways of satisfying additional or new needs.

a *Typewriter composition and photo-litho-offset printing.* Advances in printing techniques have now been quite widely applied both to speed the process of dissemination and to reduce costs. As a replacement for traditional hot-metal typesetting and printing it is now quite common to find in-house typewriter composition combined with photo-litho-offset printing ('cold type'), pioneered for scientific journals by the American Society of Civil Engineers in 1953 and by *Physical review letters* in 1958. Going one step beyond this, more recently the ASCE and the American Institute of Physics began to make use of author-typed copy photographed direct on to the printing plate. Savings are not in fact as great as might be expected because fewer words can fit on a page, and some editors have expressed concern about the deterioration of quality in the writing and lack of control over the appearance of the printed page. Indeed there has been a reaction among some journals and they are now striving to re-assert the editorial function in areas where it has all but disappeared.

b *Microfiche.* An even simpler solution is to reproduce the authors' typescripts on microfiche exactly as they stand, eg *Wildlife diseases* appears in no other form. It does indeed look as if part of the solution to the problems of the journal, at least in the immediate future, will be sought by turning increasingly to the cheapness and ease of production (and reproduction) of microforms. This will be bad news for many readers: when biological societies were asked by the Biological Information Review Committee for their views on the future of the periodical, many stated that microfiche would not be acceptable in any role.

Already mentioned is another innovation, the miniprint, made by reducing author-supplied text to about a quarter or a third of its original size in linear terms. On paper of course, not film, miniprint is readable in theory without special equipment; in practice when tried experimentally by the American Institute of Aeronautics and Astronautics in 1971 complaints of eyestrain and headache were received.

c *Photocomposition and computer processing.* The setting of text by photocomposing machines in the late 1960s and early 1970s was a further step, but some editors and publishers maintain that the future
174

lies in using the computer to drive the photocomposer. Already a number of major journal publishers have converted to computer photocomposition, eg the American Institute of Aeronautics and Astronautics, the American Chemical Society, and *Science*. One step beyond this is the use of an optical character recognition device rather than a keyboard to provide input: this is feasible, but OCRs that are versatile enough to read more than a small range of type faces are very expensive.

It is expense, indeed, in the context of the typical small-scale operation of the individual scientific journal publishers, that prompted the suggestion of a common Editorial Processing Centre. Acting as a kind of co-operative publishing office, this would centralize all the functions of journal publishing that could be programmed for computer processing, whilst leaving the editors in full control of the actual content of their individual journals. In the US the National Science Foundation has funded a series of experiments after a feasibility study found that substantial savings should be possible. In the UK a similar study was conducted by ASLIB.

Author-input to the computer is perhaps some way in the future, but there is at least one experiment to show that it is possible even with the present technology. What is bringing it nearer is the advent of word-processors. These accept input from a typewriter-style keyboard but store the text in a memory; the visual display unit then allows the text to be corrected, amended, or edited as often as desired without the necessity for complete retyping. Once the final version is word-perfect the stored text can then be typed out automatically. As it is machine-readable it can also, in theory, provide input for a computer-driven photocomposer. In practice there are still problems of compatibility between many word-processors and some computers, and as what are basically office machines most of them are unable to deal with the mathematical symbols and Greek characters (and of course diagrams, illustrations, etc) common in much scientific writing. But these obstacles will be surmounted in time: the American Institute of Physics is planning to encourage authors to submit their manuscripts in machine-readable form. It has been forecast that by 1985 75% of the world's journal literature will be prepared by word-processors (though not necessarily by the authors themselves) and 60% will be composed by computer.

But will authors desire such 'access to parts of the publishing process previously outside their reach'? As we have seen, they are conservative creatures. They have noted that such a change does transfer to them some of the production costs previously assumed by the publisher, in the same way as does the requirement for camera-ready copy, which they resent. The well known concern expressed by

175

some printing trade unions about this 'new technology' is another difficult aspect of what is basically a social problem.

d *Integrated computerized systems.* Once the whole of the *primary* text of a journal is in machine-readable form the computer can be called upon for far more than simple composing: the whole system can be so ordered that a secondary data base is generated almost as a by-product (see chapter 11). This could take a variety of forms, eg author, title, or keyword indexes, advance alerting services, etc. A trail-blazer in this field of computer-based integrated systems of primary and secondary journals was the Current Physics Information programme of the American Institute of Physics, the world's largest producer of primary journals within a single discipline with over 100,000 pages a year. To pick but one simple instance of the improved service thus made possible they can now produce and print an annual index fast enough for it to be bound in with the last issue of the year.

Finally, the existence of such a machine-readable data base opens up the possibility of making it available for searching by users either in batch mode or on-line if required (see chapter 11). In the case of the machine-readable version of the sixteen American Chemical Society primary journals totalling 35,000 pages a year experiments have now shown that such on-line searching is indeed feasible.

e *The 'electronic journal'.* The logical next step is to make all the machine-readable text available in this way to the reader at his own desk terminal. A Swedish attempt at crystal-gazing predicted that by 1991 storage and retrieval of documents will be replaced by on-line information recovery on a large scale, and by 1994 half of all new scientific literature in the world will be available full-text in computer-readable form. In 1975 the National Science Foundation sponsored a more detailed look into the future by J W Senders and a research team from the University of Toronto. Basing their arguments largely on cost-effectiveness they proposed the complete replacement of the whole scientific output in English by an 'electronic journal' of 47 million pages a year. Papers would be 'published' by being made available for retrieval in an open data base, and subscribers would read them on their own terminals connected to the computer over a telecommunications link. They calculated that such a journal would actually be cheaper than paper journals. At a conference on scientific journals in 1975 a speaker from the National Science Foundation warned 'In this electronic vision, libraries would play a substantially different role than today. Journal holdings, perhaps even many reference materials and handbooks, would disappear, to be replaced by a set of terminals for use by persons not privy to these devices in their offices and laboratories'. But the Toronto investigators were initially more cautious, admitting candidly that 'Unfortunately, there are some reasons —
176

mainly connected with the vagaries of human nature – which might make the implementation of such a system difficult'.

Later they showed more confidence: Senders claimed that 'the electronic alternative offers the possibility of keeping the good things of the print system and eliminating most of the bad ones'. 'Yes, it will happen', was the answer they gave to the question 'Will an electronic journal be published?' But even then they conceded: 'Perhaps users' acceptance will have to await the retirement of older scientists'. F W Lancaster has few reservations: in his 1978 survey on the impact of a paperless society on the library of the future he concluded that both primary and secondary literature in printed form are doomed to extinction. 'The future', he claimed, 'lies with electronic distribution'.

But as May Katzen of the Primary Communications Research Centre points out, 'the long existence of a paper-based communication system has fostered the development of subtle, complex and flexible patterns of assimilating information from printed texts, such as rapid scanning, browsing, or using several texts at once. These reading processes are not easy to undertake when the transmission system is an on-line screen'. Any new medium that bids to *replace* the scientific journal must fulfil all its main roles: current awareness, alerting readers to new work; the archival function, recording the results of research for future consultation; providing a field for browsers; and by no means least, the social functions of establishing priority, controlling quality, and allowing scrutiny by a scientist's peers.

Not unexpectedly, publishers are against the electronic journal, mainly for reasons of finance. Even if their total income were to remain the same, the radical change from the present system of subscriptions paid in advance to charges payable for use would create severe cash flow problems. They also point to the 'practical convenience of conventional print which can be read anywhere, any time, without machinery, and without charge for reading time'. More profound are the legal problems, such as copyright, and the fact that in many countries telecommunications are under state control, even when the printing press is not. Scientists in particular are also concerned about what they see as a basic moral issue: the restriction of access to those in possession of the requisite technology, mainly in the advanced industrial countries.

In the meantime experiments continue. The British Library has funded a three-year investigation at the Universities of Birmingham and Loughborough, and a completely paperless trial electronic newspaper on the subject of human computer factors has been started. A sad failure was the *Mental workload journal*, an experimental electronic journal funded by the National Science Foundation at the New Jersey Institute of Technology in 1979. The problems were of many kinds, but primarily human: in the course of a year nothing got submitted, nothing got published, and there were no subscribers. In the words of J W Senders, 'We have visited the future and it doesn't work!' His conclusion was that 'A system must do what people want it to do or it will be a failure'.

As so often the main obstacle to innovation is not technical but social. The difficulty is not the solving of the practical problems of transmitting information in new formats but in persuading the users to accept them (and pay for them). The Royal Society survey of working scientists reported in 1981 found a majority against the electronic journal.

Lists of periodicals

Over the last fifty years or so a great amount of effort has been expended, largely by librarians, on bibliographies of periodicals. Bibliographical control of the contents of periodicals through indexes and abstracts will be discussed in the next chapter: here attention is concentrated on the first stage of control — the identification and location of periodicals in science and technology.

Whilst relying heavily on the general bibliographies of periodicals (with which the student will already be familiar) and especially those lists like *New serial titles – classed subject arrangement* and David Woodworth *Guide to current British journals* (Library Association, second edition 1973) — both arranged by the Dewey Decimal Classification — which can be used as subject lists, the librarian can also turn to a range of bibliographies devoted to scientific and technological periodicals.

The most useful lists of current titles are the holdings lists of the major scientific and technological libraries, eg *List of serial publications in the British Museum (Natural History) library* (1975) with some 17,000 titles in its three volumes, British Museum *Periodical publications in the National Reference Library of Science and Invention* (1969-71) in three volumes, Science Museum *Current periodicals in the Science Museum Library* (ninth edition, 1965). Holdings lists of general libraries known to have strong scientific and technical collections are equally useful, eg the annual British Library Lending Division *Current serials received* with over 54,000 titles, and the two-volumed Cambridge University Library *Current serials* (Cambridge University Press, 1980). There are holding lists of non-current serials too: the counterpart to the latter title is *Non-current serials: a select list of holdings* (Cambridge University Library, 1978). The most extensive holdings list of all is now the BLLD *Keyword index to serial titles* (1980-) on microfiche with a replacement set every quarter which
178

includes 140,000 titles, current and defunct. As such, of course, they also locate copies (in the case of the BLLD, lending copies) of the journals listed. Keeping abreast of newly published periodicals in science and technology (and of the frequent changes of title, amalgamations, fissions, etc, to which current periodicals are so liable) is greatly aided by the Science Reference Library monthly *Periodicals news* (1970-). The standard retrospective list, with almost nine thousand titles, is H C Bolton *Catalogue of scientific and technical periodicals, 1665-1895* (Washington, Smithsonian Institution, second edition 1897). This is usefully supplemented by S H Scudder *Catalogue of scientific serials . . . 1633-1876* (Cambridge, Mass, Harvard University, 1879), which includes the transactions of learned societies (omitted by Bolton). It excludes technology, but interestingly arranges its 4,400 titles by country and town of origin.

All these lists are comprehensive in intention, but there are a large number of deliberately selective lists using various criteria of selection, eg in *British medical periodicals* (British Council, fifth edition 1977) and *Directory of Indian scientific periodicals* (New Delhi, INSDOC, second edition 1968) the titles are chosen on a national basis; *Commonwealth specialist periodicals* (Commonwealth Secretariat, second edition 1977) and UNESCO *List of scientific and technical periodicals in thirty two countries of Africa* (Paris, 1972) are obviously international; B L Emery and R T Bottle *Gratis controlled circulation journals for chemical and allied industries* (New York, Special Libraries Association, 1971), Midwest Inter-Library Center *Rarely held scientific serials* (Chicago, 1963), Science Museum Library *Periodicals on open access* (1973) and Royal Society *A list of British scientific publications reporting original work or critical reviews* (1950) illustrate four other bases of selection. *Technical and specialized periodicals published in Britain* (Central Office of Information, 1972) is unusual in that it is confined to titles throught likely to be of interest to readers overseas.

Commonly encountered are the many union lists of library holdings: these may be local, eg Sheffield Interchange Organization *Union list of periodicals* (second edition 1970) with over 5,000 titles; or national, eg *Union list of scientific serials in Canadian libraries* (Ottawa, Canada Institute for Scientific and Technical Information, eighth edition 1980) in two volumes, listing 48,000 titles in 250 libraries. The *World list of scientific periodicals* (Butterworths, 1963-5) does indeed live up to its name in including periodicals from all over the world, but as a *union list* it is a national not a world list as it only indicates holdings in British libraries. However, like all union lists this work has a bibliographical value over and above its use as a location tool. These examples remind us that the routine bibliographical task of compiling a union list is well-suited for mechanization: the Sheffield list was compiled with the aid of punched cards and the Canadian work was computer-produced. A by-product of bibliographical mechanization have been the schemes to devise a distinctive abbreviation (or more precisely, code) for each periodical title using as few letters as possible to facilitate the computer operation. Best known are the American Society for Testing and Materials four- and now five-letter codes, which, with the addition of a fifth (or sixth) check letter, are known as CODEN. Although little more than the titles are included, the lists of these codes and their accompanying alphabetical and keyword indexes can serve as a sort of bibliography: ASTM *CODEN for periodical titles* (Philadelphia, third edition 1970) with supplements in 1972 and 1974, for instance, includes 109,000 titles. Responsibility for the CODEN system was transferred to the American Chemical Society in 1975. The *International CODEN directory* (Columbus, Ohio, Chemical Abstracts Service, 1977-) is now issued in microfiche format with supplements twice a year, and covers the more than 150,000 CODEN assigned since 1954.

One of the most far-reaching developments here is the International Serials Data System which it is hoped will do for periodicals (and ultimately for the individual article) what the International Standard Book Number is doing for books. An International Centre for the Registration of Serial Publications has been set up at Paris within the UNISIST framework. The International Standard Serial Number (ISSN) is a seven-digit number plus a check digit. Some forty member states have set up their own national (or in some cases, regional) centres: from 1974, for example, the UK National Serials Data Centre at the British Library has been assigning ISSNs to all new and changed UK serials. Titles recently recorded by these centres are listed in the *ISDS bulletin*, which appears six times a year with cumulating ISSN and keyword indexes. The complete ISDS register is available on microfiche. Furthermore, ISDS will furnish for each journal that wants it an official standard abbreviation for its title in accordance with the ISO standard. ISO has in fact proposed that its standard be withdrawn in favour of an international list of title word abbreviations to be maintained by the ISDS centre at Paris. The effect of all this activity in the last few years means that, for example, the two journals *Inorganic chemistry* and *Canadian journal of botany* now have three officially approved and internationally recognized abbreviations, as follows: the word abbreviations are Can J Bot and Inorg Chem; the CODEN are CJBOAW and INOCAJ; the ISSNs are 0008-4026 and 0020-1669. The International Council of Scientific Unions Abstracting Board has recently published a double concordance providing for conversion of a known CODEN to the corresponding ISSN, and *vice versa*.

The results of modern technology can also be seen in the London

School of Hygiene and Tropical Medicine *Serials catalogue* (Boston, Hall, 1965), a reproduction by the photo-litho-offset method described earlier of the 6,000 periodical title cards in the library catalogue. As this example indicates, there are separate lists of periodicals in the special subject fields within science and technology. A useful guide to these and similar works (a bibliography of bibliographies, in fact) is M J Fowler *Guides to scientific periodicals: an annotated bibliography* (Library Association, [1966]). The student will find that such lists fall into the same categories as those of wider subject scope described above, namely, lists of a library's current holdings, eg *Mathematical and statistical periodicals held by the Science Reference Library* (1977) and the similar lists on other subjects, eg astronomy (1976), geology (1978), etc; or retrospective lists, eg Royal Botanic Gardens *List of periodical publications in the library* (Kew, 1978); or selective lists, eg P C R Mason *A classified directory of Japanese periodicals: engineering and industrial chemistry* (ASLIB, 1972); or union lists, eg ASLIB Textile Group *Union list of holdings of textile periodicals* (third edition 1962). Some of the most useful current lists are those issued by abstracting services to let their users know which periodicals they can scan, as for instance *Publications indexed for engineering* from *Engineering index*, or *Serial sources for the BIOSIS data base*, both annual. One of the largest and best known of these used to be the Chemical Abstracts *List of periodicals abstracted*, but this has now been replaced by the two-volumed Chemical Abstracts Service *Source index, 1907-1979 cumulative*, which includes 50,000 chemical titles (not merely those current) together with locations (including back files) in 400 libraries in the US, Canada, and 26 other countries. It is kept up to date by quarterly supplements, and is also available for leasing in magnetic tape form. A consolidated list of all the titles covered by the abstracting and indexing services that are members has been issued by the International Council of Scientific Unions Abstracting Board: *International serials catalogue* (Paris, 1979), with over 30,000 titles.

Not all lists of periodicals are published separately: a number of valuable examples have appeared in periodicals, eg M M Rocq and others 'Petroleum periodicals' *Special libraries* 36 1945 376-91. Similarly, the lists of periodicals abstracted by many (probably most) services appear only in the abstracting journal itself, eg in the January issue each year of *International abstracts of biological sciences*.

Further reading

B C Vickery 'Bradford's law of scattering' *Journal of documentation* 4 1948 198-203.

R H Phelps and J P Herlin 'Alternatives to the scientific periodical' *UNESCO bulletin for libraries* 14 1960 61-75.

H V Wyatt 'Research newsletters in the biological sciences' *Journal of documentation* 23 1967 321-7.

A A Smailes 'The future of scientific and technological publications' *ASLIB proceedings* 22 1970 48-54.

'Symposium on the primary journal' *Journal of chemical documentation* 10 1970 26-46.

'Record of the conference on the future of scientific and technical journals' *IEEE transactions on professional communication* 16 1973 49-189.

J Aitchison *Alternatives to the scientific periodical: a review of methods reported in the literature* (Office for Scientific and Technical Information, 1974) [Report no 5190].

L C Cross 'The primary scientific literature in the next few years' *ASLIB proceedings* 26 1974 425-9.

M B Line and D N Wood 'The effect of a large-scale photocopying service on journal sales' *Journal of documentation* 31 1975 234-45.

'Record of the 1975 IEEE conference on scientific journals' *IEEE transactions on professional communication* 18 1975 85-308.

K Subramanyam 'The scientific journal: a review of current trends and future prospects' *UNESCO bulletin for libraries* 29 1975 192-201.

M Line and B Williams 'Alternatives to conventional publication and their implications for libraries' *ASLIB proceedings* 28 1976 109-15.

Trends in scholarly publishing (British Library Research and Development Department, 1976) [Report no 5299].

A M Woodward *Editorial processing centres: scope in the United Kingdom* (British Library Research and Development Department, 1976) [Report no 5271].

A M Woodward *The electronic journal: an assessment* (ASLIB, 1976) [BL R&D report no 5322].

E van Tongeren 'The effect of a large scale photocopying service on journal sales' *Journal of documentation* 32 1976 198-206.

'Record of the 1977 IEEE conference on scientific journals' *IEEE transactions on professional communication* 20 1977 49-140.

J Senders 'An on-line scientific journal' *Information scientist* 11 1977 3-9.

M Gordon *A study of the evaluation of research papers by primary journals in the UK* (University of Leicester Primary Communications Research Centre, 1978) [BL R&D report no 5495].

M Katzen *Trends in scholarly communication in the United States and Western Europe* (University of Leicester Primary Communications Research Centre, 1978) [BL R&D report no 5496].

C Balog 'Distribution of offprints: a case study' *Scholarly publishing* 11 1979 73-8.

A J Meadows *The scientific journal* (ASLIB, 1979).

R J Millson *An experiment in synopsis publishing in the field of mechanical engineering* (British Library Research and Development Department, 1979) [Report no 5498].

H P Tseng 'The ethical aspects of photocopying as they pertain to the library, the user and the owner of copyright' *Law library journal* 72 1979 86-90.

Commission of the European Communities *The impact of new technologies on publishing: proceedings of the symposium, Luxembourg, 1979* (Saur, 1980).

EURIM 4, a European conference on innovation in primary publication: impact on producers and users, Brussels, 1980 (ASLIB, 1980).

P Hills *The future of the printed word: the impact and the implications of the new communications technology* (Pinter, 1980).

A A Manten 'Possible future relevance of publishing primary scholarly information in the form of synopses' *Journal of information science* 1 1980 293-6.

A J Meadows *New technology and developments in the communication of research during the 1980s* (University of Leicester Primary Communications Research Centre, 1980) [BL R&D report no 5562].

J F B Rowland 'Why are new journals founded?' *Journal of documentation* 37 1981 36-40.

Chapter 10

INDEXING AND ABSTRACTING SERVICES

Once their short period of currency is over and they have become back files or bound volumes, most journals would remain closed books were it not for those keys to their contents, the indexes. Most of the more responsible periodicals with material of lasting value in their pages do attempt to publish an index to their contents, commonly once a year (or once per volume, if this is not yearly), although there is a great variety of forms to be observed. Probably the most convenient is for the index to be included with the final issue of the year (or volume), printed either as an integral part or separately. Most journals need longer to compile their indexes and have to adopt the practice of including them with a subsequent issue. More frustrating is the index which is not automatically supplied to subscribers but has to be ordered specially, and sometimes paid for as an extra. Each periodical is a law unto itself in this matter, although a number of the bibliographies of periodicals do indicate the practice of individual journals where known, eg *Ulrich's International periodicals directory*, and there is one bibliographical tool designed specifically for this problem: *Directory of title pages, indexes and contents pages* (British Library Lending Division, 1981), covering some nine thousand titles, not all in science and technology of course.

Even more valuable are the cumulated indexes, covering not just one but ten, fifty, or even more years of a periodical, eg *Analyst*, volumes 1-20 (1877-96) and every ten years since; *Scientific American cumulative index, 1948-78*; *Engineer*, 1856-1959; Institution of Mechanical Engineers *Brief subject and author index of papers ... 1837-1962*; *American scientist*, volumes 34-61 (1946-73), a continuation of the index to the first 33 volumes. Keys of this kind can turn a run of a journal into a valuable reference source, and a helpful aid is C M Devers and others *Guide to special issues and indexes of periodicals* (New York, Special Libraries Association, second edition 1976), listing over 1,200 US and Canadian titles.

There are many journals, however, which do not provide any index at all, and many more which provide no more than an author index. When queried, in some cases their editors will maintain that the effort (and the expense) is not justified; in others the point will be made that their particular periodical has a current value only. In some ways more disturbing, because more misleading, are inadequate indexes. Far too common is the index which through half-heartedness, technical incompetence or misguided policy, serves as a guide to only part of a journal's contents. The danger is of course that the searcher will assume that if what he is looking for is not in the index it is therefore not in the journal. As Ronald Staveley has said, 'Some indexing services could certainly be dispensed with, and the work of others would be lightened, if journals were themselves indexed fully and systematically. It is surprising that so many editors continue to neglect this elementary duty.'

Indexing services
For many years, among the most important bibliographical tools for controlling the periodical literature of science and technology have been those indexes which have analysed the contents not just of one but of a wide range of titles. The student can gain a valuable insight into the operation of typical current services of this kind by an examination of two parallel but contrasting monthlies in the same subject field, *Biological and agricultural index* and *Bibliography of agriculture.* The former, founded in 1916 and known for its first 49 years as *Agricultural index*, is an obviously very selective alphabetical subject index to about 180 periodicals. Published by the famous H W Wilson Company, the world's largest publisher of indexes for libraries, it is a characteristic example, with the monthly issues cumulating quarterly, annually (and in the case of the earlier *Agricultural index* every three years as well). Following regular Wilson practice the titles indexed (all in English, and mainly from the US) are chosen by the subscribers themselves in a poll conducted at intervals by the publishers, and price is determined according to 'the service basis method of charge, based on the principle that each subscriber should pay in proportion to the amount of service used'. *Bibliography of agriculture*, on the other hand, was formerly an official US Government publication of the National Agricultural Library. Starting in 1942, it has attempted to index not merely periodical articles but all literature, both domestic and foreign, received by the library. Arrangement is in classified *ad hoc* order, with geographic, author, corporate author and subject indexes in each issue. An average of ten thousand citations per issue makes it well over twice the size of *Biological and agricultural index*. Comprehensive coverage of the subject field is attempted with over 1,200 periodicals and thousands of other publications in over forty languages indexed.

Until a few years ago, it would have been true to say that these two services, despite their many differences, were straightforward conventional indexes. The last twenty years, however, have seen a revolution in the application of machines to indexing. In *Bibliography of agriculture* the changes that were introduced in the 1960s and 1970s are a demonstration of this revolution in practice. In 1962 an internal Task Force was set up in the National Agricultural Library to study how mechanization could meet the anticipated growth of the *Bibliography*, and as a result year by year a series of steps was taken towards an automated system. Starting with the monthly author index, and the annual cumulation, and progressing to the subject indexes, computer sorting and print-out based on input from optical scanning of typewritten text became a reality. By 1970 the effect of all this was to replace the printed indexes altogether within the library itself. The publication of the monthly *Bibliography* for 'routine public needs' was taken over from the NAL by a commercial publishing company using their own software on the magnetic tapes supplied by the library: the data base itself was available monthly in magnetic tape form as CAIN (*Ca*taloging-*in*dexing). And since 1973 on-line search facilities have been provided via the Lockheed DIALOG system or the System Development Corporation ORBIT system. In 1976 the data base was named AGRICOLA (*Agric*ulture *on*line *a*ccess) and currently search facilities are available from a number of commercial and other on-line service suppliers.

And time does not stand still. In future the clear comparison that it has been possible to make may not be so instructive. In the first place, the H W Wilson Company has recently established a Computerized Bibliographical Services Department which is 'engaged in the development of an online input and publication system for the printing of Wilson indexes and in the creation of an online retrieval service'. In the second place, we have had since 1975 a third contender in the ring in the shape of *Agrindex*, the printed monthly indexing journal derived from the data base of the Agricultural Information System for Agricultural Sciences and Technology (AGRIS) sponsored by the Food and Agriculture Organization of the United Nations at Rome. It furnishes an interesting contrast with the two US systems: its deliberately decentralized structure takes input in over forty languages from 76 countries and 7 international or multinational organizations.

Of course the best known and most thoroughgoing mechanization of an indexing service is to be seen at (US) National Library of Medicine. The MEDLARS (*Med*ical *l*iterature *a*nalysis and *r*etrieval *s*ystem) story is too well known to be told again here, but the heart of the system was the complete computerization in 1964 of *Index medicus*, founded in 1879 and the world's largest index in any subject field. With (at that time) some 13,000 citations per month, it was obviously ripe for
186

mechanization. At a cost of about $3 million and 30 man-years of programming labour the computer took indexed citations in the form of paper tape, and within five days (compared with over three weeks by manual methods) produced via the phototypesetter the 600-odd printed pages of the *Index*. Since then of course, the system has been refined and extended, MEDLARS II has been running for some years and MEDLINE is the most extensively used on-line service in the world. An upgraded MEDLARS III is in prospect.

The student will be aware that mechanization in this context (and up till now in the context of any other indexing service apart from the primarily experimental) relates to the clerical process of indexing, eg arranging, sorting, cumulating, printing-out. The intellectual process of indexing each paper for MEDLARS still takes between ten and fifteen minutes of the time of a graduate literature analyst. Gerald Salton of Cornell University wrote in 1970 about mechanized indexing 'All the available evidence indicates that the presently known text analysis procedures are at least as effective as more conventional manual indexing methods . . . there are no obvious technical reasons why manual document analysis methods should not be replaced by automatic ones'. But real progress in applying machines to the intellectual tasks involved in indexing is very slow, despite intensive research.

Characteristic of indexing services for science and technology (but not of abstracting services) is the attempt to cover the whole field. According to R T Bottle, 'The dream of a comprehensive multi-disciplinary bibliographic index is probably almost as old as librarianship itself, yet the ever increasing flood of literature of all types and subjects must doom any such project to failure'. One cannot deny that the path is strewn with the remains of failed indexes of this kind (eg *Index of technical articles, Cleaver Hume technical article index, Pandex: current index to scientific and technical literature*), but we do have the example of *Applied science and technology index*, which under its earlier title *Industrial arts index* has been covering a very wide range of disciplines since 1913. Apart from its highly professional approach (it is one of the oldest of the H W Wilson subject indexes), it probably owes its success to the care taken to respond to the needs of its users in choosing the three hundred periodicals to be indexed. Obviously, these make up only a fraction of the total journals, and all are in English (mostly published in the US), but librarians can testify to its value in libraries of all kinds.

Although established only since 1962 the corresponding *British technology index* has many features of interest to the student. Like a number of other indexes it has been fully computerized in stages over the last few years. Once again, of course, it is only the clerical procedures that have been mechanized: 'The production of *British technology index* is *human* based, and there is no intention in the foreseeable future to try to dispense with the human intellectual effort which we believe to be necessary to achieve the standard of retrieval performance which we have set ourselves'. Even more significantly, however, it has served as a practical focus for a second revolution in indexing to parallel the computer revolution mentioned above. This of course is the great surge of interest in subject-indexing theory over the last decade or so. Compared with earlier examples, BTI uses a very sophisticated classificatory method of indexing, with headings consisting basically of a string of subject terms, together with inversion cross-references, and synonym and relational cross-references. This approach contrasts interestingly with the *Index medicus* method, also designed in the light cast by the latest research into indexing theory: subject headings and cross-references are drawn from MeSH (Medical subject headings), a 16,000 term thesaurus published annually in two volumes: *Annotated bibliographical list* and *Tree structures*. It is relevant here to recall Cyril Cleverdon's view that for searching a mechanized data base controlled-language indexing of the usual kind is irrelevant: 'searching on the natural language terms as found in the title and abstract is at least as efficient as searching on controlled language terms assigned by an indexer'.

British technology index changed its title in 1981 to *Current technology index*. No mere cosmetic or capricious substitution, this coincided with a major rethink, for users were told that 'Changes are taking place in the following areas: references, in particular the form of entry; terminology; and page layout'. It is perhaps too soon to assess the effects yet, and in any case more changes are planned, but there are indeed markedly fewer references (ie cross-references, not entries), the terminology has been brought 'more into line with that used in the literature', and the pages have been rearranged 'to draw the user's attention to blocks of entries and blocks of references in a "clean", attractive manner'. The list of titles indexed has been expanded by a further twenty and consideration is being given to others, in particular non-UK titles in English, feature articles in newspapers, and certain 'popular' journals.

In fact the indexing of popular, non-specialist journals in science and technology has been neglected territory until recently, when since July 1978 we have had monthly from H W Wilson *General science index* with the familiar quarterly and annual cumulations, covering 89 English-language periodicals; it is designed particularly for the needs of the non-specialist, the student, and the general public library reader. Less obviously 'professional' but well worthy of study is *Clover information index* quarterly since 1975 with annual cumulations, an admirable attempt on a shoe-string budget to provide a subject index to about a hundred 'popular' magazines of the mind regularly found in small libraries but often unindexed elsewhere. Distinctly practical in nature, most of the titles have a scientific or technical content, eg *Motorcycle mechanics, Sky and telescope, Practical computing, Everyday electronics, Flight.* A similar and now well established Australian example is *Pinpointer,* published six times a year by the Libraries Board of South Australia. This likewise concentrates on some fifty 'popular' periodicals, mainly Australian, not indexed elsewhere, eg *Aeromodeller, Electronics Australia, Practical motorist, Your garden.*

Some indexes, of course, confine themselves to the periodicals taken in a particular library, but where that library is a major collection in its own special field the index may have a use far beyond the library's own or potential users, eg the quarterly *Anthropological index to current periodicals in the Museum of Mankind.*

To the aid of the human indexer the computer brings nothing more than greatly increased manipulative power. It is this power which allows him to dispense with much of his clerical help and still to produce his index in a tenth of the time. But it also permits him to consider methods and processes that he would otherwise dismiss as far too tedious and time-consuming for even the most junior human hand. This is the explanation for the revival of interest in rotated indexes and citation indexes. Both have been known for many years, but are exceedingly laborious to compile. In a rotated index (used as long ago as 1864 by Crestadoro in the printed catalogue of the Manchester Public Library) each of the keywords in a title serves as the subject heading in its turn, the title being printed as many times as there are keywords: in its commonest computerized form, the printout centres these keywords down the page with the rest of the title (still in the same order) 'wrapped round'. Its great advantage is that its compilation needs no 'intellectual', ie indexing, effort at all, and some of its recent success is probably due to shrewd publicity combined with a brilliant choice of acronym in KWIC (keyword in context). Much criticism is heard about the actual physical difficulty of using such indexes, and more basic doubts are frequently expressed on the reliability of the title alone as an indicator of a paper's content, but so speedily can they be produced that they are now widespread. A typical example to study is *Chemical titles.*

Citation indexes

A citation index is a list of cited articles, under each of which is a further list of documents where they have been cited within the period covered by the index. This system has been used in the field of law for many years in the US (*Shepard's citations*), and was used in the name index to *Physics abstracts* from 1898 to 1924, when it was abandoned probably because the manipulation of the increasing mass of material had become too burdensome by manual methods. All that is novel, therefore, is the use of computer processing. Again, the great advantage of this method is that its compilation is a purely clerical operation, and no subject indexing as such is done. In effect the author of a paper serves as his own indexer: each time he provides a citation in his paper he is indexing some aspect of his own work and re-indexing the scientific literature. The success of a citation index depends of course on the extent to which items cited in the bibliographies to papers reflect the contents of those papers. In searching a special technique is necessary: before he can begin, for instance, the user must have a starting point; he must know at least one article on the subject of his search. By far the best known is *Science citation index*, covering some 2,400 journals containing over half-a-million articles a year. These 'source' items generate something over three million cited references. To manipulate this data manually would be a superhuman task, but the computer produces not only the quarterly issues of *Science citation index* itself but the companion *Source index* (authors and titles of citing articles), *Corporate index*, and *Permuterm subject index* (a pre-coordinated index showing the permutations of all possible pairs of terms derived from the title of each article).

An interesting by-product here for the student to note is the mass of data that citation indexes furnish for bibliometric analysis (see chapter 9). Examination of the patterns of citation within particular subject fields is providing valuable new insights into networks of communication among scientists, and giving us new yardsticks to measure, for instance, the eminence of individual workers within a discipline. Of course such data has to be used with care, but Harriet Zuckerman found that in the 1960s a typical Nobel prize winner would receive at least two hundred citations in the year *before* the award. William Stuckey used citations to predict the physics award to Arno Penzias and Robert Wilson. Not the least curious of the findings from citation analysis studies is that successful drugs leave 'bibliometric tracks' quite different from unsuccessful drugs.

Retrospective indexes
An indication of the importance attached by scientists to the periodical literature and its bibliographical control is the effort expended on retrospective indexing, giving science advantages still not shared with most other disciplines. Outstanding is the Royal Society *Catalogue of scientific papers, 1800-1900* (Clay, 1867-1902; Cambridge UP 1914-25), covering in its 19 volumes over 1,500 nineteenth century periodicals.

This is an author index only, however: the complementary *Subject index* (Cambridge UP, 1908-14) was abandoned incomplete after only three of the projected 17 volumes. Also abandoned was the successor to this venture, the ambitious *International catalogue of scientific literature* (1902-19), designed 'to record the titles of all original contributions since Jan 1, 1901'.

The fact that this last title included books as well as papers in journals reminds us that many indexes, current and retrospective, cover more than just periodicals. As we have noted (chapter 8) they combine the features of an index (ie of periodical articles) and a bibliography (ie of books), eg the *Annual index to IEEE publications*, covering not only some 45 Institute of Electrical and Electronics Engineers journals, but papers presented at over a hundred conferences, standards, technical reports, etc, totalling 100,000 a year, the annual *Zoological record*, B Dean *Bibliography of fishes* (New York, American Museum of Natural History, 1916-23), with 35,000 titles, 'all the published references to fishes', in three volumes. And occasionally to be encountered is the composite book index, eg L J Fogel *Composite index to marine science and technology* (San Diego, Cal, Alfo, 1968), a collation of the indexes of 30 books. More recently, from the publishers of *Science citation index*, we have had *Current book contents*, the first attempt on a regular basis to analyse in detail the individual contributions to 'non-journals'; the contents of such multiple-author books are now analysed in the series of half-dozen weekly 'current contents' journals from the same publishers (see below).

Abstracting services

To fit the definition of an indexing service, all that is necessary is for the citation to provide sufficient bibliographical information about each item to enable it to be identified and traced. In practice this means that the scientific or technological information content of a reference is limited to the subject-heading chosen and the title (or to the title alone in title-based indexes such as KWIC). It is probably true that titles in science and technology are on the whole more informative and explicit than in other fields; some indexes do 'enrich' titles with words and phrases added by the indexer; and descriptors accompanying a citation can sometimes be pressed into service as a 'skeleton' summary of subject content. Nevertheless, as sources of information, indexes have obvious limitations. Abstracts, however, are archetypal secondary information sources: comprising not merely citations but also summaries of the content of publications or articles, they manifestly 'organize the primary literature in more convenient form'. As a current-awareness tool for the scientist or technologist the abstracting service is dual-purpose: not only does it alert him (as an indexing service does) to newly-published work that the law of scattering has so dispersed that he would without its aid miss completely, but it can often obviate the actual perusal of the original documents. Faced with the steadily rising tide of primary publications, research workers not surprisingly grasp eagerly at services which save them time. Even where the user judges that access to the original is essential the abstract in the meantime will serve as a provisional substitute, particularly in those frequent cases where the full document is difficult to obtain, or in an obscure language, or published in condensed form only (as are some conference papers). A 1980 survey of the several abstracting journals published by the Commonwealth Agricultural Bureaux found that they would cease to be of value to over half the users if the abstracts were dropped and the entries were to contain bibliographical citations only.

Retrospectively, as a repository in summary form of the literature in its field, an abstracting service permits retrieval of specific information. Indeed, many writers lay great stress on what might be called this 'encyclopedic' function, eg 'The index of an appropriate abstracting journal should be used whenever information is being sought on any subject, and it is likely to lead to information of more value than would be obtained from many other books of reference' (ASLIB *Handbook*). And of course one must not forget the browsing value of an abstracting journal.

A distinction is commonly drawn between an *indicative* abstract ('a brief abstract written with the intention of enabling the reader to decide whether he should refer to the original publication or article'), and an *informative* abstract (which 'summarizes the principal arguments and gives the principal data in the original publication or article'). In practice there are many abstracts published which do not fall clearly into one or other category, and many services (a quarter of 130 surveyed in 1962) publish both, eg *Horticultural abstracts, International petroleum abstracts*. In any case it is probably unwise to rely solely even on very long informative abstracts: any serious worker would also consult the original. Most abstracts, whether indicative or informative, are merely descriptive, with no attempt at evaluation. Indeed, abstractors are commonly instructed to avoid such an assessment, eg 'An abstract should be impersonal . . . and should not take critical form . . . the only acceptable criticism is that of giving very little space to the abstract, or of ignoring the article entirely if it is definitely lacking in value'. *Chemical abstracts* strives for even greater objectivity: 'Judgement as to the importance of papers should be left largely to the present and future users of the journal'. The subtitle of *Psychological abstracts* reads 'nonevaluative summaries of the world's literature'. Of particular interest therefore are those few services which do attempt evaluation,

192

either in the text of their abstracts, eg *Computing reviews, Applied mechanics reviews*, and a number of medical abstracting services; or by their policy of selecting only 'some of the more important and interesting recently published papers', eg *Chemistry and industry, Food manufacture*. Of course within a particular organization or institution, in private or internal abstracting services, evaluation and criticism are sometimes expected by the users.

Form of abstracts
In contrast to indexing services, a large number of abstracting services are not separately published but appear as a feature within a particular journal, eg *Vacuum, Journal of the Institute of Brewing, Ultrasonics, Glass technology*. This, of course, is a practice far older than the pure abstracting journal, of which the earliest authentic example appeared in Germany in 1714, but it seems to be on the decline. A number of well-known journals that used to carry abstracts have recently abandoned them, eg *Journal of the science of food and agriculture, Journal of applied chemistry* (now *Journal of chemical technology and biotechnology*). Occasionally abstracts can be found appearing in review journals, eg *Metron: SIRA measurement and control abstracts and reviews*.

But the most obvious form for an abstracting service is a journal devoted to abstracts, eg *Dairy science abstracts, Lead abstracts, Computer and control abstracts*. Some of them aim at comprehensiveness, trying to abstract every publication or article appearing in their subject field which contains valuable or original material, eg *Fuel abstracts, British ceramic abstracts*. Others are deliberately selective, and include only those items which they think are major contributions to knowledge or are likely to be of use to a particular class of reader, eg *Plastics abstracts* confines itself to British patents; *Patents abstracts bibliography*, issued twice a year by the National Aeronautics and Space Administration, is even more closely confined to patents obtained to protect NASA's own inventions, but which are nevertheless available for use under licence. *Core journals in pediatrics* abstracts articles only from those journals that bibliometric studies have identified as providing a high proportion of significant papers. A variant of the abstract journal is the abstract annual, issued in the form of a bound volume, eg *Energy index*.

A format very common on the Continent of Europe but less frequently encountered in Britain and America is the abstracting service on cards. A trifle unusual in that it uses 6in by 4in cards rather than 5in by 3in is *Training abstracts*; even more striking is the American Society of Civil Engineers *Publications abstracts* with removable cards

193

in journal form; each of the pages making up the now suspended *Building science abstracts* could be cut into four A6 sized cards, and each of the pages making up *Water resources abstracts* can be cut up into three 5in by 3in slips. To help in information retrieval some services use 80-column punched cards, and *Instrumentation abstracts* uses edge-notched cards. A number of services provide cards as well as a conventional journal, eg *Metal finishing abstracts, Apicultural abstracts, Engineering index* (an abstracting service despite its title). Several journals provide the wherewithal to make your own cards by printing their abstracts on one side of the paper only ready for mounting, eg *Analytical abstracts*, (and in some cases already gummed), eg *Journal of animal ecology*. In the case of *Packaging abstracts* the reader is offered the choice of a journal, cards or paper. *RAPRA abstracts* from the Rubber and Plastics Research Association is available as a journal, or on cards, or as computer printout either on cards or on fan-fold paper. Gummed paper is often used for the so-called 'homotopic' abstracts (again more common in Europe): these are usually printed on a sheet inserted near the front of a journal and comprise abstracts (often by the author himself) of the actual articles in that issue, eg *Glass technology*, where the abstracts are in French and German as well as English. Loose-leaf services are also occasionally encountered, usually with each abstract printed on a separate sheet, eg *Quality control and applied statistics abstracts service, Operations research/Management science abstracts service*. A number of services will provide subscribers with specific sections of the regular journal, eg the INSPEC Tailored Abstracts, which permit the individual user to choose any of the 55 chapters from *Electrical and electronics abstracts* and *Computer and control abstracts*.

Frequency of publication obviously can vary from daily upwards, with perhaps monthly being the commonest. *Abstract journal in earthquake engineering* is a rare example of an annual. Cumulations (of the actual abstracts, as opposed to the indexes), are rare, but not unknown: a very well-known instance is *Engineering index*, a monthly which cumulates annually. Cumulations covering narrower subject areas within the main field are produced by *Solid state abstracts*, eg crystal physics. There are isolated examples of abstracts which do not appear currently at all: *Abstracts of microbiological methods* (New York, Wiley-Interscience, 1969) is a book containing a collection of abstracts from the literature of the previous 25 years.

Categories of abstracting services
There are probably about two thousand abstracting and indexing services currently available in science and technology: 1,855 of these were listed in 1963 in *A guide to the world's abstracting and indexing services in*
194

science and technology (Washington, National Federation of Science Abstracting and Indexing Services), and the British Library Lending Division has published *A keyword index of guides to the serial literature* (1974) with over 1,500 titles. Another holdings list is A Mukherjee *Abstracting and indexing periodicals in the Science Reference Library* (1975) with some 1,500 titles. A national guide is *Inventory of abstracting and indexing services produced in the UK* (British Library Research and Development Department, 1978) with some 350 services, mainly in science and technology; this also includes computerized services. The student could never hope to familiarize himself with more than a fraction of these, but like periodicals, for purposes of study abstracting services can usefully be grouped according to their source of origin, as follows:

a *Learned societies and professional bodies*: eg *Photographic abstracts* (Royal Photographic Society); *Mineralogical abstracts* (Mineralogical Society of Great Britain); *Physics briefs* (American Institute of Physics); and *Applied mechanics reviews* (American Society of Mechanical Engineers) and *Mathematical reviews* (American Mathematical Society), both of which are abstracting services despite their titles.

b *Academic institutions*: eg *Abstracts on hygiene* (London School of Hygiene and Tropical Medicine), *Geomechanics abstracts* (Royal School of Mines), *Petroleum abstracts* (University of Tulsa). And as with periodicals (see chapter 9) in some instances the actual publisher is a commercial house, eg *Ergonomics abstracts* (University of Birmingham) from Taylor and Francis.

c *Governmental bodies*: the state is often responsible for abstracting services issued by its various organs such as ministries, departments, commissions, boards, etc, eg *Industrial ergonomics abstracts* (British Steel Corporation), *Health and safety laboratories abstracts* (Health and Safety Executive), *Monthly review of technical literature* (British Rail), *HRIS abstracts* (US Highway Research Information Service). International governmental bodies too produce abstracting services, eg *Euroabstracts* (Commission of the European Communities). Some of the best known are the series of over two dozen abstracting journals from the Commonwealth Agricultural Bureaux, which represents 26 Commonwealth countries, eg *Plant breeding abstracts, Weed abstracts.* An example involving even more countries is *INIS atomindex*, the twice-monthly computer-produced journal of the International Nuclear Information System of the 120-nation International Atomic Energy Agency. Formerly, as its name implies, an indexing service only, from 1976 it began to include full abstracts, superseding the long-standing US service, *Nuclear science abstracts.*

More commonly in Britain, state support of abstracting services has taken the indirect form of a grant-in-aid to bodies such as industrial research associations, which often publish abstract journals as part of their function eg *Automobile abstracts* (Motor Industry Research Association), *Journal of abstracts* (British Ship Research Association). Some titles in this category, however, are restricted to members, eg *British Leather Manufacturers Research Association journal*, others used to be but have now been made more widely available, eg *Steel castings abstracts* (Steel Castings Research and Trade Association). A feature of recent years has been the number of cooperative ventures in abstracting, eg *World textile abstracts*, a combined operation by five research associations in the textile field; *Food science and technology abstracts*, a computer-produced journal prepared jointly by the Commonwealth Agricultural Bureaux (UK), Institute of Food Technologists (US), Centrum voor Landbouwpublikaties en Landbouwdocumentatie (Netherlands) and Gesellschaft für Information und Dokumentation (Federal Republic of Germany).

The student will have noticed that a number of these governmental abstracting services are what is called 'mission-oriented' as opposed to 'discipline-oriented'. It is a fact that the organization of abstracting services reflects the organization of science itself. Large-scale public funding (eg, defence, space, pollution, energy) has produced a situation where much scientific and technological research is 'directed to the solution of complex problems in a society rather than to the advance of knowledge in an academic field'. Abstracting services oriented to missions are characteristically interdisciplinary.

In certain countries it is found that the state itself operates specialized abstracting organizations, eg Centre National de la Recherche Scientifique (Paris) compiles *Bulletin signaletique*; VINITI (Moscow) compiles *Referativnyi zhurnal*; the Indian National Scientific Documentation Centre (Delhi) compiles *Indian science abstracts*.

The last title reminds us that a specific national orientation is often to be found in abstracting services. Coverage may be deliberately restricted to the literature produced within their own country, eg *Korean scientific abstracts* (Seoul, Korea Scientific and Technological Information Center), or the literature relating to their own country, eg *Thai abstracts: series A Science and technology* (Bangkok, Thai National Documentation Centre). Such abstracts are often produced in an English-language version specifically for distribution abroad, eg *Abstracts of Rumanian technical literature, Scientific abstracts* (People's Republic of China).

d *Independent research institutes*: these are not common, and are usually confined to topics of current interest to the institute, eg *Selected Rand abstracts* (Rand Corporation).

e *Commercial publishers*: so laborious and thankless a task is abstract compilation that most publishing houses steer clear, but there

196

are exceptions, eg *Acoustics abstracts* (Multi-Science), *Oceanographic abstracts and bibliography* (Pergamon), *Genetics abstracts* (Information Retrieval), *International petroleum abstracts* (Heyden).

f *Industrial and commercial bodies*: abstracts issued by individual firms are usually produced by their library or information departments for internal consumption only, but many do circulate more widely, often free of charge, eg *Nickel bulletin* (International Nickel Co), *Sugar industry abstracts* (Tate and Lyle Ltd), *Titanium abstract bulletin* (ICI Ltd, Metals Division). Trade development associations are active in this field also, eg *Cadmium abstracts* (Cadmium Association), *International copper information bulletin* (Copper Development Association), *Zinc abstracts* (Zinc Development Association).

g *Specialized abstracting services*: a number of the larger abstracting services are run as independent organizations, sometimes incorporated. Theoretically self-sufficient they may on occasion receive financial support from the state, eg *Biological abstracts* (BioSciences Information Service), *Energy abstracts* (Engineering Index, Inc).

h *Libraries and information services*: as the student will have observed in the examples noted, the libraries of learned societies, government departments, research associations, industrial firms, etc, frequently take or share responsibility for the abstracting service issued by their parent bodies, eg *Current information in the construction industry* (Property Services Agency Library Service), *Department of Environment and Transport library bulletin.*

It would be inappropriate to the purpose of this textbook to describe in detail any particular abstracting services, but it is essential for the student wishing to appreciate their role in scientific and technological communication to devote some time to the actual physical examination of major services in their basic printed form, eg *Chemical abstracts, Biological abstracts, Science abstracts* (comprising *Physics abstracts, Electrical and electronics abstracts, Computer and control abstracts*). From this he should then move on to a broader study of the recent radical advances made by these services. To some extent these developments have been forced by the increasing inadequacy of manual methods in the face of the ever-growing flood of publications, and typically take the form of a total computerized system, not only producing the basic printed abstracting journal but offering a range of other services also, eg CAS (Chemical Abstracts Service), BIOSIS (BioSciences Information Service), and INSPEC (International Information Services for the Physics and Engineering Communities) (see chapter 11).

A descriptive account of some 140 services is D B Owen and M M Hanchey *Abstracts and indexes in science and technology* (Metuchen, NJ, Scarecrow, 1974).

197

Coverage of the literature

Estimates prepared at the National Lending Library (now the British Library Lending Division) indicate that the 26,000 scientific and technological periodicals current in 1965 contained approximately 850,000 authored articles. King Research estimated a total of 420,000 scholarly refereed scientific and technical articles in the US alone in 1980; A A Manten's estimate of the world total of scientific articles in 1975 was 2,500,000. Other estimates have ranged as high as 3.5 million, but none have approached 7.5 million, which is a reliable estimate of the number of references produced annually by the indexing and abstracting journals. Clearly duplication must be widespread. In a classic series of statistical tests carried out for ASLIB it was shown that 47% of a sample of 3,420 references were abstracted more than once, and in some subjects 22% were covered four times. More recently in an investigation of the journals abstracted by BIOSIS, CAS, and *Engineering index* it was found that 27% of the 14,592 journals monitored were abstracted by two of the three services. Overlap between BIOSIS and CAS was 40% and between *Engineering index* and CAS was 43%. Of course journal overlap does not necessarily indicate article overlap because not every article in every journal is abstracted, but a follow-up study of the actual articles abstracted during a sample twelve months showed an overlap between BIOSIS and CAS of 48,856 articles and between *Engineering index* and CAS of 21,583 articles.

It is sometimes advanced as justification for such overlapping that each abstract is made from a particular viewpoint, and a summary of an article prepared for a chemist, for instance, will not serve the needs of a biologist. There is some truth in this, and the Commonwealth Agricultural Bureaux, for example, responsible for over two dozen abstracting services, often prepare three or four abstracts of the same paper deliberately, each with a different subject-slant. Similarly, in the USSR the All-Union Institute of Scientific and Technical Information (VINITI) finds that as many as 15% of the abstracts it prepares appear in more than one of the various sections of the *Referativnyi zhurnal*, but only 3% are used without modification: 'Experience of single abstracting has shown that it is very difficult to obtain a universal abstract. The sectional editorial boards prefer to get their abstractors to match their abstracts completely to the various branches of science involved'.

But the operation of such deliberate differential abstracting policies is not a sufficient explanation for so much duplication: the ASLIB tests showed very little evidence of genuine slanting in duplicate abstracts, and an earlier US investigation of papers covered twice or more in nine abstracting services showed that most of the abstracts were in fact written by the authors of the papers themselves, and in a

198

very large number of cases they had appeared with the original papers. Duplication is still very evident among the new computerized systems, and in 1974 fourteen major science abstracting services, all except one members of the National Federation of Abstracting and Indexing Services, agreed to participate in an overlap study sponsored by the National Science Foundation. The investigators confined themselves to the journal literature. 25,902 journals were scanned by the 14 services in 1973, of which 10,560 were scanned by two or more services. Only 5,466 of them, however, were found to have at least one article selected for abstracting or indexing by two or more services. Actual *article* overlap was found for only 23.4% of the articles for this group of 5,466 titles, but 108 articles were covered 6 times and 9 articles were covered 7 times. Within a specific subject field a 1979 study of the three major world agricultural data bases already referred to in this chapter (AGRICOLA, AGRIS and CAB) found a 20% duplication of periodical titles between AGRIS and CAB; 25% between AGRICOLA and CAB; 24% between AGRICOLA and AGRIS; and 10% among the three.

A far more disturbing problem of coverage is omission. The ASLIB tests showed that 21% of the 3,420 references were not covered by the abstracting services, supporting an earlier US survey that had concluded: '. . . other techniques must be combined with the judicious use of abstract journals and indexes for the greatest possible efficiency'. In 1964 an examination of coverage of the electrical engineering periodical literature found similarly that 'The published abstracts journals do not abstract material completely enough . . . to make our own [ie internal] indexes unnecessary'. In 1962 a writer on the 'Information crisis in biology' painted an even gloomier picture: 'The biological literature which is abstracted and indexed is less than one quarter of that published'. It is probable that the new computerized systems in both these fields have improved coverage, but there is still no solution yet to one interesting problem discovered by the ASLIB team: '. . . we have been unable to identify any general reason why material is not covered by abstracts services. There is certainly no evidence that the material not abstracted is irrelevant or of lower quality.'

The student should also be careful to distinguish when examining abstract journals between nominal and actual coverage: a number of services cover their listed journals only selectively, not to say haphazardly. Helpful to the user here is the practice adopted by INSPEC of systematically listing separately those of the journals scanned that are so central to the subject field and of such a high quality that they are abstracted completely. The student should also remember that some services exclude some forms of literature, eg books, patents, theses, research reports, conference papers, etc. Coverage from year to year may not be consistent: the reasons may be financial rather than scientific, but on occasion they may be merely capricious. Some abstracting services are deliberately selective as a matter of policy, eg the series of *Key abstracts* issued by INSPEC: they choose references only from the most important journals. And of course Bradford's Law of Scattering (see chapter 9) is highly relevant here: one could scarcely wish for a more precisely defined subject than turtles, yet we are told that the literature on turtles appears in 600 journals.

Author abstracts
The editors of many journals now insist that authors provide abstracts for their papers: physics journals in particular lead the field here, but biology, for example, has been slower to follow suit. Normal practice, of course, is for such author abstracts to be printed for the convenience of the reader at the head of the article below the title, and as we have noted, they are frequently repeated verbatim in the abstracting journals. Indeed for a quarter of a century the Abstracting Board of the International Council of Scientific Unions has been active in encouraging this trend, and in showing how by careful construction an author abstract can be made more usable. In 1968 UNESCO published a *Guide for the preparation of author's abstracts for publication.* That such efforts are bearing fruit is evident from a *Chemical abstracts* report in 1973 that 'during the past few years we have seen a marked improvement in the quality of the abstracts that accompany a primary journal article'. The widespread acceptance of such author abstracts by respected abstracting services does indicate that subject-slanting is not always necessary, and that it is possible to produce a useful 'neutral' abstract. INSPEC uses the author abstract whenever possible provided it is 'an adequate and reasonably unslanted summary'. The scientists surveyed for the 1981 Royal Society report on the UK scientific information system were quite willing for author abstracts to be used unchanged in abstracts journals. And who better, many would say, than the author to distil the essence of his own paper?

On the other hand, many professional abstractors would maintain that their craft demands special training and experience, and the major abstracting journals impose the most rigorous standards on their abstractors. In *Chemical abstracts*, for instance, there has been a marked shift towards full-time, on-site specialists away from volunteer, part-time, abstractors, down from 3,245 in 1967 to 1,029 in 1979. H E Kennedy of *Biological abstracts* is of the opinion that 'It is generally difficult for authors to avoid the bias or emphasis conditioned by their closeness to the data'. Following a British Library-funded study in 1975 standardized notes on abstract writing were produced for inclusion
200

in the 'Instructions to authors' sections of primary journals. Supporting information, incorporating a check-list of elements potentially valuable in an abstract was provided for editors to submit to authors. These guidelines and the back-up information were tried as an experiment by some fifty UK and overseas journals in agriculture. It is instructive to see what the Commonwealth Agricultural Bureaux abstracting journals have to say about their policy towards the author abstract. Sometimes the CAB abstract is reprinted unaltered from the abstract accompanying the paper; more often it is changed 'due to elimination of superfluous material and inclusion of important information absent from the original abstract. In some cases it will have been necessary to correct errors of substance where an original abstract does not correspond with the paper to which it is attached'.

In view of the increasing concern for the protection of their proprietary rights expressed by authors and their publishers in recent years it is worth remembering that author abstracts are copyright, and it would therefore seem that indexing and abstracting services should not make use of them in their own publications without permission. It must also be noted that this is disputed by some publishers of abstracts, notably Dale B Baker of *Chemical abstracts*. It has even been said by some that this fear of infringement has led a number of services deliberately to abandon the use of author abstracts entirely. Of course this whole issue is no more than one aspect of the increasingly bitter confrontation between the publishers of the primary literature, particularly journals, and the secondary services, mainly indexing and abstracting services, that arise from them.

Indexing of abstracts
Most indexing services are self-indexing by reason of their usual arrangement in alphabetical order of subject. Most abstracting services, on the other hand, have adopted another arrangement, commonly a fairly broad *ad hoc* classification, eg *Weed abstracts*. Some follow a recognized general scheme like the Universal Decimal Classification (which is very popular), eg *Apicultural abstracts*; some use a special faceted classification, eg *Occupational safety and health abstracts*. Since 1977 INSPEC has employed the new *International classification for physics and related fields* specially developed for the purpose under the aegis of ICSU/AB. Other arrangements are less studied: *ICE abstracts*, published by the Institution of Civil Engineers, is in 'no discernible order', according to Walford. Obviously, if such services are to fulfil their role as retrospective retrieval systems they must be provided with indexes to permit specific subject access. The user is normally content to scan each current issue as it appears in order to keep himself alerted, but if

he is obliged to search back numbers for specific information he demands a subject index.

This is highly relevant to coverage: what is not indexed (even though abstracted) will not be retrieved in a retrospective search. This is well understood by the major services: the now defunct *Nuclear science abstracts* used to warn its staff that 'A collection of abstracts is only as good as its indexes'. Close on half of *Chemical abstracts'* full-time professional manpower and a quarter of the operating budget are devoted to the indexing effort, producing some six index entries per abstract. *Index medicus* employs a hundred indexers.

The more frequently an index is produced, the fewer unindexed issues will require page-by-page searching in the course of a typical search. Annual indexes are probably the commonest still, but mechanization has allowed several services the luxury of more frequent appearances, and it is now possible for very little extra effort to have subject indexes in every issue (although most of these are usually no more than keyword indexes). Variations are to be found of this standard pattern: *Metals abstracts* and *Metals abstracts index* are quite separate publications, although appearing simultaneouusly; *Engineering index* (an abstracting journal in spite of its title, of course) is already arranged alphabetically by subject and therefore needs no subject index. The manipulative power of the computer also permits author indexes, patent number indexes, corporate indexes, etc, from the same data base. The other side of this particular coin, however, is the multiple sequences facing the searcher pursuing an exhaustive search. Cumulated indexes are obviously of value here, eg *Gas chromatography abstracts, 1958-63, Index to Apicultural abstracts, 1950-72,* the nine-volumed *Cumulative index, 1973-1977* to *Engineering index* with its 450,000 entries; and the immense decennial and now five-yearly *Chemical abstracts* indexes (35 volumes and 75,000 pages to cover 1961-71, with 2,203 authors surnamed Smith!). The 9th collective index for 1972-76, though covering only five years rather than ten, is even larger with 96,000 pages and 20 million index entries in 57 volumes. CAS believe this is the largest printed index ever published. The 10th index will exceed 80 volumes.

The student will know that the efficiency of subject indexes is one of the topics most actively investigated at the present time. He will know that no index is 100% perfect, and in the words of E J Crane, for many years Director and Editor of *Chemical abstracts*, 'Even in the use of the best subject indexes the user must meet the indexer part way for good results'. And there will still remain some references undisclosed. The ASLIB tests reported that a searcher would be 'unlikely to find more than three-quarters [of abstracts on the subject of his search] via the subject indexes, and he is unlikely to be able to find more than half without the

202

exercise of considerable ingenuity or a good knowledge of the subject'.

Tracing an appropriate abstract is usually only the first stage in a bibliographical quest, particularly if it turns out to be merely indicative, and access to the original article is commonly required. This, of course, is the task of the library, but some abstracting services do offer help in a variety of ways, eg lending the original from the library (*Zinc abstracts*); offering a photocopy (*Chemical abstracts*, through its world-wide Document Delivery Service, or *World textile abstracts*, which includes tear-out application forms in each issue); seeing to it that copies are available from certain libraries (*Physics abstracts*, which makes known the fact by printing advertisements in its issues for the British Library Lending Division, or *ICE abstracts*, which offers not the facilities of its own library at the Institution of Civil Engineers but recommends users to the Science Reference Library). To provide access to items covered by *Agrindex* there has been established within AGRIS a world-wide network named AGLINET, comprising sixteen major agricultural libraries and one international centre.

Co-operation in abstracting

The dilemma of the abstracting services is summarized with crystal clarity by the ASLIB investigators: 'The search product is the available portion of the indexed portion of the abstracted portion of the total relevant literature'. Part of the solution lies in co-ordinating current effort: as Wilfred Ashworth says, '. . . there is little doubt that enough energy is already being used which would, if properly applied, give complete coverage of all literature'. In 1978 the Biological Information Review Committee urged co-operation on the three hundred services in the life sciences. They found particularly hampering the incompatibility between indexing practices, eg controlled vocabularies *versus* natural-language indexing terms. Centralization would be one method of standardizing citation and indexing practices as well as eliminating overlap and enabling gaps to be more easily seen: such is the system operated since 1952 at the national level by VINITI in the USSR. By contrast, the USA relies on voluntary co-operation co-ordinated by the National Federation of Science Abstracting and Indexing Services, founded in 1958, and at the international level since 1952 there has been the Abstracting Board of the International Congress of Scientific Unions. Many services of course are members of both. One major contribution that ICSU/AB has made to the timeliness of many abstracting services is persuading a number of the most important primary journals to airmail proofs or advance copies to the abstracting organizations. *Chemical abstracts* has estimated that this saves them as much as two months. An extended account of the efforts made towards better co-ordination of services would be inappropriate here, but the student can observe some positive results in the literature, eg *Metals abstracts*, formed in 1969 by a merger of *Metallurgical abstracts* of the (UK) Institute of Metals and *Review of metal literature* of the American Society for Metals; already mentioned is *World textile abstracts*, a collaboration under the aegis of the Textile Research Council by the research associations for the cotton (including silk and man-made fibres), hosiery and wool industries. An even more interesting merger between the public and private sectors is *Aquatic sciences and fisheries abstracts*, formed by amalgamation in 1971 of the Food and Agriculture Organization *Current bibliography for aquatic sciences and fisheries* and the Information Retrieval Ltd *Aquatic biology abstracts*; it is published by Information Retrieval under contract to FAO, and since 1978 has divided into two parts.

The computer casts its long shadow here also, making possible and even stimulating several examples of national co-ordination and international co-operation. There are now a number of international agreements under which individual countries abstract their national literature to provide decentralized input to international systems in return for the use of the centralized data base. One of the most successful is the International Nuclear Information System, referred to above, based at Vienna: the Atomic Energy Research Establishment is the body responsible for British input. An alternative kind of international arrangement divides the work among participants, eg the International Food Information Service, where Britain provides the input, Germany undertakes the processing, and the United States markets the product. The first of these international co-operative systems designed for computer operation was the International Road Research Documentation System which works slightly differently: there are three centres operating in Britain, France, and Germany, each member contributes its output to one of them, and then the centres exchange the material among themselves.

This is perhaps the most appropriate place to refer to UNISIST, which is not an acronym, but an invented code-name best explained by its official subtitle: 'Intergovernmental programme for co-operation in the field of scientific and technological information'. A joint ICSU/ UNESCO feasibility study was followed by a conference in Paris in October 1971 attended by delegates from 83 nations where approval was given for the launching of the scheme. Its basic philosophy is the promotion of a global network by voluntary co-operation among existing information services rather than the establishment of a single centralized system. Its five original primary objectives include two of particular relevance here: improving the tools of systems interconnection, and improving information transfer. Evidence of duplication with the programme of UNESCO's Division for Documentation, Libraries

204

and Archives (DBA) led in 1977 to a merger within the broader General Information Programme (PGI). From the start emphasis was laid on the needs of developing countries, but pressure from UNESCO members has produced a surge of effort to help such countries gain broader access to scientific and technical information and to correct the balance in access between them and the industrialized countries. UNISIST II, a second major conference at Paris in May-June 1979, confirmed this change in priorities. Currently it has an extensive programme of activities with budgeted funds for 1981-3 of $19 million. Concrete instances of the progress that has been made are the various UNISIST international information centres that have been set up in recent years, eg the three referred to above: the International Information Centre for Terminology, in Vienna, the Clearinghouse for Thesauri and Classification Schemes, in Warsaw, and the International Centre for the Registration of Serial Publications, in Paris. Others include the International Referral Centre for Information Handling Equipment, at Zagreb, and the International Information Centre for Bibliographic Descriptions, in London. UNESCO has now decided to concentrate all its activities relating to scientific and technological information and documentation within the UNISIST programme.

'Current awareness' journals

What the computer has not yet been able to do is to write an abstract. This of course is an intellectual process, and what John Martyn wrote in 1967 still remains true: 'It is not yet possible to produce an abstract which will summarize a document, in sentences not found in the document, by any other than human means'. R E O'Dette of *Chemical abstracts* is even more down to earth: '. . . scholars of information science continue to announce new ideas and techniques for automated abstracting and indexing. While these reports make interesting reading, for the weekly or monthly production of a real world secondary service containing more than a handful of abstracts, abstracting and subject indexing remain as human endeavours'. After extensive research into automatic abstracting at CAS it was concluded in 1975 that 'It will not be possible to produce abstracts of manual quality by computer without a great breakthrough in linguistics, especially in the area of semantics'. As an example of what is possible the student should study the computer-simulated abstracts in journals like *Biological abstracts/ RRM*, called more modestly 'content summaries'.

Because abstracting and indexing are still basically human activities, they are also slow processes, responsible for much of the delay between the publication of an article and its appearance in an abstracting journal. Delay hampers the repository role of an abstracting service hardly at all, but it can cripple its alerting function. It is a telling indication of the pace of scientific and technological discovery and of the growth of the literature that over the last twenty years or so a number of services have been started to bridge this gap between the publication of an article and its abstract, eg *Current chemical papers*, no longer extant but the first (1954) publication designed specifically for 'current awareness'. The fact that a number of these have been produced by the abstracting services themselves, eg *Current papers in electrical and electronics engineering*, 'congruent in its coverage of the literature with its corresponding abstracts journal', demonstrates the truth of R T Bottle's statement that 'Production delays and the time required to produce subject indexes for abstracts have almost eliminated their one-time function as a news-giving service'. The 1964 survey mentioned above of electrical engineering abstracts discovered that six months after publication only 17% of a particular sample of papers had been abstracted, and only 66% after twelve months. As alerting services the abstracts were clearly failing.

These newer services are able to appear more rapidly because most of them are merely title *announcement* lists, arranged in broad subject classes, but not indexed, eg *Current papers in physics*. A number are indeed title *indexes*, but depend for their promptness on mechanization, eg *Chemical titles*, a KWIC index produced by *Chemical abstracts*. Commonly they claim to include papers within a month of receipt, but they often concentrate on a limited number of core journals: *Chemical titles*, for instance, covers 725 out of the 14,000 monitored by the parent service; *Current physics titles* concentrates on no more than 70.

An even simpler form of title announcement service is the contents list, requiring a minimum of preparation. This is simply a transcription of the contents of current issues of the journals within a particular subject, and is a development of the common habit of learned journals of printing the contents pages of their contemporaries. Modern technology has given us in recent years a revival of this form of alerting service, using actual reproductions of contents pages printed by photo-litho-offset and sometimes with computerized indexes, eg *Current contents in marine sciences*, the half dozen weekly titles from the Institute for Scientific Information, eg *Current contents: physical, chemical and earth sciences*. Unusual insofar as it confines itself to the journals issued by one publisher is the American Chemical Society twice-monthly listing, *Single article announcement* service which reproduces the contents pages of some twenty titles.

Such current awareness services, rightly used, have no permanent value, for the individual papers listed are in due course covered by the abstracting services. But for current awareness, there is now some evidence to show that abstracts do not offer significant advantages over

206

such simple title lists. Furthermore, recent research indicates that the titles of scientific papers are becoming more informative, in two ways: by including more keywords, and simply by being longer. What we are observing in the literature is a common enough transmutation in life — differentiation and specialization.

Use and neglect

Many user studies reveal a surprisingly low usage of abstracts and indexes, particularly among technologists. 53% of mechanical engineers do not see any abstracting or indexing journal regularly. In the electrical and electronics industries, one survey showed that only 38% of technologists are aware of abstracts in their special field, and only 31% claim to make use of them. An ASLIB survey of technical library use found abstracts the least productive in yielding source material, with 7% compared with 40% for periodicals and 19% for textbooks. Investigations of items requested from the NLL (now the BLLD) showed that (with the exception of *Chemical abstracts*) 'abstracting journals when regarded as individual sources may be relatively insignificant': out of 9,182 references only 43% had come from abstracting journals. It is true that academic users and pure scientists make more use of such services, but it is still surprisingly light. The guide to the literature published by the Geological Society, mentioned in chapter 2, claims: 'The mint condition of abstracting journals in most geology libraries is only one proof of our neglect of these aids'. The inherent paradox here can be seen openly admitted in the recent survey of scientists carried out for the Scientific Information Committee of the Royal Society: they 'valued highly the abstracting and indexing publications, while not actually making very frequent use of them'.

Further reading

F A Tate and J L Wood 'Libraries and abstracting and indexing services — a study in interdependency' *Library trends* 16 1967-8 353-73.

'Published indexing and abstracting services' Jack Burkett *Trends in special librarianship* (Bingley, 1968) 35-72.

R L Collinson *Abstracts and abstracting services* (Oxford, ABC-Clio, 1971).

Primary publications and secondary services: partners in information flow (Paris, ICSU/AB, 1974).

T C Bearman and W A Kunberger *A study of coverage overlap among fourteen major science and technology abstracting and indexing services* (Philadelphia, National Federation of Abstracting and Indexing Services, 1977).

207

B M Manzer *The abstract journal, 1790-1820: origin, development and diffusion* (Scarecrow Press, 1977).

H East 'UK abstracting and indexing services – some general trends' *ASLIB proceedings* 31 1979 460-75.

D B Baker and others 'History of abstracting at Chemical Abstracts Service' *Journal of chemical information and computer sciences* 20 1980 193-201.

Chapter 11

COMPUTERIZED DATA BASES

As the student will have noticed while reading the preceding chapter any discussion of present-day abstracting and indexing services is next to impossible without frequent references to mechanization. At its simplest, the computer is seen as the *only* method of coping with the continued expansion of knowledge: to take but two instances, there are now over a thousand new chemical compounds reported in the literature every working day; from its foundation in 1940 to 1972 it had never been found possible to produce a subject index to *Mathematical reviews*, but in 1973 the computer permitted this for the first time. But many see in the application of machines to an abstracting service an opportunity to effect a complete reconstruction. Some grasped the chance in the 1960s and transformed their services (often with substantial aid from public funds), producing not merely a mechanized abstracting service, but adding a new dimension: 'In a computer-based system, information selected in a single intellectual analysis of the source documents, an analysis combining both abstracting and indexing, is put into a unified machine-manipulative store through a single keyboarding. Then from the unified bank of information, material appropriate for special-subject alerting and retrieval publication can be drawn, largely by computer programs'. Thus the computer brought in its train two immediate improvements: firstly, it speeded up the production of the traditional printed indexing and abstracting services, cutting the average time needed to process an item for *Chemical abstracts*, for instance, from 28 to 14 weeks, and then to 10 weeks; secondly, from the computerized data base it began to produce new forms of printed product previously impossible without disproportionate effort or cost, for example:

a Current awareness bulletins, eg *Chemical titles*, the very first (1961) computer-produced journal (CAS).

b Regular bibliographies on specific subjects, eg *Artificial kidney bibliography* (quarterly from MEDLARS).

c Specific bibliographies resulting from file searches, eg 'Rubella or German measles' (*Biological abstracts*); 'Battered child syndrome' (MEDLARS); 'Black holes' (INSPEC).

d 'Package' alerting services (also called macro-profiles or standard interest profiles), eg *Topics*, the weekly INSPEC service on cards, in any of 73 different subjects, such as road traffic control, colour television, crystal growth; *CA selects*, bulletins twice a month from the Chemical Abstracts Service in any of well over a hundred subject areas, such as forensic chemistry, corrosion; *Bioscans* are similar bulletins from the BioSciences Information Service and *Ascatopics* in over four hundred fields are from the Institute for Scientific Information. A listing by M D Bonham *An index to standard interest profiles in science and technology* (Bloomington, Indiana University, 1979) is itself computer-produced.

Each of these is produced with a minimum of extra effort simply by 'exploitation of the machine record'. Produced with even less effort, almost as by-products of the regular service, are derivative publications such as *Abstracts of mycology* (from *Biological abstracts*), *Epilepsy* (from *Excerpta medica*), *Potato abstracts* (from the Commonwealth Agricultural Bureaux data base), *Bioengineering abstracts* (from *Engineering index*), *Key abstracts* (covering selected portions of the INSPEC data base, eg electronic circuits, communication technology). The computer can also work in the opposite direction and combine a number of separate printed abstracting services into one, eg the PIRA data base corresponds to *four* journals: *Paper and board abstracts, Printing abstracts, Packaging abstracts* and *Management and marketing abstracts.* We have seen too how extra or more frequent (and indeed more detailed) indexes are also quite feasible.

The computer can also eliminate the typesetting bottleneck at the printing stage: many of the major services now use computer-driven photo-composition machines in preference to hot-metal typesetting. Indeed this was one of the original objectives of mechanization.

But speed and 'spin-off' of this kind were only the first of the benefits brought through the vast manipulative capacity of the computer. We were to witness in the 1970s a classic instance of the operation of the Law of the Instument. Put in its most homely form this Law states: 'Give a small boy a hammer and he will find a large number of objects that require hammering'.

Realization soon dawned that the transmutation of the printed indexing and abstracting journal into a machine-readable bibliographic data base made it feasible to use this new tool, the computer, actually to undertake an individual's literature search for him, at least so far as concerned the manual rather than the intellectual component. We have seen (chapter 8) that a searcher's bibliographical needs are of two kinds: current, keeping abreast of what has recently been published; and retrospective, searching either for any suitable literature on the topic of interest (the everyday approach) or for all relevant literature (the exhaustive approach). Bespoke or custom searches to meet these needs soon began to be offered to scientists and technologists in a variety of ways.

'In-house' searching
Many institutional subscribers such as libraries, information units, laboratories and research departments acquired for themselves by purchase or lease the actual data bases on magnetic tape instead of or in addition to the printed versions. There are now hundreds of such data bases on the market, most of them in the fields of science and technology, and many of them covering quite narrow or specialized topics, eg ASFA (*Aquatic sciences and fisheries abstracts*) ABIPC (*Abstract bulletin of the Institute of Paper Chemistry*). Many of them have evolved from a long-established printed abstracting or indexing service which is still as a rule the most widespread fruit of the producer's endeavour and the source of 90% of his income. Sometimes the tape version is available only to subscribers to the printed journal, eg META-DEX, the *Metals abstracts* tapes.

In some instances the whole of the data base (ie citations, abstracts, indexes, etc) can be acquired in machine-readable form, eg COMPEN-DEX, which is *Engineering index* on magnetic tape. More common perhaps is the service offering the citations and indexes only, eg META-DEX. In effect, therefore, these are computerized *indexing* services like MEDLARS and *Science citation index* rather than abstracting services. INSPEC (*Science abstracts*) subscribers have a choice: INSPEC-1 carries the full abstracts; INSPEC-2 provides only the bibliographical references, classification and subject indexing. Other producers offer selected packages, either on a subject basis, eg POST, a twice-monthly *Chemical abstracts* service covering polymer science and technology, and including abstracts; or, like *Chemical titles*, on the basis of significance (the 725 most important chemical journals). Available too are a number of data bases confined to particular categories of publication, eg the machine-readable version of *Government reports announcements and index* (see chapter 14); various tapes from INPADOC (International Patent Documentation Centre), the largest computerized patent data base in the world (see chapter 15); or the computerized version of *Comprehensive dissertation index* (see chapter 19). One stage beyond these are those data bases available only in machine-readable form, for which there is no printed equivalent; potential producers were not slow to see that a data base did not have to derive from a printed original.

Selection is helped by a whole range of bibliographies, some of which are described at the end of this chapter. Subscriptions are often higher for machine-readable versions, and they normally only cover leasing not outright purchase; there is sometimes a usage fee as well. And of course in-house processing costs are extra. But once the tapes have been acquired the institutions are free to manipulate them at will on their own computers to provide their users with individual search services.

a *Selective dissemination of information* (SDI), as the current-awareness service has come to be called, is the computerized version of the long-established library practice of drawing to the attention of individual scientists and technologists any recently published material on their known subject interests. Readers have always shown appreciation of this traditional bespoke alerting service, but it can make excessive demands on a librarian's time and skill when performed manually. By devising more sophisticated techniques for describing an individual's interests in a 'search profile', and then employing the computer to match that profile regularly in a batch with others against the latest week's or month's tapes, items of interest to the individual reader can be automatically selected, printed out on cards or slips or continuous computer stationery, and forwarded to him. Commonly they will arrive ahead of the printed journal, often by as much as several weeks, since the printing process itself cannot commence until the tapes are ready.

The best SDI systems provide for the user to report his degree of satisfaction with the references drawn to his attention, thus enabling his search profile to be refined, if necessary. It will be obvious to the student that the success of SDI depends upon careful and detailed indexing of the literature and so currency is likely to be less than in those printed current-awareness services based on title keyword indexing methods (see chapter 10). In both, indeed, the reader is obliged to accept a compromise between immediacy and retrieval quality.

b *Retrospective searches* can likewise be provided from in-house data bases, although it should be remembered that the data bases extend back only to the advent of computerization as a rule and manual searches are still required for searches covering more than the most recent years. The computerized search process is not dissimilar to that employed for SDI searches: the search analyst, or information officer, or librarian, first devises a search 'strategy' based on the enquirer's statement of his problem, which is often provided in written form. The question as put is then translated into a search statement which is usually fairly formal in structure and is written in language compatible with the indexing terms used in the computerized data base. The day's or week's searches are then run as a batch, often overnight using cheap off-peak computer time. As the computer prints out the bibliographical references the search strategist normally scans them to ensure than they are relevant to the problem before forwarding them to the enquirer.

212

The integration of such 'bought-in' services into existing libraries and other local information systems has provided a whole new dimension of economic, technical and managerial problems for the librarian and information officer. Many libraries with computer facilities have naturally made use of them to improve and develop their own internal services and there are now many successful examples of the use of external tapes together with internally generated data in an integrated information service.

Searching by data base producers

Of course not all potential users, whether institutional or personal, have access to a computer, and to meet the needs of such searchers some of the data base producers will undertake retrospective searches for customers, at a charge, and many will also provide an SDI service, eg the American Geological Society offers both from its GEOREF data base, as also does the American Society for Metals from the METADEX data base. The Institute for Scientific Information offers a weekly SDI service known as ASCA (*A*utomatic *s*ubject *c*itation *a*lert) from the *Science citation index* tapes, but not a retrospective search service; *Petroleum abstracts*, on the other hand, will undertake retrospective searches only.

Though there are many other similar examples some data base producers have decided not to offer direct services at all. Chemical Abstracts Service is a good example: searches of the data base are available only from other agencies of which there are about forty world-wide. In the UK, for instance, the Royal Society of Chemistry acts as exclusive agent through its United Kingdom Chemical Information Service (UKCIS) based at Nottingham University, which has done much to develop search services from the variety of CAS data bases. Of course such decentralized provision is found with other data bases also, eg the University of Loughborough offers search services from the COMPENDEX data base and UKCIS also provides a range of search services from the BIOSIS tapes. MEDLARS has employed decentralized output from the earliest days, with overseas stations in many countries, eg British Library Lending Division. It is worth noting that traffic is sometimes two-way: UKCIS is responsible for *input* to CAS of UK-published chemical literature; BLLD performs the same service to MEDLARS for the medical literature.

Searching by service suppliers

It is clearly not a large step for a library, university, or other institution to broaden its role beyond that of simple agent for the data base producer. Neither is it surprising to find a number of institutional subscribers to data base tapes wishing to offer their in-house search facilities to those outside their own walls, for a fee. What we have in fact seen emerge is a new kind of information provider, the service supplier, intermediate between the data base producers and the ultimate users. Variously described as data base processors, distributors, vendors, or hosts, they are in effect information retailers using their own computers to repackage the bibliographical data on an *ad hoc* and often eclectic basis for those users who lack the equipment to process it themselves, or those who do not need the whole of a data base, or those for whom the search services of the data base producers are either inadequate or lacking altogether. The data base producers, indeed, have been more than willing to assist this development and many have concluded licensing (as opposed to leasing) agreements with the service suppliers permitting them to offer search services more widely.

The kinds of institutions that act as retailers of search services in this way are so many and various that it is hardly possible to give a typical example. Commonly they are universities: a well-established instance is the Georgia Information Dissemination Center at the University of Georgia at Athens. A pioneer of a library-based service is the Royal Institute of Technology (KTH) in Stockholm which has been providing computerized search services to the general public since 1967. What is certainly an outstanding example of a service supplier operating on a nation-wide basis is the Canada Institute for Scientific and Technical Information at Ottawa. CAN/SDI is the current awareness service matching some three thousand profiles regularly against some twenty data bases; ULSS (*U*nified *l*iterature *s*earch *s*ervice) is the retrospective service carrying out each year close to a thousand comprehensive subject searches and well over two thousand shorter searches, drawing not only on the most relevant of a hundred data bases but CISTI's large collection of printed abstracting and indexing services.

One service supplier that operates in a single subject field is the Centre for Agricultural Publishing and Documentation (PUDOC) in the Netherlands: it provides a highly successful SDI service based on four data bases — AGRICOLA, CAB, AGRIS, and FSTA (*Food science and technology abstracts*) — which are processed in-house, with additional data bases from BIOSIS and CAS being processed by other agencies. A variation on this kind of provision is the establishment of small cross-disciplinary systems, typically mission-oriented, formed at a particular processing centre by combining selected parts, or 'sub-sets', of a number of large data bases, eg the Batelle Energy Information Center in Columbus, Ohio, has assembled in a single machine-readable file selected material on energy from over thirty computerized data bases.

214

Searches of the kind so far described, both current and retrospective, are undertaken for obvious reasons of economy and efficiency by the computer in batches, as we have seen, and the enquirer has to wait until it is time for the next computer run for his query to be processed in his absence. The result of his search, in the form of computer print-out, has then to be delivered to him, perhaps by post, days and sometimes weeks later. An immediate response to an enquiry, dialogue with the computer, and adjustment of the enquirer's requirements in accordance with the progress of the search (or *vice versa*) is scarcely possible. Such batch searches are still carried out in quite respectable numbers, but they have been largely overshadowed by 'live' searches undertaken by direct interrogation of the computer (which may be many miles away), often by the scientist or technologist in person, with virtually instantaneous feedback allowing man to interact with machine, using 'on-line' access.

On-line searching
In the late 1960s and early 1970s three simultaneous technological advances brought on-line bibliographical searching into the realm of the practical: firstly, the development of random-access computer memories on disk with greatly increased storage capacity; secondly, the availability of simple and cheap acoustic couplers and the more reliable modulator-demodulators (modems) to convert analogue signals to digital (and *vice versa*) and thus to allow computer terminals to be linked to the regular telephone system; and thirdly, improvements in cable, microwave and satellite telecommunications which allowed greatly increased amounts of computer data to be transmitted faithfully over greatly increased distances. Such on-line facilities permit the individual user to conduct his own bibliographical search of a distant computerized data base using a two-way telecommunications link from a teletype terminal (which is like an electric typewriter with more keys than usual and is sometimes fitted with a visual display unit) in his own laboratory or office or library. It should be explained that the terminals can be connected to the computer by private lines or dedicated lines leased from the telecommunications authorities, or the connection can be made by the user dialling up the computer as required, over the regular telephone system. Increasingly, these methods may use the facilities of data (as opposed to voice) telecommunications networks, designed especially for computer traffic. This shows a great saving in cost, particularly if the data is transmitted by the 'packet-switching' technique, ie interleaved automatically on a time-sharing basis with other packet-based traffic. Would-be users normally take out a subscription with a service supplier; they are then allocated a unique and

215

confidential 'password' which allows them to identify themselves to the computer and thus gain access to the data base they wish to search. On-line bibliographical searches have increased enormously in numbers in recent years, particularly in the United States and Western Europe. They are particularly widespread in the libraries and information units of industrial firms, especially those that are science-based or are deeply involved in research and development.

Compared to batch searches their advantages are manifold: they are obviously faster, and much more convenient, with the time spent at the terminal being of the order of fifteen or twenty minutes, and there need be no waiting. A 1976 survey for the Commission of the European Communities found that over half of on-line users had adopted that method of searching because of the speed of the response. On-line searches are often cheaper, and as they can be carried out in person they are likely to be more effective. But this boom in on-line searching cannot be explained merely by reason of its superiority to batch searching. Its growth is due to the advantages it also offers over manual searches in many instances, though by no means all, as we shall see. Not the least of its immediate attractions is its much wider potential availability, wherever there is a telephone line in fact. All predictions are that this growth is to continue: it is estimated that on-line bibliographical searches will exceed 4 million in North America by 1982 and 1 million in Europe; by 1985 this could be 15 million and 4 million respectively.

On-line access is now possible to well over two hundred bibliographic data bases containing over 70 million records, mostly in science and technology. It has been claimed that 90% of all significant machine-readable bibliographic data bases are already on-line. Currently the largest, most successful and most heavily used on-line data base in the world is MEDLINE, the on-line version of MEDLARS, with over a million searches a year, including half of all US searches and an even greater proportion of searches in Europe. Supported by substantial government funding, it became operational in the US in 1971, a pioneer once more like MEDLARS I in 1964. A number of significant advances are being planned for MEDLARS III. Normally only part of the data base is scanned in a direct search — the current year and the previous four years, amounting to some 1.5 million citations from 2,600 journals — but the back files to 1964 are available for batch searching on demand. There are plans to make the remainder of the data base available on-line as a series of separate files. The introduction of on-line access in 1971 evoked an amazing leap in demand to several times the level of batch-processed searches. A data base containing only the current month's citations — about 20,000 in number — called SDILINE, provides a current-awareness service, available several weeks before the
216

corresponding issue of *Index medicus*. The National Library of Medicine is also responsible for TOXLINE, another pioneering venture and now one of the most widely used composite data bases which covers the field of toxicology by merging relevant entries from MEDLARS, *Chemical abstracts, Biological abstracts* and a number of other sources.

Other services were quick to follow the lead given by MEDLARS, eg BIOSIS started to offer on-line access to its PREVIEWS data base in 1975; after trials with partial access the full *Science citation index* data base was also made available as SCISEARCH. But quite soon many data base producers saw the economic sense of not attempting themselves to provide an on-line retrieval service on a single data base. Increasingly they directed the would-be searcher to the service suppliers with whom they negotiated licensing agreements.

On-line service suppliers
Thus we saw emerge yet another category of information provider, the big-league service supplier, sometimes called system operator, concentrating exclusively on on-line provision and offering access to data bases on an international scale over a variety of telecommunications links. The market leader here with its five thousand subscribers is DIALOG Information Services at Palo Alto, California, a subsidiary of the Lockheed Corporation. Lockheed Missiles and Space Company was closely involved in the mid-1960s with the development of on-line systems for the National Aeronautics and Space Administration and what was then the US Atomic Energy Commission. DIALOG provides what Lockheed has called 'fingertip access' to well over a hundred data bases with a total of more than 45 million citations or abstracts. This access is by dialling direct, or by leased line, or via TYMNET or TELE-NET which are major commercial data communications networks mainly serving North America but with nodes around the globe. The second major US-based supplier is the Search Service of System Development Corporation at Santa Monica, California, a subsidiary of the Burroughs Corporation. Using the ORBIT retrieval system, SDC offer over sixty data bases for searching. They undertook much of the development work for MEDLINE at the National Library of Medicine and also hold the distinction of being the first bibliographical search service to make use of a satellite communications link on a regular basis. As with DIALOG, access is by dialling direct to the computer, or by leased line, or by dialling the TYMNET or TELENET node nearest to the user's terminal. In the UK users of both services dial their nearest node to connect with the British Telecom International Packet Switching Service (IPSS). The third major US supplier is Bibliographic Retrieval Services (BRS) of Scotia, New York, with some thirty data bases.

In the UK the largest supplier is BLAISE (*British Library automated information service*) which commenced in April 1977 under the auspices of the British Library but using a rented computer. On-line search facilities in science and technology are mostly concentrated on MEDLINE and the related data bases. Access over the public telephone system is via some half-dozen nodes strategically sited across the country. Access from the Continent of Europe and the United States is via the IPSS network. Linked with this development is MERLIN (*Machine-readable library information*), a new integrated computerized system being developed by the British Library to support bibliographical and cataloguing services and other library house-keeping activities for its own purposes as well as for external users. In addition to bibliographical search facilities BLAISE already offers a range of services specifically directed at the library cataloguer, but the comprehensive MERLIN system is currently described as 'in cold storage' on account of cuts in government expenditure.

The largest service supplier in Europe is the Information Retrieval Service of the European Space Agency (IRS-ESA) at Frascati, near Rome. Formerly the Space Documentation Service and developed from the batch service set up in 1965 at Darmstadt, Germany, originally with the NASA data base only, it converted to on-line in 1969 using the system developed for NASA by Lockheed. Currently IRS-ESA offers some thirty data bases with about 18 million references. For some years now it has been marketed in the UK by the Department of Industry Technology Reports Centre at Orpington under the name DIALTECH but the future of this arrangement is in doubt in the light of the developments at a European level to be described later. There are several other important on-line service suppliers in Europe, including a number of British suppliers, commercial and state-supported, but the most distinctive feature of European provision is its international character deriving from the political presence of the European Communities, as will be seen.

The student will have noticed one obvious additional factor in the information transfer equation for on-line searching that is absent with batch searching: the telecommunications companies. As we have seen, they may indeed be commercial companies, but in many countries telecommunications are operated by the government alone. In either case they obviously require payment for their services. Costs for on-line searches therefore comprise four elements:

a *Data base producer's charge*: normally in the form of a royalty related to actual use and charged to the service supplier who passes it on to the user.

b *Service supplier's charge*: designed to cover computer costs and other operating expenses and normally charged to the user at so much

218

an hour (though a subscription may also be charged).

c *Telecommunications charge*: covers the cost of the phone call to the computer (or to the node plus the network charge). It is normally based on time and distance (sometimes on traffic carried) and is charged to the user direct by the telecommunications company.

d *Print charge*: arises when the user requires the results of his search printed out for him to retain. For a small number of references this can be done on-line at the terminal, but is normally done off-line by the supplier to save on-line costs and is then sent by post. The supplier usually charges the user at so much per reference, or sometimes by the page.

The institutional user must obviously meet the costs of his own terminal, whether purchased or rented, the costs of any other telecommunications equipment, eg telephone line, staff costs and associated overheads. And since death and taxes are the only two things in this world of which we can be certain, as Benjamin Franklin reminded us two hundred years ago, the user also has to pay those taxes levied on his search that his government has seen fit to levy, in Europe mainly Value Added Tax.

As with batch services the bibliographical data stored in the system comprises in most cases the citations together with the indexes, which are usually in the form of keywords or descriptors. Some data bases store the abstracts as well, eg BAT (*Biological abstracts* on tape) contains the abstract texts for the citations in BIOSIS PREVIEWS. There have also been experiments by INSPEC with ultrafiche (see chapter 21) display of abstracts of the citations retrieved in the on-line search. The ultimate would be reached with a data base that made the full text of each document available for searching and retrieval. As mentioned in chapter 9 this has already been achieved experimentally by BRS with certain American Chemical Society journals. Backfiles can normally be searched off-line, ie by the service supplier in batch mode overnight on the user's behalf, though the search results can sometimes be displayed on-line the following day.

On-line SDI is technically feasible, as the MEDLARS SDILINE shows, but as Peter Leggate of the University of Oxford Experimental Information Unit predicted in 1975, 'The continuous character of the SDI service reduces the value of on-line provision; the researcher will probably prefer to receive output by mail at regular intervals rather than have to remember to go to a terminal and reactivate his profile. The likely compromise is that SDI profiles will be negotiated and refined on-line but run in the batch mode'. A number of suppliers now offer this service, eg BLAISE, with its AutoSDI, allowing users to store their search profiles and then have them run automatically whenever the appropriate data bases are updated. A similar service is available

219

for all appropriate DIALOG files, and SDC have their SDI search facility.

But it is not only the individual user with access to a terminal who has benefited from the on-line revolution. Many of the service suppliers offering current awareness and retrospective searches in batch mode (as described above) have themselves taken advantage of these technological advances to speed up their services and now undertake some or all of their searches for their clients by interrogating the distant computer of an on-line service supplier rather than running a batched search of a leased data base on their own computer.

On-line search service brokers
This has led logically and inevitably to the emergence of yet another category of information provider in the shape of the 'broker' (or service intermediary), whose role it is to provide or sell on-line search services on demand to those who do not have access to a computer terminal. The student should note that unlike service suppliers such brokers do not necessarily require their own computer facilities or data base tapes: all they need is a terminal providing on-line acccess to distant computerized data bases, together with some experience in searching them.

They fall into two categories. First is the wide variety of public or semi-public institutions, already providing some kind of information services, such as universities and colleges, hospitals, learned societies, research associations, and of course many libraries, including public libraries, eg University of London Central Information Service, St Bartholomew's Hospital, Commonwealth Forestry Bureau, International Food Information Service, Sheffield City Libraries. Many such institutions have now begun to offer on-line search services to all comers without restriction, though usually for a fee. The searches they undertake are of course of data bases distributed across half the globe, but the dimension of service they provide is essentially domestic or local. While they would obviously not refuse to entertain requests by post or 'phone, their services are basically for their own clientele or the enquirer who visits them in person. Indeed the enquirer is often encouraged to be present at the search, eg Institution of Electrical Engineers, Paint Research Association, Science Museum Library. In some cases he is *required* to be present, eg British Library Lending Division, Royal Society of Chemistry, Cheshire Information Service. Sometimes he may actually use the terminal himself if he wishes, eg Hatfield Polytechnic, Science Reference Library. Occasionally the enquirer is not encouraged to be present, but he is usually asked to pay a second visit or the results are posted to him, eg Huddersfield Polytechnic, Leicestershire Libraries, Metals Society, ASLIB; thus even though the search

220

may indeed be carried out on-line, it appears in its results similar to a batch search so far as the user is concerned.

The second category comprises the increasing number of on-line brokers operating as commercial enterprises, eg Global Access Ltd, Information Services Ltd. Obviously they always make a charge for their services, sometimes by combining a fixed annual subscription with a fee that varies with each search. The continuing rapid spread of such on-line facilities does raise the question of what is to happen to service suppliers providing only batch services. In 1976 Gordon Pratt of ASLIB warned that they were 'not exhibiting the growth characteristics displayed by the on-line interactive systems and all indications point to a stagnation or decline of the services offered'. Roger W Christian went further: 'Unless these "first generation" information centers make their wares available on-line in the reasonably near future, they may not have a future, given the apparent preference of users for the flexibility and immediacy of on-line searching'. The BLAISE Postal Service offers MEDLARS and other search services in this way, but there has been a steady fall in demand ever since users could search MEDLINE themselves from their own terminals.

Information networks
It seemed logical to many that linking information users and all the various information providers (data base producers, suppliers, brokers, etc) in an on-line network would be an obvious next step, certainly quite feasible with current technology. The aim would be to make the information in any one system available to as large a group of users as possible. Such a development is very much in line with the original objectives of UNISIST (see chapter 10).

For some countries of Western Europe just such a data transmission network has been established under the direct sponsorship of the Commission of the European Communities, which stated that its policy was 'to incorporate in the European network all information centres, systems or other useful institutions, which exist or are being set up in the Member States, and link them together'. Agreement was reached in 1976 by the postal and telecommunications authorities (PTTs) of the nine EEC countries on the establishment of EURONET, the actual international telecommunications network. The ensemble of information services available to users via this network has itself been christened DIANE (*D*irect *i*nformation *a*ccess *n*etwork for *E*urope) and was opened in 1980. The President of the Commission, Roy Jenkins, described it as 'a new highroad on which to transport a key resource — information'. As his metaphor implies, EURONET/DIANE is not a service supplier but a decentralized network linking the computers of a range of service suppliers. Currently about 40 'hosts' (as EURONET/DIANE calls them) are connected: some offer only a single data base, eg the Institut Textile de France in Paris, with TITUS; but others offer several, eg IRS-ESA in Frascati, with about thirty. In all access is provided to almost three hundred data bases, and the network is not exclusively for the ten Community countries: by special agreement users in Switzerland also have access.

A problem brought very much to the fore in such a system, sponsored as it is by an international community committed to the concept of equality among its several official languages, is multilingual access to the data bases. It is regarded as a matter of principle that users should be able to put queries to the system in their own tongues. This has led to much study and experiment, with one line in particular being pushed enthusiastically: the development of multilingual thesauri (see chapter 4). In the longer term, help may be found from machine translation (see chapter 17): SYSTRAN is already being used experimentally to translate some DIANE data bases.

Benefits of computerized searching

The student should appreciate that a computerized data base is more than a mere automated version of its printed counterpart, and that the benefits it brings to the searcher are more than mere speed and convenience. Computerization of an abstracting and indexing service is a 'value-added' transaction inasmuch as it brings in its train a number of other benefits that did not exist before:

a The capacity to make highly intricate searches involving many terms in complex logical relationships.

b Printout when required, sometimes with abstracts, obviating the need to copy out citations or take notes.

c Considerably more access points to each document. On the subject side, for example, in *Index medicus* the average article appears under three headings, but in the MEDLARS tapes it is tagged with twelve or more for computer retrieval; it has been estimated that the effect in the case of MEDLINE is that the computer will find roughly twice as much on a concept as a manual search of *Index medicus*. As for searching fields other than the subject, in addition to the author, which one expects, it is often possible to search by date, language, country of origin, address of author, document type, report number, ISSN, CODEN, etc, all items that never appear in the usual conventional printed index.

d Elimination of the physical handling of multiple volumes — no light matter, as any *Chemical abstracts* searcher will readily confirm.

e A continuously updated file with improved timeliness: not only

222

does the single file replace the sequence of issues that one has to search with a printed journal, but almost without exception the machine-readable version is available before its printed counterpart.

f Remote access, virtually from any service point with a telephone, given the appropriate equipment.

g Access to data bases, eg TOXLINE, which have no printed equivalent.

The Director of the Publications Division of the American Institute of Physics has confessed 'it is very difficult for us to imagine using the pre-computer secondary data bases of abstract journals and card files for many of the purposes now considered essential'.

There has also been one benefit of an unanticipated kind that seems to have no basis in logic but was noticed in a 1975 study of on-line users: 'Engineers, scientists and researchers more readily accept the results of on-line searching than they do the results of manual searching, even though the quality of the search is obviously still determined by the adequacy of the search prescription formulated by the librarian'.

There are also further benefits to come. Already feasible is the use of a microcomputer at the terminal (the so-called 'intelligent' terminal). This can reduce costs and inconvenience while at the same time improving searching by formulating the search statement off-line with editing by the computer before transmitting automatically at maximum speed. SDI profiles and other regularly used searches can be stored for re-use as required.

Problems of computerized searching

The advent of the computer has not been without its particular difficulties. While few would deny the benefits just listed, others that are claimed have been challenged. Perhaps the two most controversial issues are cost and effectiveness: of course they are intimately linked.

On the face of it at least, measuring the cost of an on-line search in order to compare it with the alternative, ie a traditional manual search, should be a simple straightforward matter. Yet investigators appear unable to agree on a methodology: each of the scores of research studies that have been reported seems to provoke immediate rebuttal, one camp arguing that on-line is cheaper, the other retorting that when properly costed it is not. What is clear is that it is a more complex matter than it might seem at first: even direct costs such as charges that are itemized in a bill can fluctuate widely for a variety of reasons, as we shall see. And as just one example of how variable are some of the less obvious costs, such as human time, A J Harley in discussing MEDLARS searches makes the point that 'If the person using the output was a professor of medicine, it was almost always cheaper to use the computer. If he was a research student it was cheaper to make him use *Index medicus* in the library'. For a reasonably objective view the student might consider the findings of a literature survey sponsored by the British Library in 1979 which found the costs of manual and on-line searches to be approximately equal. It also found that 'Some of the early literature was overtly or covertly partisan for on-line techniques, tending to produce high, and non-comparable, manual costs to indicate the advantages of the new approach'.

But cost comparisons alone prove nothing: what is really important is the effectiveness of the search. Comparisons here are even more difficult. Applying the usual measures of recall and precision may not be helpful because of differences, for example, in the nature of the data base and the mode of searching, though the student should note that reports in the literature usually indicate effectiveness measures for on-line searches of less than 50%, ie around 50% recall and 50% precision. Jason Farradane has told us 'I am beginning to question very seriously the value of online services when any attempt to achieve completeness or lack of noise is required. The sad fact is that users of online services seem to look only for the easiest way to get *something*, and to be unaware of what they fail to retrieve, and even not to care'. That is a stern judgement, but the rapidity of response and the deceptive ease with which references can be called forth seems to have beguiled many who should know better into accepting a compromise over retrieval quality, as with batch SDI services. Some would argue that we might have to get used to compromise: a standard text on the subject published in 1980 explains that 'The case for on-line systems is not primarily that they are a more cost-effective way of carrying out our present activities; it is, rather, that a new expense will give the average user a completely new level of contact with information in his subject area'.

Whatever the truth about the relative cost and effectiveness of on-line and manual searches the comparison to be fair can only be drawn for those searches that are judged appropriate in the first place for an on-line search. The student must not make the mistake of assuming that this is so in all cases. There is widespread agreement that for the less intricate everyday and background searches, for exhaustive searches covering many years, and for browsing, manual methods are not only more economical but also more effective. Experience has also shown that it is usually misguided to embark on an on-line search with a vague or imperfectly-formulated question.

A decided limitation in some data bases, as has already been mentioned, is that retrospective coverage extends to no more than a few years, the start of the run normally coinciding with the computerization of the corresponding printed service, eg *Chemical abstracts*

224

from 1967, INSPEC from 1969 (the printed versions date from 1907 and 1898 respectively). Scarcely any are to be found covering the period before 1960, though one or two services have made sterling efforts at retrospective conversion: the continuing importance for geologists of the older literature, referred to in chapter 9, has no doubt prompted GEOREF to extend its North American coverage back to 1785, taking advantage of the excellent printed *Bibliography of North American geology, 1919-1970* (Washington, USGPO, 1931-73) and its predecessor J M Nickles *Geologic literature on North America, 1785-1918* (Washington, USPGO, 1923-4). Understandable though this lack of coverage may be, it does mean that in many cases a retrospective search to be exhaustive must combine manual and on-line approaches.

One perturbing effect of computerization has been the dropping from the subscription lists of some libraries of certain printed abstracting journals once they become available on magnetic tape or on-line. What is worse, it has been reported that the first UK public library to go on-line sold off long runs of its abstracting journals such as *Engineering index* and the various INSPEC services. Even more fatal is the case of those abstracting journals no longer produced at all because of the availability of machine-readable alternatives, eg *Road abstracts*. Similar problems are caused by those producers who have allowed the machine-readable version of their service to outstrip the printed journal, eg from 1980 the PSYCINFO data base included 25% more citations than the printed monthly *Psychological abstracts*.

There are those who would agree that this may be reducing rather than increasing access to information. What is likely is that the practice will spread. UKCIS, for instance, makes no attempt to hide this: 'Though conventional abstracting and indexing services are still very much alive and kicking it is now possible to foresee a time when they will be largely or entirely replaced by combinations of electronic and photomicrographic systems of great power and flexibility'. Of course the information formerly available in printed form will continue to be accessible via magnetic tape or disk-pack or 'bubble' memory (or whatever their technological successors might be) for current or retrospective search – but at a charge, for the introduction of computerized services has prompted the adoption by many libraries and information units of a policy of cost recovery. The implications will be discussed later. At the moment those bibliographic data bases that have no printed counterpart are in the minority, but an increasing number of new data bases exist in no other form, giving a new twist to the problem, eg TITUS.

Then there are a whole series of what might be called house-keeping problems, inherent in the nature of the equipment. For one thing, service is rarely available round the clock, or even every day of the

225

week: BLAISE, for instance, is accessible only during office hours, Monday to Friday; DIALOG is accessible longer (114 hours a week) but not yet 24 hours a day. And intercontinental link-ups have to take account of time zones, national holidays, etc. Physical access too is limited, to one searcher at a time at each terminal and a limited number of ports at the computer. Equipment malfunctioning is by no means uncommon, both at the computer and in the telecommunications link, and for services so totally dependent on technology can be disastrous. Similar are the sometimes only temporary inconveniences suffered with some data bases and some supplier systems, eg file inconsistency, lack of promptness in updating, inaccuracies, etc. While by no means to be condoned these are the routine kinds of faults that have always been encountered in the less efficient services whether printed or computerized. The frequent changes that are made in systems and the data bases they provide can also cause difficulties or at least temporary confusion. It is to be hoped that most of these are intended as improvements that will benefit the user, but some are mere commercial adjustments that may not.

A number of much more fundamental problems stem from the fact that many data bases, as has been described above, were not created primarily for on-line searching: they usually originated as by-products of a publication system, consequent to computerization. As J J Pollock explains, they are 'essentially electronic copies of printed products and it would be surprising if an information medium designed for one purpose turned out to be ideal for a different one'. This has two far-reaching effects, on content and on access. A E Cawkell has pointed to the first: 'The low cost of tape used in this way permits the recording of rubbish together with useful information . . . Now when it comes to running on-line services, where storage and access costs are high, what do we do? We use the same tapes containing the same "noise", much of which will never be used by anybody'. As for the effect on access to the content of the data base, it has to be admitted that the standard of indexing in many cases is poor. There is now considerable evidence that the methods of indexing so carefully developed over many years for manual searching are less appropriate for machine searching and nothing of comparable quality has yet taken their place. The student must not fall into the trap of blaming the computer for this; the fault lies with the indexers and their inadequate grasp of the nature of information. Farradane's opinion of the search method necessary with the inverted file method of indexing is characteristically forthright: 'the descriptor plus Boolean logic methods being applied in machine systems are really primitive and need to be greatly improved, if not entirely recast'. Hans Wellisch is even more blunt: 'Computerized means of frustration and electronically coded ineptitude are no better than the

226

manually produced variety. Indeed they are more dangerous'. Research continues into improved search methods, particularly in the direction of searching the texts of natural language data bases using techniques such as variety generation, for example. It is highly significant in this context that in recent years the National Science Foundation Office (later Division) of Science Information Service has become so convinced that the march of technology has outstripped the halting progress of the underlying disciplines of information science that it no longer supports individual information services. Its efforts are now concentrated on more fundamental studies of human communication and information transfer. The view expressed by Lee G Burchinall of NSF was that 'We are building services and networks without understanding the basic processes involved'.

On one matter there does seem to be no doubt: on-line searching does save time. Even if one includes preparation time an on-line search can be done in a third or even a fifth of the time taken for a manual search. Perhaps the best approach for any potential searcher to take is the commonsense one of regarding on-line searching as a perfectly normal activity, to be used or not used as the case may be. For some searches it will be the only satisfactory way of proceeding, just as there will be others that can only be done manually. Many searches can be carried out either way, and as we have seen some must be undertaken using both methods. Even when the decision has been taken to search on-line, many would still agree with the conclusions of a team of lubrication engineers who compared CA CONDENSATES and ASCA: 'one cannot totally dispense with some hand searching and following-up of references to be reasonably sure of good coverage'.

Proliferation

In his contemplation of the search services of DIALOG, SDC, BRS, IRS-ESA, and others the student will notice a considerable degree of duplication among the data bases offered: INSPEC, COMPENDEX and AGRICOLA are available from all four suppliers, for example; dozens more are available from two or more. CAS data bases are on offer from at least six on-line suppliers in the UK alone. Scores of the data bases on the host computers in EURONET/DIANE overlap with those offered by US commercial service suppliers. More disturbingly, they duplicate and even triplicate each other within Europe. The student will be told that the way each supplier loads them on to his computer is unique and distinct and allows different search facilities to be provided, but he may conclude that what he is observing is simply the effect of free market competition. In practical terms, to searchers in the UK the big US service suppliers offer advantages in the larger number of data bases they make accessible and in the flexibility of the system facilities, whereas the use of local suppliers, either based in the UK or the Continent of Europe, means lower telecommunications charges.

Nevertheless, the student can be forgiven for feeling confused by the proliferation of providers of information. And it does not make for clear understanding to find particular kinds of providers playing a variety of roles at the same time: service suppliers may also be data base producers, producters may also be suppliers, etc. Libraries, with which the student will be particularly concerned, in addition to providing in-house search services for their own readers may also be data base producers (as the National Library of Medicine with MEDLARS), or agents for data base producers (as the University of Loughborough with COMPENDEX), or service suppliers (as the Royal Institute of Technology at Stockholm), or on-line search brokers (as the Science Reference Library).

Amid such disarray perhaps the best advice that can be offered to the perplexed is to suggest that they fix their attention on the functions being undertaken rather than the institutions that are undertaking them at any particular point in time; it is to be hoped they can then keep distinct in their minds the role of producers, of suppliers, of brokers, and of users, while setting aside for the moment the confusing reality that a library or a university or a research department may in fact fulfil at various times any or all of these functions.

The use of search intermediaries
As has been mentioned, on-line searches can be made by the scientist or technologist in person: indeed this is claimed as one of the main advantages of on-line service. Determined efforts were made from the earliest days by some data base producers, some service suppliers and some subscribing institutions to encourage the enquirer himself to learn how to search. *Excerpta medica*, for instance, stressed that its service is 'designed for access by the ultimate user and not by a documentalist intermediary'. At the Science Reference Library on-line services were offered to searchers on condition that they operated the terminal themselves after initial guidance by the library staff. Put at its simplest, the case for the scientist or technologist undertaking his own search is that he alone knows exactly what he wants and he is familiar with the subject field. If he could but add to that the skill and experience of the expert searcher there would be no argument.

But a bibliographical search has never been an easy task even for professional searchers and the complexities of on-line have added to the difficulties. We have already noted the prerequisite in each particular case of deciding first of all whether an on-line search would be appro-

228

priate; we have also seen that in many instances an on-line search has to be combined with a manual search to be fully effective. These are not decisions to be made lightly and require more than simply a knowledge of the subject field. They demand a detailed acquaintance with alternative sources of information, their content, arrangement and accessibility, their strengths and weaknesses, and not least their comparative costs. Even when an on-line search has been determined on all is not plain sailing. No one can deny that many of these computerized systems lack what has been called 'transparency', letting a searcher use the service without first becoming an expert in the complexities of its structure.

To what is basically the intellectual barrier of search formulation and strategy the advent of on-line has added a further hazard, immediacy, and has erected a quite new barrier, the command language. Both are direct consequences of the interactive nature of the search process. In the first place, because the search is undertaken by the computer while the searcher is actually at the terminal, he has to be prepared to respond immediately to the progress reports that it makes; in other words his search strategy needs to be flexible. In the second place, because instructions to the computer on how to proceed with the search have to be individually given in each case, the searcher has to be familiar with the range of commands that may be needed. The student should be careful to distinguish these commands about *how* to search from the descriptors used in the search statement which indicate the subject of the search, ie *what* to search for.

It is still true, by and large, that the searcher has to learn a different command language for each system he interrogates, but there is some progress to report. Already developed in Europe is a DIANE Common Command Language enabling users to search data bases offered by different service suppliers by using the same set of commands. At the moment only half-a-dozen or so have implemented it; the others remain to be persuaded. One critic has warned that 'it may be utopian to expect them to agree to use it'. A number of studies, for example at the University of Manchester Institute of Science and Technology, have examined how microprocessor-assisted terminals could provide automatic translation between different command languages or could allow the use of simplified dialogues between the searcher and the system. The fact remains, however, that the number of times the average working scientist or technologist might need to use an on-line system would not give him sufficient practice to develop real skills. Even the hurdle of the teletype keyboard has proved a stumbling block for the surprising number who have never used a typewriter, though the actual amount of typing needed for a single search is really very little.

The Science Reference Library advises its readers that 'effective

229

searching . . . requires careful prior preparation involving the consultation of manuals, thesauri and term lists'. For we have seen blossom a quite new form of reference tool, the on-line user's guide, often called manual and commonly in loose-leaf format. Such works usually confine themselves to one data base, eg *AGRICOLA online users guide*, *BIOSIS search guide*, but sometimes several data bases may be covered in a single work, such as those offered by a particular service supplier, eg the *DIALOG guide to databases* in four volumes. Sometimes there are two or more manuals for the same data base, eg *Engineering index COMPENDEX online user's manual for the Lockheed DIALOG Information Retrieval Service* and *SDC COMPENDEX user manual*. We also have at least one bibliography of such guides: *Online reference aids: a directory of manuals, guides and thesauri* (San Jose, Cal, California Library Authority for Systems and Services, 1979). In addition to the various thesauri needed to formulate search statements for particular data bases (see chapter 4) a number of libraries have also found it useful to have copies of other reference works by the terminal, such as dictionaries, and even *Roget's thesaurus*. Many of the large on-line service suppliers also issue regular newsletters giving details of changes in the system, newly added data bases, altered times or charges, and general advice to searchers, eg *DIALOG chronolog*, *SDC searchlight*, *BLAISE newsletter*, *EURONET DIANE news*, *MEDLARS technical bulletin*, *News and views* from IRS-ESA.

One insight into just what it takes to be an on-line searcher was given in a paper read recently to a library research group: 'A medium-sized information unit typically will have about a 10-foot run of manuals concerned with online, its staff will have spent many man-hours on courses run by the hosts or data base suppliers, and they will spend a proportion of their time each week reading the large number of "online" newsletters which detail introductions, changes, modifications and other information essential for online searching'.

Apart from offering courses the service suppliers exert great efforts to make on-line systems easier to use. One tool that has been devised is the 'data base selector' or 'master index', such as DIALINDEX from DIALOG or DBI (Data base index) from SDC. The aim of such aids, which the searcher actually consults on-line, is to indicate which of the many data bases on offer are the most appropriate for a particular search topic. This they attempt by indicating the number of postings of any desired term or combination of terms in any specific group of data bases. Personal assistance too is not lacking: BLAISE has a Help Desk; SDC maintain Customer Action Desks; DIALOG have their own librarians to provide advice over the 'phone.

Among the data bases themselves, however, uniformity is still notably lacking in indexing terminology and methods, though there

230

have been attempts at harmonizing CAS and BIOSIS indexing practices and allowing the use of their data bases in common. Similar experiments by the American Institute of Physics and *Engineering index* were aimed at making their two data bases interchangeable by devising computer software to convert the indexing terms, format, etc, one into the other. There is a basic dilemma here as V Stibic has indicated: 'All categories of users want a friendly system . . . Unfortunately, the requirements of experienced users on the one hand, and of the beginners or incidental users on the other, are contradictory'. This is a classic instance of taking one step forwards and two steps backwards. H S White has explained that 'by reducing file structures and access software to a common denominator, we back away from the optimum capability which sophisticated file design and search strategies for uniquely structured data bases frequently offer us'. Conversely, the provision of more sophisticated search facilities, which is one method used by suppliers quite naturally desiring the maximum use of their services, means further elaboration and complication, and more difficulties for the untrained searcher. The challenge to the system designers for the future is to provide the variety of levels of response to enquirers that the human reference librarian almost instinctively offers. But the truth of the matter still remains, as Peter Vickers of the ASLIB Research Department reminded us in 1980, 'we are still miles away from designing systems that anyone would find easy to use . . . The most abbreviated list of common commands is over a page long. The actual operating manuals are the size of encyclopaedias'.

In fact it had become very clear even in the earliest days of batch searching that there was a major problem, familiar to reference librarians the world over: the need for some kind of skilled and experienced intermediary between the users and the system, to explain, advise, teach and indeed to search on their behalf. In his pioneering 1968 evaluation of the MEDLARS demand search service F W Lancaster found that 'The greatest potential for improvement in MEDLARS exists at the interface between user and system'. Referring particularly to the failures, he concluded that 'A significant improvement in the statement of requests can raise both the recall and the precision performance'. Evidence from the early British experience with MEDLARS at the University of Newcastle showed that searches formulated after discussion with liaison officers showed an 8% improvement in recall and a 9% improvement in precision.

The plain statistical facts of the matter are that ever since on-line searching was introduced the great majority of searches have been undertaken by intermediaries such as librarians, information scientists, search analysts, liaison officers, interface specialists, in-house strategists, etc. The view expressed by DIALOG, certainly the largest and probably the most experienced supplier of on-line services, is that librarians are finding a profession within a profession in learning to search a multiplicity of data bases on-line. It was reported in 1975 that over three-quarters of all on-line bibliographic searches in the US were being carried out by librarians or other skilled intermediaries. Similarly in Europe KTH reported in 1974 that almost all the searches were carried out by an intermediary, one of the library's documentalists. A 1976 survey for the Commission of the European Communities found that in 95% of the cases the terminal was operated by a trained intermediary. SDC found in 1980 that over two-thirds of their US users actually held degrees in library science. A 1980 UK survey by the ASLIB Online Information Centre reported that of the 289 libraries and information units searching on-line in only 12 did the ultimate user make the search.

Indeed, one unanticipated by-product of the coming of on-line was the number of organizations, anxious to exploit the potential of the new service but lacking librarians or information staff, that made such an appointment for the first time. Even the Science Reference Library softened its previously hard line described above and now states that 'searches will be carried out for you by trained intermediaries'. What are in some ways even more remarkable are findings such as those by Judy Wanger of SDC: her study of ten on-line services in the US revealed that almost half of the libraries and information centres conducting searches had never before performed *any kind* of literature search for their users. Far from eliminating the middleman the arrival of the computer had created the need for an intermediary where none existed before.

It is not claimed that any of these barriers in the path of the lay searcher are insuperable, and searching is certainly no occult art inaccessible to ordinary mortals. So far as the preliminaries are concerned, using a terminal is much easier than driving a car and command languages far simpler to remember than the *Highway code*. Achieving the necessary familiarity with data bases together with a grasp of the intellectual complexities of subject searching offers a greater challenge, but is not beyond the reach of any scientist or technologist who is anxious to develop new skills and is prepared to devote sufficient time to the task.

Here we can discern the flaw in the reasoning of those who argue that the ideal to aim for is every scientist and technologist making his own search. We have little evidence that they wish to do so. H S White indeed has written with some scorn of 'the myth that the ultimate users of information enjoy the process of search and discovery in the literature, and they would prefer to do this themselves, given the option'. Recognizing the professional satisfaction they themselves obtain from literature searches, some librarians and information workers
232

find it hard to credit that chemists, for example, may prefer to do chemistry, or engineers to practise engineering. Yet scientists and technologists have for years been giving quite clear signals as to what they require. In 1974 the reported experience of KTH, home of the largest scientific and technical library in Sweden, was that 'almost no customer has any desire to learn all the intricacies of the system since he is more interested in acquiring information than in the acquisition process'. A National Science Foundation study carried out by industrial engineers reported in 1975, when the on-line world was a much simpler place than today: 'It needs a full-time skilled librarian to know of all the various information retrieval systems available ... there must be a severe training problem when a user tries to learn more than one system'. Four years later the Agriculture Information Review Committee, composed mainly of scientists, found with regard to the three major data bases in its field (AGRICOLA, CAB and AGRIS) that 'it is becoming increasingly difficult for non-specialists (who do not have a detailed knowledge of the differences in subject coverage, level of input, entry points and costs of the various databases), to select the best combinations of databases for each search, both in terms of subject coverage and search costs'. And this last point reminds us that not least of the considerations to be borne in mind is the observed fact, not overlooked by the scientists and technologists, trained observers all, that an experienced intermediary can cut searching time by half. This weighs particularly with those enquirers who know they are going to receive a bill for the search, whether they carry it out themselves or not; for computer use is not cheap and telecommunications charges are far from trivial and both are time-related, as we have seen.

Yet, against all the indications, some remain unconvinced. A speaker at the 1979 Online Conference in London stoutly maintained that in the future the ultimate users of information will do their own searching, once we have more 'user-friendly' systems. As recently as September 1980 Georges Anderla, Director of Information Management for the Commission of the European Communities, promised as the next major development 'changes in the systems to make them usable by unskilled users'. It may well be that they are right to persist; perhaps it is solely the lack of time that accounts for the reluctance of busy scientists and technologists to learn a new skill. If on-line searching does become child's play some day — though it is difficult to see how the intellectual component ever could — perhaps they will come to enjoy it.

But the problem of user education may have deeper roots than many suspect. As we have seen in chapter 10, most user surveys show widespread under-utilization of abstracting services, with up to 20% of scientists rarely if ever referring to them, many more using them only irregularly, and technologists consulting them even less frequently than scientists. The coming of computerized services has done little to change this pattern, at least so far as scientists searching for themselves is concerned. Interest has certainly been evoked by on-line, but some of this is explainable as the 'new-toy syndrome'. It may well be that A J Meadows is right when he suggests that behind this reluctance is the 'fundamentally important point that scientists are oriented towards the literature less as recipients than donors. Most are more concerned with publishing their own work, than in following up the work of others. Therefore bibliographical aids of all sorts, and not only abstract journals, are used by many scientists only when they cannot avoid it'.

What is becoming ever clearer as each new study reports its findings is that the best results come from the collaboration of the scientist or technologist and the professional searcher. If knowledge of the subject and expertise in the search cannot realistically be looked for in one being, the best alternative is to seek them in a partnership. A detailed investigation of the potential of on-line searching in the Ministry of Agriculture, Fisheries and Food concluded that searches would best be conducted by appropriately trained personnel, preferably the local librarian or information officer, but the search strategies would be best formulated in close co-operation with the enquirer. It has been reported that routine experience in the years since has confirmed the wisdom of that conclusion. At the library of KTH where all the searches are carried out by intermediaries, as we have seen, the method so elegantly described by Roland Hjerppe could well serve as a model for others: 'The query requester is always invited to participate in the search and he usually does so if he has the possibility. The documentalist will in that case, in a manner of speaking, be the pilot of the system while the requester is the navigator'. Such collaboration often brings with it a bonus denied the enquirer searching for himself: assistance during the pre-search interview in clarifying in his own mind exactly what it is he is searching for. All reference librarians will confirm that it is by no means certain for enquirers to be certain of that when they commence their search. Results from a 1977 research study of on-line searching at the University of Manchester Institute of Science and Technology showed that 'the most successful searches were those in which the intermediary knew sufficient about the background of the user's query to ask pertinent questions, which would draw full answers about his requirements from the user'.

Data base economics
One of the least anticipated effects of the on-line revolution, and still hardly recognized by many library and information workers, is a funda-
234

mental change in the economic structure of the secondary information sources, in particular abstracting and indexing services. When they first appeared on the market in the 1960s computerized data bases were normally sold, like their printed counterparts, on subscription, being regarded by their publishers/producers as an extra service to their subscribers, bringing in a modest but very welcome additional revenue. The extension of their use outside the original purchasing institutions in the way described above provoked a sharp switch to leasing or licensing arrangements. Few of the major data bases can now be purchased outright: a subscriber may only lease them for a period, and he first has to sign an agreement governing the use that can be made of them. Some producers impose restrictions on who can use such leased data bases, others limit the geographical areas where services can be offered. For those institutions wishing to offer services outside their own walls special licensing arrangements are necessary. Furthermore, some producers also charge a royalty or usage levy over and above the leasing or licensing fee. Charges can obviously become very complicated and may vary considerably for the same data base.

On-line searching brought a whole new dimension to the issue: indeed for the secondary information services it marked a watershed. They realized, to their horror, that the new technology that they had been exploiting to mechanize their indexing and abstracting publications had also made feasible, at least in theory, an information system where 'a single copy of a tape can be acquired centrally, stored in a direct access memory, and then be accessed by thousands of researchers all over the country at remote terminals'. Even more startling, further advance also makes possible 'a network in which only a very few copies of a computer-readable service would be technologically adequate to service the entire world's needs'. They found they were witnessing the emergence of the concept of a global library in which the objective of 'holding' was likely to be superseded by the objective of 'accessing'. At this point some 90% or more of the income of the data base producers was still derived from the sale of the printed versions of their services. Thus it was clear that if subscriptions began to drop as a result of the availability of on-line services the producers would have to take steps to ensure an income sufficient to cover the costs of compilation, which of course are the same irrespective of the number of copies sold.

Following studies by the Commission of the European Communities and the International Council of Scientific Unions Abstracting Board, T P Barwise of the London Business School reported in 1979 that bibliographic data base royalties accounted for between 20 and 40% of the direct costs (ie excluding staff costs and overheads) of on-line searching (item a above), the remainder comprising computer charges

235

paid to the on-line service supplier (item b above) and telecommunications charges (item c above). Evidence shows that the data base producers had in fact adopted an age-old commercial tactic by deliberately holding down royalties in order to allow on-line searching to become established. In other words they used the revenue from their printed abstracting and indexing journals to subsidize the new service. Now that on-line searching has become virtually indispensable to its users the argument for subsidy has ceased to apply, while the financial risks of continuing it have increased. Barwise warned that on-line royalty charges may double by 1985. One straw in the wind is the sharp change in pricing policy of SCISEARCH, imposed in 1980 by the Institute for Scientific Information. On-line users who also subscribe to the printed version, *Science citation index*, have had their charges more than halved; non-subscribers on the other hand have had them almost doubled, to four times the subscribers' charge.

It should be said, however, that even in the absence of such financial penalties there is increasing evidence that the switch to on-line does not automatically bring about the cancellation of subscriptions to the printed versions. It was reported in 1980 that less than 1% of current or prospective users of the Commonwealth Agricultural Bureaux data base have indicated that they would cancel existing subscriptions. It does seem that in many cases the two modes of use are complementary. Indeed the previous year R J Rowlett of the Chemical Abstracts Service observed that on-line services had opened up many new markets for information without significantly diminishing the demand for more traditional publications. This may not of course influence at all the data base producers' commercial decision to increase their income from on-line searching, but how long the two modes will continue to co-exist is an open question.

The student will have gathered that the information market-place, like many others similarly buffeted by technological change, is a far from orderly or even sensible forum. Indeed a study of the trade in on-line services in particular reveals a scene of extraordinary chaos, with what seems like everyone taking in each other's washing. As Roger W Christian described it, 'Data base publishers who market their wares directly, as well as through vendors, are competing with their own most important customers. Further, the data base distributors — particularly libraries who pass the service benefits through to individual users free or on a cost-recovery basis — compete with their own suppliers, as well as having a devastating advantage over commercial "retail" outlets. Moreover, to the extent that machine-readable data bases are substitutes for the same information in print media, the data base publishers are even competing with themselves. Everyone, of course, is competing for the same limited supply of money available for bibliographic search of whatever kind'. And that is not the end of it: Carlos Cuadra warned recently that 'Some online services are making tentative steps toward the development or ownership of databases'. Dimly understood from the very outset, the economics of computerized data bases is now that of the madhouse. Of its 1980 introduction of the markedly differential charges mentioned above, the Institute for Scientific Information admitted 'We are all just groping'. The very real prospect often held out by the service suppliers, as computer operators, of unit costs falling year by year as a result of technological advance combined with economies of scale has been markedly dimmed by the cruel commercial truths that data base royalties seem certain to be raised well above their present artificially low level, and that tele-communications charges in the hands of the state monopolies are unlikely to be responsive to market forces.

User charges

When the question of charges for on-line searching is being discussed the student should be careful to distinguish between the 'user' who is actually the institutional subscriber and the 'user' who is the person who needs the bibliographical information. So far as the service supplier is concerned, the 'user' is the institution (or much more rarely the individual) to whom the bills are invariably sent. Whether or not the cost is then recovered by the institutional user from the ultimate user (or the 'end-user' as he has come to be called), or whether a nominal charge or a flat charge is made, or no charge at all, is a matter of institutional policy. As the student will have seen there is considerable scope for variation in costs from one search to the next, but this is as nothing compared to the actual *charges* made to the institutional user. There are many reasons for this, one of the most significant being the degree of support from public funds, which can range from nil to 100%, and may apply to any of the four elements itemized above. There are often differential charges for special categories of subscriber, eg educational institutions, or for different times of the day, and there are many discount plans and special offers and price changes, up and down, not to mention alterations in the rate of VAT and other taxes.

The question of passing some or all of these charges on to the ultimate user has been a matter of prolonged debate in libraries from the outset, and nowhere more than in public libraries, which have traditionally offered their services free of charge. Much concern too has been expressed in academic libraries, which although not free to all comers in the same sense, have not previously made charges for service to their regular users. It was scarcely to be expected, however, that libraries would stand aside to let the tide of technological progress flow past,

237

just at the point when bibliographical searching by computer became possible. And after some initial doubts, libraries in general have welcomed the on-line terminal. Early government-funded experiments in public libraries made two significant findings: many enquirers were people who had not used the library before; and when fees were charged after an initial period of free searches demand dropped considerably. Roger Summit has disclosed that 'several hundred' public libraries are DIALOG subscribers. The majority of the major university libraries are now 'on-line', and many of the rest have run trials. Since 1978 ASLIB has been operating its Online Information Centre with government support as a source of help and advice to librarians and others. We have seen that some libraries have developed into major suppliers of batch services; similarly there are now examples of libraries operating as major on-line search service brokers, eg the Science Reference Library with access to over fifty data bases.

Some public and academic libraries have set their faces firmly against charging for on-line services, although most do restrict searches to cases where the librarian judges it appropriate. Their argument is that such a service is no different in principle from the provision of printed abstracting and indexing services, or indeed that 'personal assistance to individual readers in pursuit of information' which for almost a hundred years has been known as reference work. Like on-line searching both of these services can also be very costly. To make a charge for a bibliographical search, merely because it is via a computer, is, they maintain, as illogical as to charge for consultation of the library catalogue when it is computerized. This policy has not gone without critical comment from other libraries, particularly in those instances where 'free' searches have been funded in part by cancelling subscriptions to the printed abstracting journals.

An agreed professional philosophy has yet to emerge, but in the meantime every variation imaginable can be found: simply taking the first five in a recently published list, Aberdeen City Libraries make no charge; Aberdeen University Library costs each search and charges the user; ASLIB makes a basic charge of £25 to include up to 20 minutes on-line and up to 50 printed citations; Bexley Central Library charges the cost of the search and adds 50% for search preparation; Birmingham Public Libraries charge £1 per minute plus print charges. The American Library Association is currently conducting a survey of how on-line services are financed in publicly-supported libraries and non-profit organizations and institutions.

As we have seen, a common experience when computerized services are tried experimentally in libraries is for volunteers to flock around when free searches are offered, only to drift away as soon as a charge is made. The MEDLARS station at Stockholm experienced a halving of

238

demand when users were asked to pay, and there is at least one example of a service that collapsed when it began to make charges. No one is more aware than the providers of computerized information services that alternative means of access to their information exist; part of their dilemma is that in retaining the manual printed services they are shoring up the redundancy in the system.

The imposition of charges on users accustomed to free access to the printed abstracting and indexing services in their libraries is often a profound shock, and not only to their pockets. A study by C I Birks for the British Library found that 'private individuals almost never use services which charge'. But scientists too, for whom most of the searches would be in the course of their daily work and who in many cases would expect their institution to pay, find it difficult to adjust to the idea of paying for information. Unlike businessmen, for whom information almost always has a definite commercial value, scientists identify it with 'knowledge' or 'truth', and expect it to be free.

Access to the literature
A well-nigh universal experience when computerized bibliographical search services are made available to scientists and technologists is a greatly increased demand for access to the literature. One great advantage possessed by service suppliers that are library based, eg KTH, National Library of Medicine, is that they are often able to provide copies of the documents cited, for these of course are what searchers really want, not bibliographical references or even abstracts. Increased library use generally, in fact, has accompanied the growth of computerized search services, inasmuch as information-seekers now turn to the library first, instead of, as so often in the past (see chapter 1) as a last resort. Many of the new clients, indeed, have not previously been library users at all. This new demand in its turn has increased pressure on library interlending services because of the inability of all but the largest libraries to stock more than a small proportion of the documents cited in the data bases.

It does perhaps need pointing out that what enquirers ultimately want, to be quite precise, is not documents as such at all, but the facts or ideas that they contain. In this strict sense, as has been stressed earlier more than once, it is misleading to use terms like data base and information system if what is being manipulated is not data or information but bibliographical citations, abstracts, and documents. This is a point which greatly exercised the members of the Agriculture Information Review Committee: they felt that integrated and evaluated information was what many agriculturalists really wanted, and they

239

made a number of detailed proposals to increase such provision in the form of reviews (see chapter 12).

But even the simple provision of documents poses serious problems for non-library service suppliers. As we have seen (chapter 10) sometimes the data base producers themselves have already made provision for 'document delivery', as it has come to be called: the arrangements made by *Zinc abstracts, Chemical abstracts, World textile abstracts*, for example, have already been described. For users of *Science citation index* the publishers have been offering for several years their Original Article Tear Sheet Service (OATS) which furnishes either the actual articles torn from their journals or photocopies. So valuable a timesaver has this become that several other data base producers and providers of search services now routinely rely on OATS as the primary source of their full-text copies. We have also seen that some services enter into formal or informal arrangements with major libraries to supply documents, eg *Agrindex, Physics abstracts, ICE abstracts*.

In fact all such current schemes sooner or later have to fall back on library collections. Here, for once, the UK is in a uniquely strong position. In the British Library Lending Division users have available, virtually by return of post, the great bulk of the world's scientific and technical literature, a service without rival in the world. Indeed about 18% of its three million requests come from 120 different countries outside the UK: the library of KTH, for instance, uses BLLD as a matter of course for items it cannot supply; the US National Library of Medicine requests over a thousand photocopies a month from BLLD. The document access systems plan proposed for the US in 1975 by the National Federation of Abstracting and Indexing Services was aimed to ensure that copies of all documents cited by member services would be obtainable from a number of selected research libraries. Users would be charged for documents provided: this would recompense the libraries and provide a royalty for the publishers. A later alternative was the proposal for a National Periodicals Center, already mentioned (chapter 10), but now effectively shelved by the US Congress.

The coming of on-line simply aggravated the problem, which has rightly been described as 'a painful but unavoidable corollary of high-quality secondary service provision'. Once again those suppliers or brokers that are library-based are in a particularly fortunate position: the Science Reference Library, for example, already holds the vast majority of items that are likely to be retrieved in the form of citations from the fifty data bases it searches. The big on-line service suppliers too have recognized the problem of document delivery: SDC offers its searchers its Electronic Maildrop Service in the form of full-text copies of items cited in some of the data bases they market; DIALOG has a

240

similar service. Supply is by commercial entrepreneurs, who frequently obtain the copies from libraries. BLAISE has its Automatic Document Request Service which permits a subscriber who wishes to see the full text of an item he has located in his own on-line search to transmit his request via the BLAISE computer to BLLD by typing in a simple command and the serial number of a prepaid BLLD loan coupon. It was reported in 1980, however, that use had remained disappointingly low.

Electronic ordering of documents, therefore, we already have, but the actual copies are still brought to our doors by the non-electronic postman. One alternative, at the moment waiting on technology despite its invention in the mid-nineteenth century, is facsimile transmission. Currently available equipment is either too expensive or too slow: machines already in use at BLLD transmit at an average rate of three to four minutes a page, or no more than a dozen articles a day per machine. For some the future is seen to lie in combining electronic ordering (such as we already have) with electronic delivery, perhaps transmitting the texts by satellite, which would allow greatly increased transmission rates. Studies undertaken for the Commission of the European Communities suggest that in Europe the answer may lie in document digitalization whereby the text is coverted to digital form by an optical character recognition device so that it can be stored on the service suppliers' computers and then transmitted via the EURONET link, probably during off-peak times, at speeds of up to a thousand pages an hour. The user's document delivery terminal would then reassemble and print out the text of the documents at high speed. It was reported that such an overnight document delivery system is both technically and economically feasible; it has even been given an ingeniously chosen acronym, ARTEMIS (*A*utomatic *r*etrieval of *t*exts on the *E*uropean *m*ulti-national *i*nformation *s*ervice), the Greek name for DIANE (or Diana as the Romans called her).

But even for the library-based services there are storm signals warning of two separate dangers ahead. The first, looming ever larger, is the scale of the demand. Some years ago we were warned from what was then the National Science Library of Canada: 'as data base subscriptions increase internationally it is impossible for any one institution to carry the burden of documentation supply'. The other cloud on the horizon is legal and financial: in 1979 A J Harley of BLLD reminded us that government costing policies and stricter legislation on copyright could easily reduce the effectiveness of their document supply service. The legal problems are multiplied when national boundaries have to be crossed, as the planners of EURONET/DIANE were to discover. They had wisely anticipated that 'Apart from the provision of access to online data, clearly the provision of documents is a key factor for success. Nothing is guaranteed more to frustrate a user than to give him a reference to a document he cannot obtain. A satisfactory on-line service can only be obtained through a blend of computerized and non-computerized systems and services. Extensive co-operation with libraries is therefore a prerequisite for a successful network'. However, once they got down to studying the problem they were confronted by the harsh realities of copyright law, which in the form of author's rights another arm of the Commission of the European Communities was striving to protect! The law applies as much to reproduction by electronic means as to reprography and all the arguments rehearsed about library photocopying from journals (see chapter 9) were heard again about documents stored in a computer and transmitted in digital or facsimile form. In 1979 the publishers sternly warned the CEC: 'it is not permissible for the EURONET system or any systematically operated delivery service for copyrighted documents to operate without the express consent of the copyright owners'. They are very concerned that any scheme that is established should not be subsidized by the CEC or member governments, and they have made firm proposals for substantial royalty charges to be paid by the user in addition to any photocopying or other charges. The CEC report on document digitalization and teletransmission just mentioned was frank enough to confess 'such critical issues as copyright (intellectual property rights) are still to be resolved'. No answer has yet emerged, though CEC officials have spoken of 'a pragmatic, internationally harmonized solution within the Nine Countries such as an appropriate system of cross-royalties', and later of 'a voluntary licensing system which could be freely negotiated between all parties concerned'.

The politics of computerized information

The imposition of charges on the ultimate user, at least in public libraries and often also in academic libraries, is as much a political as an economic decision. The problems of access to information, including 'trans-border flow' as it has come to be called, are also basically political, for information is the path to knowledge, and 'Knowledge itself is power', as Francis Bacon, a politician, told us nearly four hundred years ago. The implications range far beyond the world of scientific and technical literature.

In 1977 Eric Moon, as incoming President of the American Library Association, warned that the next few years would see the most impressive network of bibliographical control the world had ever seen, and as a basis for a national policy the principle of free access to all should be affirmed. More recently A E Wessel has expressed the view that 'the existing inequitable patterns of access to the world's informational resources are being made still more inequitable by
242

the current information system and computer-system technologies and their methods of implementation'. L J Taylor's opinion is that 'these developments point to the closure of traditional open publication and the establishment of groups of privileged consumers'. The 1979 CEC and ICSU/AB study mentioned earlier in this chapter found clear agreement that in India, for example, on-line usage will still be minimal in 1985, and its scientists and technologists will still be almost entirely dependent on printed abstracting and indexing services. One recalls Gandhi's warning a generation or more ago: 'there should be no place for machines that concentrate power in a few hands'.

Eugene Garfield is one who has already seen his predictions on this score come true. He has further told us: 'Twenty years from now, the only serious question will be how to take care of the people who can't afford the fee-based information society'. Some have taken up this challenge. In 1981 a group was formed within the American Library Association to focus on the public's right to electronic information: 'Unless the public library becomes a home for the electronic devices for information, the information gap between the haves and have nots will become even greater'.

The Director of Information Management for the Commission of the European Communities gave a similar warning some time ago: 'In fifteen or twenty years time people will either have or be denied access to very powerful systems — sources of knowledge and enrichment. Those without access will be second-rate citizens'. There is a vigorous political thrust behind EURONET/DIANE, stemming of course from its direct sponsorship by the CEC. Its stated aim is to 'create a Common Market in scientific and technical information with all the advantages that this can confer'. This implies the provision of service 'equally and without restriction' to citizens of all ten countries, with special effort being devoted to reducing the regional imbalance in information supply. This is seen to practical effect in the flat-rate or 'distance-independent' telecommunications tariffs within Europe, putting citizens of Copenhagen, Palermo, Dublin and Berlin, for example, on an equal footing. And though it is sometimes denied there seems little doubt that there has been a deliberate attempt to reduce European dependence on private US service suppliers and telecommunications networks. The PTTs have used their state monopolies to set low telecommunications tariffs for EURONET/DIANE hosts and punitively high charges for US service suppliers. Roger Summit of DIALOG recently warned the first world-wide conference of the Special Libraries Association at Hawaii that this politically motivated pricing structure makes US-based on-line information systems five times as expensive to use as European systems. The expression 'information war' is increasingly heard.

Computerized data banks

So far this chapter has concerned itself solely with *bibliographic* data bases, ie those containing literature citations and sometimes abstracts; the on-line searching that has been discussed is for bibliographical references. As the student will remember from chapters 5 and 6, this is not 'data' in the strict sense of numeric or quantitative information. Increasingly, however, such data is found in non-bibliographic computerized data bases, which now in fact make up a substantial majority of the thousand or so data bases publicly available. The category now extends to include not only scientific and technical data but numeric data of any kind, such as financial or statistical. And increasingly to be found are data bases containing substantive factual information of many kinds, not merely numeric or quantitative. These are commonly referred to as data *banks*, or factual or factographic data bases.

It is perhaps more instructive to adopt the recently proposed division into *reference* data bases, ie those that refer users to another source for complete information; and *source* data bases, ie those that themselves contain the source information. Under this classification the bibliographic data bases treated in this chapter are obviously reference data bases, although those that contain some substantive information in the form of abstracts could be regarded as hybrids. Examples of topics in science and technology represented by source data bases are electronic components, weather records, patents, psychotropic (ie 'mind-bending') drugs, chemical terminology.

It is not only the nature of the content that distinguishes source data bases from reference data bases. Perhaps the first and most obvious distinction is that only in a minority of instances do they derive from a printed equivalent. Numeric data bases in particular owe their origin more to the data processing industry than to the publishing trade. To some extent as a consequence they are sold primarily to the ultimate users, with libraries and information centres being largely by-passed. As most source data bases contain economic and financial rather than scientific or technical information most subscribers come from the worlds of business and commerce. Some are available only to closed user groups; this is not uncommon in the business world even with printed information. The second and quite critical difference is that their content is constantly shifting, as is the nature of much statistical and financial data, such as the population of Hong Kong or the exchange rate of the US dollar. This contrasts sharply with reference data bases, where a bibliographical citation, once made, is final. The format of the record too is much less standardized in the source data base.

A third difference that has emerged with use is that the searchers of source data bases are less commonly intermediaries; typically the ultimate user makes his own search. This is partly a matter of propinquity:

244

if the subscribers to such services are businessmen or research scientists the terminals are more likely to be located in offices and laboratories rather than in libraries. Another part of the explanation lies in the nature of the data retrieved: end-users will obviously be more inclined to search for themselves if the search product is not a list of bibliographical references but 'hard', factual data that can be applied immediately to solve a problem. One finding of the study by C I Birks of fee-charging information services was that 'Businessmen found that the bibliographic information provided by one on-line literature search service was of no use to them because they wanted hard facts'. Similar research on the needs of scientific workers has shown that two out of three require hard data not bibliographical citations, which for them are still one step removed from the answers they want.

We have been told that 'nonbibliographic data bases are the wave of the future', and some observers have remarked growing expectations among library users that the on-line terminals they see should also be used to interrogate the growing number of source data bases. Librarians are not resisting this: the Canada Institute for Scientific and Technical Information has already added scientific numeric data bases to its list; 40% of EURONET/DIANE data bases are data banks; the National Library of Medicine already offers TDB (Toxicology data bank) for on-line search, and MEDLARS III plans to include several source data bases. Though similar, however, the searching skills required are not the same as for reference data bases. One aspect in particular librarians may find challenging: as Carlos Cuadra explains, 'Nonbibliographic databases typically permit, invite or require some degree of manipulation, in addition to the pure searching function. Therefore, to use these databases, librarians must not only learn some new types of skills but must also be willing to move into a new role'. An example of what he refers to are the programs that some systems provide for the user to undertake mathematical operations on the data. Some librarians face this with confidence: Ray Jones told a conference of the American Library Association that 'The reference function can acquire an added dimension when computer techniques for the retrieval and manipulation of numeric data are learned and utilized ... There is no question of the obsolescence of traditional reference skills. The new dimension is to be added and integrated within our profession'.

There are signs too that changes in the content of reference data bases may bring them closer to source data bases. In his 1979 address to the National Federation of Abstracting and Indexing Services D W King predicted that 'it is going to be more and more difficult to distinguish between retrieval of information used for the identification of publications and the retrieval of full-text of primary information. Since the processes and technology used for on-line search and retrieval can be used in the future for transmission of full-text the two kinds of retrieval will eventually blend into a single system'. We already have an interesting example of a reference data base and a source data base combined, though it is a general index rather than one confined to science or technology: the New York Times Information Bank derives ultimately from the printed *New York Times index*, but in its on-line version it has been reported that 85% of the queries put at the terminal are answered without having to consult the newspaper files themselves.

Videotext

This is perhaps the place to mention videotext (sometimes known as videotex), the generic name for a variety of systems linking a distant computerized data bank to the domestic television set. Pioneered in the UK with CEEFAX and ORACLE, which are broadcast systems, and PRESTEL, which is an interactive system using a telecommunication link, they have the potential to provide what a *Times* headline recently proclaimed as 'Fireside access to sum of human knowledge'. Primarily designed for the general public and the businessman, they are proving slow to realize this potential, but the scientific and technical community is watching closely. The Agricultural Development and Advisory Service is already an information provider on PRESTEL, and has devised an agricultural information routing structure or 'search tree', in effect a self-service indexing service. BLAISE has also been experimenting with PRESTEL. By the year 2000 it has been estimated there will be 40 million videotext terminals in the countries of the European Communities.

Lists of data bases

The last ten years have provided the user of computerized data bases with an embarrassment of lists, guides, inventories, indexes, directories, source books, etc. The student should take care to distinguish these from the user guides and manuals described earlier; despite the variety of titles these are all in fact bibliographies. However, it is possible to discern a number of distinct categories, as well as some titles which combine two or three categories into one.

The most obvious are those listing reference data bases available on-line, eg J L Hall *Online bibliographic databases: an international directory* (ASLIB, second edition 1981) with entries for almost two hundred, containing a total of some 70 million references. As a counterpart to such lists we now have titles such as A Dewe and M A Colyer *British information services not available online: a select list* (ASLIB, 1980) covering eighty services. Of course it must not be forgotten that
246

many data bases are available for batch searching also, some are available for batch searching only, some are also available for lease or licensing, some are only available for lease or licensing, and some are available for neither. The best guide to this intricate market is M E Williams *Computer-readable data bases: a directory and data sourcebook* (Washington, American Society for Information Science, 1979). Including both batch and on-line services, this is by far the most extensive list, with world-wide coverage and over five hundred exhaustive entries; it is itself computer-compiled from a data base on data bases at the University of Illinois Information Retrieval Research Laboratory.

As has already been noted in chapter 10, some bibliographies list printed as well as computerized bibliographical services, eg *Inventory of abstracting and indexing services produced in the UK* (British Library, 1978). A new kind of list that we have seen emerge is not a bibliography of data bases but a directory of on-line brokers: J B Deunette *UK online search services* (ASLIB, 1981) describes 88 services, including some 20 public libraries, for those 'individuals and organizations without their own computer terminals'.

There are several listings of source data bases, mostly in specific subject fields and usually combining batch and on-line services, eg *Factual data banks in agriculture* (Wageningen, Centre for Agricultural Publication and Documentation, 1980), and *Directory of computerized data files and related software* (Springfield, Va, National Technical Information Service, 1974-), which is an annual listing of US federally-generated publicly available data. As most source data bases so far are distinctly business-oriented the listings reflect this, eg *The directory of computerized business information* (Berkhamsted, Trade Research, 1979).

Some lists of course attempt to cover the whole range of computerized services, eg the annual *EUSIDIC database guide* (Oxford, Learned Information, 1978-), which includes bibliographic and non-bibliographic data bases available for batch or on-line searching, totalling some 1,400, and describes itself as 'an international directory with a European emphasis'. The list with the widest scope is A T Kruzas and J Schmittroth *Encyclopedia of information systems and services* (Detroit, Gale, 1981) and its periodic supplement *New information systems and services*. Though it excludes the traditional printed abstracting and indexing services, and (since the fourth edition) libraries unless they are also data base producers, its alphabetically arranged entries, computer photocomposed and totalling over two thousand, make it a combination of world-wide bibliography, directory, and encyclopedia.

Further reading

P Leggate 'Computer-based current awareness services' *Journal of documentation* 31 1975 93-115.

R K Summit and O Firschein 'Public library use of online biblio-graphic retrieval services: experience in four public libraries in Northern California' *Online* 1 1977 (October) 58-64.

P W Williams and J M Curtis 'The use of online information retrieval services' *Program* 11 1977 1-9.

C I Birks *Information services in the market place* (British Library Research and Development Department, 1978) [Report no 5430].

T P Barwise 'The cost of literature search in 1985' *Journal of information science* 1 1979 195-201.

Commission of the European Communities *Workshop on document delivery, Luxembourg, 1979: proceedings* (Brussels, Bureau Marcel van Dijk, 1980).

H East 'Comparative costs of manual and on-line bibliographic searching: a review of the literature' *Journal of information science* 2 1980 101-9.

W M Henry and others *Online searching: an introduction* (Butter-worths, 1980).

A J Oulton and A Pearce *On-line experiments in public libraries* (British Library Research and Development Department, 1980) [Report no 5532].

'Perspectives on on-line systems in science and technology' *Journal of the American Society for Information Science*, 31 1980 153-200.

'Planning for online search in sci-tech libraries' *Science and technology libraries*, 1 1980 3-132.

R Woolfe *Videotex: the new television/telephone information ser-vices* (Heyden, 1980).

C-C Chen *Online bibliographic searching: a learning manual* (New York, Neal-Schuman, 1981).

B Cronin 'Databanks' *ASLIB proceedings* 33 1981 245-50.

S Keenan and others 'On-line information services in public libraries' *Journal of librarianship* 13 1981 9-24.

S Ljungberg 'Euronet-DIANE — pros and cons: some observations by an outsider' *Information services and use* 1 1981 17-21.

R Winsbury *Viewdata in action: a comparative study of Prestel* (McGraw-Hill, 1981).

Chapter 12

REVIEWS OF PROGRESS

The elaborate and irreplaceable apparatus of indexing and abstracting services described in the previous chapters can unfortunately do nothing to stem the swelling tide of primary publication, doubling every ten or fifteen years with the volume of new research reported. Invaluable as these keys to information undoubtedly are, it cannot be denied that they *add* to the total amount of literature, and the scientist or technologist often finds himself in sympathy with James Thurber's friend who complained in some distress: 'So much has already been written about everything that one can't find out anything about it'.

It has been clear for some years that even the scanning of indexes and abstracts is proving too much for some workers, and there have been urgent pleas for more easily digestible forms of secondary publication. In response we have seen a remarkable revival of the review, a literature form far older than the abstract, but which has lain in its shadow for a hundred years or more. Not to be confused with the book review (of the kind found in the *Times literary supplement*), it takes the form of an evaluative summary (and possibly synthesis) by a specialist of developments in a particular field of endeavour over a given period.

Such reviews of progress are now seen very definitely to be of great importance; by some they are regarded as offering a possible pathway out of the literature jungle. In some fields they are used more heavily for literature searching than abstracts and indexes. H V Wyatt for instance considers that 'The future of biological literature lies not in classification by words but in distillation by review'. Lord Todd, Cambridge chemist and Nobel prizeman, has stated: '. . . it will, I believe, be necessary to develop review journals intensively if we are to have a really effective information system'. And of course the editors of review series are ardent apologists: the preface to the first (1962) *Advances in nuclear science and technology* refers to the 'bewildering information problem to both the expert working along its narrow crevices and the dilettantes hoping to keep abreast of the ever expanding frontiers. Clearly what is needed by both groups are well-organized review articles.' The introduction to the first (1965)*Advances in chromatography* says: 'It is clear that the individual worker, if he is to preserve even a moderate knowledge of the entire field, must rely more upon responsible surveys than on the attempt to read the avalanche of original research papers'. And neatly drawing on his own subject – a method of chemical analysis – by way of illustration, the writer goes on to explain the particular literature difficulty that the scientist has found better solved by the review than the abstracting service: 'The problem briefly stated, is one of information sorting; we wish the uses of chromatography were so universal that it could separate information – the hard core advances from the overwhelming mass of supporting evidence and data that, although necessary in research articles, quickly swamps the digestive process'. The preface to the first (1960) *Advances in computers* describes how the review is 'intended to occupy a position intermediate between a technical journal and a collection of handbooks or monographs. It is customary for a new scientific or technical result to appear first in a journal, in a form which makes it accessible to specialists only. Years later it may be combined with many other related results into a comprehensive treatise or monograph. There appears to be a need for bridging the gap between these modes of publication, by surveying recent progress in a field at intervals of a few years and presenting it in a form suitable for a wider audience.' That such reviews are seen as supplementing rather than supplanting the abstract journals, however, is well demonstrated by the USSR All-Union Institute of Scientific and Technical Information (VINITI), the world's largest abstracting organization, which publishes about a hundred different series of annual reviews, based on the abstracts in *Referativnyi zhurnal*.

As is so often the case with other types of information source, there exists an area here of considerable confusion over terminology. As we have noted, the word 'review' can also mean a book review; the dictionary definition also allows it to cover a periodical or newspaper in general, eg *Westminster review, Quarterly review*. Although the term does appear in the titles of many of the publications we are examining in this chapter, eg *International review of cytology, Annual review of physical chemistry, Review of coal tar technology, World review of nutrition and dietetics*, the vast majority of such works call themselves something else. By far the most popular titles are *Advances in . . .* and *Progress in . . .* , but other examples are *Reports on progress in physics, Survey of progress in chemistry, Recent progress in hormone research, Survey of biological progress, Record of chemical progress, Developments in food packaging, Medical annual*. There are also a number of titles which give no indication that they are review series, although sometimes their subtitles may, eg *Physics and chemistry of the earth, Vitamins and hormones, Oceanography and marine biology*, but even

250

more misleading are works with review-type titles that are really something else. *Progress in fast neutron physics* (Chicago UP, 1963) actually contains the proceedings of an international conference; *Advances in enzyme regulation* comprises the annual proceedings of a series of symposia, with each volume containing a keyed photograph of the participants; *Progress in nuclear energy* is the overall title for twelve series totalling some fifty volumes of a miscellaneous nature; *Progress in microscopy* is a monograph by M Francon; *Advances in chemistry* and *Mathematical surveys* are both monograph series issued respectively by the American Chemical Society and the American Mathematical Society; *Mathematical reviews* is an abstracting journal; and *Physical review* and *Review of scientific instruments* are both primary research periodicals.

A more serious effect of this terminological pliability is that it conceals the fact that there are two distinct categories of reviews of progress. These stem from the dilemma that always faces the editor or compiler or publisher of a regular survey of a particular subject area: should he attempt an ordered and balanced coverage of the whole field, or should he emphasize those topics of the most pressing current concern? By and large, the various review series have chosen to follow positively one path or the other: unfortunately, this vagueness of vocabulary means that users cannot determine from the title alone which choice has been made. A moment's examination of the text, however, is usually sufficient to distinguish the two types.

Comprehensive reviews
These are thorough, systematic, and condensed accounts of developments in a broad field over a narrow time interval (and sometimes within a particular geographical area). Long-established examples to study are *Annual reports on the progress of chemistry* (1904-), and *Annual review of biochemistry* (1931-). They are written by teams of specialists for fellow-specialists, and usually assume extensive knowledge of the subject. Firmly based on the literature, they provide extensive references, eg in *Annual surveys of organometallic chemistry* for 1965 the six-page survey on aluminium has 76 references; the fifteen-page account of carpets in *Review of textile progress* for 1965-6 has 114 references. They can unfortunately degenerate at times into narrative bibliographies or, even worse, into sterile 'literature reports'. The introduction to the first volume (1965) of *Advances in chromatography* refers with some scorn to 'the glorified bibliography or reference-finding system that some reviews tend to become when the author, actively engaged in research in a very restricted area, attempts a comprehensive survey'. Indeed, one of the contributions to the 1965

251

Progress in dielectrics is a 33-page bibliography! As Mellon says, 'With the facilities now available, one familiar with index serials and abstracting journals should be able to compile his own report of progress, or at least a non-critical summary'. Of course, the better series are far more than mere compilations: not only are they evaluative, but they play a very positive role synthesizing the developments newly-reported in the literature and relating them to the already accumulated knowledge in their discipline. Here we see clearly their advantage over the abstracting services, which merely report. The best of the review series go further in attempting to isolate trends and even to point the way forward.

Probably the majority of such surveys appear annually, in the form of a single bound volume, eg *Reports on the progress of applied chemistry, Progress of rubber technology, Annual review of nuclear and particle science*, although in recent years the pressure of new literature has forced both the well-known series mentioned at the beginning of this section to expand into two volumes. A questionnaire revealed that the average subscriber to *Annual reports on the progress of chemistry* claimed to read 30% but admitted to ignoring 30% of each volume: from 1968 therefore the work was made available in two separate volumes (A: general, physical and inorganic chemistry, and B: organic chemistry), and from 1980 in three, with physical chemistry achieving its own volume C. The preface to the 1966 volume of *Annual review of biochemistry* is more waggish: 'This year our *Annual review* has shown behaviour akin to a primary biological phenomenon: it has undergone binary fission'.

The Royal Society of Chemistry has gone much further than this: the field of chemistry has been divided up into over thirty sections, for each of which what is called a Specialist Periodical Report is published annually, or biennially, eg *Organometallic chemistry, volume 8* (1979); *The alkaloids, volume 7* (1979).

An alternative solution to this problem of course is to publish more frequently than once a year, eg *Reports on progress in physics* was originally a single annual volume, but by gradual stages has developed to the point where it can be obtained in monthly parts or as a bound volume each quarter.

Some surveys are published as articles in periodicals, eg 'Progress in heat transfer – review of current literature', annually in *Process engineering*; 'Annual review of the literature on fats, oils and detergents' in *Journal of the American Oil Chemists' Society*. One issue each year of *Rubber chemistry and technology* is 'Rubber reviews'.

Surveys of this kind have an obvious current appeal to the specialist in that their comprehensive nature enables him to fill any gaps in his knowledge of recent developments and their broad coverage can often

252

give him a new angle on his subject. Retrospectively, they are well-suited to serve the exhaustive approach also.

Topical reviews

These are 'state-of-the-art' reports on selected, specific topics of active current interest. Increasingly in the last two decades these have appeared collected in volumes issued as a series, eg *Progress in semi-conductors, Reviews in engineering geology.* Examples of individual reviews in such volumes are 'Jewels for industry' in *Modern materials: advances in development and applications* 6 (1968), 'Immunity to ticks' in *Advances in parasitology* 18 (1980); 'History of noise research' in *Advances in electronics and electron physics* 50 (1980); 'Dehydrated mashed potatoes' in *Advances in food research* 25 (1979).

They are specifically designed to be intelligible to the non-specialist, and while not 'popular' in approach are aimed at all levels of readership from the student to the director of research. One particular aim they have is interdisciplinary cross-fertilization, and their target is the worker in related fields of science and technology anxious to remain in touch with the more significant developments outside his immediate area of interest. The preface to the first volume of *Advances in micro-bial physiology* (1967) writes of the 'premium on the availability of review articles with which biologists, who are researching in one area, can become acquainted with progress that is being made in areas out-side their immediate sphere of interest'. The gravity of this problem is emphasized by Lord Todd: '... it is difficult to get information conveyed between one science and another, but this is usually because the practitioners of different sciences don't quite know how to express what they are looking for in comprehensible terms or in a way that will register with each other. Yet such interchange is vital to development in the borderlands between established disciplines and it is in such borderlands that the growing points of science are frequently found. I believe that only the properly written review journal, coupled with personal contacts, can solve this problem.' That such reviews of progress are achieving some success is evidenced by the reaction to the first (1960) *Advances in computers* volume, 'felt by many readers as a welcome antidote to the ever-growing specialization of technical fields'. The standing instructions to authors issued by *Science progress* are that 'Articles should be reviews of a particular region of research, not for specialists in the field but for professional scientists in quite different disciplines and for undergraduate students'. And some have even more specific aims: *Surveys of progress in chemistry* is intended to improve the transmission of new material mainly to the college chemistry teacher. The first volume of *Advances in lipid research* (1963) tells us

253

that 'new insights may be obtained into one's own work by reading concise, critical expositions of advances in areas of tangential interest'.

Although written by specialists, like the comprehensive surveys, these topical reviews are seen by their editors as something much more flexible. The authors writing for *Advances in botanical research* are asked 'to express opinions freely and to speculate as widely as they dare on future trends and future developments'. The author contributing to *Polymer reviews* is 'encouraged to speculate, to present his own opinions and theories to give a more "personal flavour" than is customary'. Likewise, the contributions to *Problems in biology* 'differ from other reviews in that their accent is on exposition. They are not intended to present a high-powered balanced account of the current factual information, instead the author has been encouraged to be selective in the choice of material and, if necessary, to present a personal view of the subject.' They are naturally (and indeed deliberately) far less 'bibliographical' than the comprehensive type of survey: they have been characterized as 'subject-oriented' rather than 'literature-oriented'. They are not limited in coverage to a particular span of time or geographical area. The editors of *Viewpoints in biology* instruct prospective contributors that their 'broadly-based reviews . . . should not only summarize the state of the subject but also indicate the direction in which progress may be expected, and stress unsolved problems. While putting a cogent, well-argued point of view the authors will, however, not necessarily be asked to give exhaustive documentations of all work in the subject'. Even more firmly the preface to the first (1963) volume of *Progress in nucleic acid research* states: 'We do not wish to sponsor an annual or fixed-date publication in which the advances of a given period of time are summarized, or a bibliographic review or literature survey. We seek rather to encourage the writing of "essays in circumscribed areas".' The philosophy of Annual Reviews, Inc, publishers of some two dozen titles each year, is even more specific: 'Colorless synoptic summaries or annotated bibliographies of all relevant papers published in the preceding year are discouraged'.

This same freedom of approach permits the author of a typical review if he so wishes to write at greater length than is usual in a conventional periodical article, without having to produce a full-blown monograph. However, if a contributor feels he must produce a work of such substance, there is evidence that the review format is hospitable enough to accommodate him: the fifth (1966) *Advances in marine biology*, until then a conventional review series, is entirely given over to a 435-page work by T C Cheng 'Marine molluscs as hosts for symbioses'. An examination of J P Hirth and G M Pound *Condensation and evaporation* (Pergamon, 1963) soon reveals that it is also known as volume 11 of *Progress in materials science*. But this is not really the best way to make use of this particular manifestation of the literature of science and technology. The ideal review, in the words of the preface to the second (1961) *Advances in computers*, should be 'long enough to introduce a newcomer to the field and give him the background he needs, yet short enough to be read for the mere pleasure of exploration'.

Topical surveys of this kind can obviously be published in a variety of forms, but it is the burgeoning review series such as *Advances in chemical engineering, Progress in optics, Recent progress in surface science, Macromolecular reviews*, which have been responsible for the spectacular rise to its present prominence of this form of scientific and technical literature. Each of these new series follows more or less the same pattern, with separate volumes containing half-a-dozen or more review articles, appearing at intervals. The titles chosen often reflect their selective, topical, flexible character, eg *Topics in stereochemistry, Modern aspects of electrochemistry, Essays in biochemistry, Modern trends in accident surgery and prevention*, and some even show a touch of imagination, eg *Vistas in astronomy*. Perhaps the most descriptive kind of title is *Current topics in developmental biology*.

Freed as they are from the necessity of surveying a particular period in time, they are far less likely to appear regularly every single year: *Advances in radiation biology* comes out every two years; between the first and second volumes of *Currents in biochemical research* ten years elapsed. Some appear more frequently than once a year: the annual *Chromatographic reviews* changed in 1967 to twice or three times a year; *International review of cytology* comprises four bound volumes issued in sequence each year; belying their titles, a number of *Annual reviews* now appear quarterly, eg *Annual review in automatic programming*. With subjects chosen for their topicality *ad hoc* for each volume there can obviously be no claim that the whole field is systematically covered, but editors do take care to cast their net widely, and some even have elaborate schemes of cycling to ensure total coverage within a measurable period. For instance, *Progress in ceramic science* explains that 'within a single volume no attempt will be made to associate topics with each other, but it is hoped that in the first few volumes most of the important parts of the field will be reviewed'. The introduction in the first volume (1958) of *Advances in petroleum chemistry and refining* outlined its plan to produce 'at annual intervals a collection of progress reports written by leading authorities on particular subjects . . . In the course of several years, the series of Advances will have touched upon all parts of the far-flung industry, and the set of volumes will assume the character of a reference book. Cumulative indexes . . . will tie the material together.' By the tenth volume (1965) the grand design must have been completed, for its preface described it as 'the final volume of the series'. *Progress of rubber technology* uses another

255

method to try to get the best of both worlds: though it is basically an annual review of the comprehensive type, in alternate years it also includes one or more topical reviews 'selected on account of notable advances'.

Reviews need not be published in collected volumes: they can be issued separately as are the paperback Sigma science surveys, deliberately limited to 5,000 words in length and issued at the rate of four a month. Some appear in both forms: each article in *Progress in materials science* is also published separately to make it available quickly, and the preface to the 1966 *Advances in applied mechanics* announced its intention 'to publish at least the next volume of "Advances" in the form of several successive fascicles, in order to present the material as rapidly as possible'. Some publishers have issued selected reprints from their annual volumes as a collection, eg *Annual reviews reprints: immunology, 1977-1979*, chosen from six separate titles in their *Annual reviews* series. *Water purification in the European Communities: a state-of-the-art review* (Oxford, Pergamon, 1980) is a 474-page volume.

A popular way for 'state-of-the-art' surveys to appear is in the form of papers read at conferences, and these may be later published separately in a periodical or collected in a volume of conference proceedings (see chapter 13). A number of scientific societies try to include such review papers in their programmes as a matter of deliberate policy. Occasionally the whole conference may consist of reviews or review-type papers: the series *Advances in the astronautical sciences* are the proceedings of annual and other meetings of the American Astronautical Society, and *Progress in astronautics and aeronautics series* is based on papers read at symposia of the American Institute of Aeronautics and Astronautics.

A little exploited source of state-of-the-art reviews is the thesis or dissertation (see chapter 19). In many cases a review of the literature to date is a required preliminary part of the work.

But by far the oldest form of publication for a review is as an article in a regular scientific or technological journal, eg *Analyst, Biochemical journal, Journal of chemical education, Endeavour*. A typical example would be the 6-page article 'Structure and evolution of the moon' in *Nature* for 13 September 1979 when the tenth anniversary of the manned Apollo missions furnished 'an appropriate opportunity to review the current state of our knowledge'. Such reviews appeared in the earliest journals like the *Philosophical transactions of the Royal Society*, and a survey published in 1964 indicated that more than half of all reviews appeared in primary research journals, compared with 7% in annuals and other special types. In 1975 *Tetrahedron*, a twice-monthly organic chemistry research journal started a regular feature,

'Tetrahedron reports', which are in fact specially-commissioned topical reviews. The April issue of *Analytical chemistry* is a special annual reviews issue. About half of each issue of Institution of Electrical Engineers *IEE proceedings part A* is given over to reviews, eg 'Airborne electrical displays' in May 1981.

Review journals

For many years (and increasingly of late), there has been a special category of periodical solely devoted to review articles, eg *Science progress, Chemical reviews, Biological reviews, Quarterly review of biology, Contemporary physics*. Apart from their format and frequency, these review journals are often indistinguishable from the review series discussed above: the editorial policy of the *Review of modern physics* is that 'The best papers in the journal should be milestones of physics, embodying the intellectual contributions of hundreds of others whose work appears in the original literature'; the objective is to publish 'perspectives and tutorial articles in rapidly developing fields of physics as well as comprehensive scholarly reviews of significant topics'. CRC Press has recently initiated a series of journals, eg *CRC critical reviews in analytical chemistry*.

Bibliographical control

The ways in which reviews of current progress serve the scientist and technologist are obvious. Currently they enable him to remain aware of the major advances outside his own particular area of activity; retrospectively he finds the bird's eye view by a perusal of a good review article is often the ideal way to orientate himself in a field comparatively unfamiliar. As for the exhaustive approach, the first thing an investigator looks for is a review article or a bibliography, preferably both. Indeed, the review holds the remarkable distinction of being just about the only source of information that scientists appear to welcome and happily read. Recent surveys of use have demonstrated its popularity again and again. The major survey of 3,021 physicists and chemists carried out by the Advisory Council on Scientific Policy demonstrated the almost universal use of reviews: over 90% of the sample had read or consulted a review within the previous month, well over half rated them as the most useful source for current aware-ness (a higher proportion than either abstracts or conferences) and between 46% and 55% would like more of them. A more recent survey of 2,702 mechanical engineers (as a group far less literature-conscious than pure scientists) showed that reviews are used consistently by workers in all types of activity (eg management, research, design, testing, sales, production, etc) and far more frequently than abstracts and indexes by all except those in research. Over half thought it would be useful if more review articles were published.

So the demand is there. The attempt by librarians to satisfy it soon brings home the fact that even for the resources which already exist bibliographical control although improving still has some way to go. A useful aid to identify appropriate collected reviews is the UNESCO *List of annual reviews of progress in science and technology* ([Paris], second edition 1969), with some two hundred titles in subject order; the British Library Lending Division has produced lists which include review journals as well, eg *Some current review series* (1964) and *KWIC index to some of the review publications in the English language* (1966). The most comprehensive list is A M Woodward *Directory of review serials in science and technology, 1970-1973* (ASLIB, 1974), with about five hundred titles, thought by the compiler to amount to some 80% of the total in this format. Many of these series take particular care over indexing their contents to enable individual reviews or topics to be located: not only is it common to find each volume indexed by subject (and often by author), but regular collective indexes are frequent, eg every five years for *Reports on progress in physics*. Cumulative indexes are even more useful, eg to the first 70 volumes of *Chemical reviews*, and the first 46 years of *Annual reports on the progress of chemistry*. As an alternative to separately published indexes of this kind, several series include collective and cumulative indexes in the volumes themselves: an author and title index of volumes 30 to 35 is included in volume 35 (1966) of *Annual review of biochemistry*; contents lists for the preceding volumes are included in each issue of *Advances in heterocyclic chemistry*; volume 29 (1967) of *Advances in enzymology* contains a cumulative index to all the 29 volumes. But this is merely the tip of the iceberg: as we have seen, only a minority of review articles appear in such collections. The remainder, in their thousands, are scattered in the main throughout the regular periodicals, and it is a review of this kind on a specific topic such as waves, metal solutions, ignition and combustion of solid rocket propellants (rather than a review series) that is most commonly demanded. One survey in the fields of biology and medicine showed that less than a quarter appeared in review publications, and over two-thirds were to be found in the primary journals; the estimated 1972 output of 22,000 reviews in science and technology was made up of 10,500 in primary journals, 4,500 in review serials, 4,000 in conference proceedings, and 3,000 in books and reports. Of course they are included (although not always abstracted) in the indexing and abstracting services with other periodical articles, and can be traced in the same way: it is estimated, for instance, that 6% of the entries in *Chemical abstracts* are for review

258

articles. But like bibliographies (see chapter 8) there is no certain way to ensure that reviews on a topic emerge at the beginning of an investigation. They are embedded in the literature and it can take an exhaustive search to prize them loose.

A further handicap is that they are not always easy to recognize: out of 8,601 reviews in the field of chemistry it was found that only two contained the word 'review' in the title. Woodward found it impossible to identify reviews conclusively without consulting the articles themselves: '. . . there appears to be little prospect in the immediate future of identifying reviews by any means other than objective assessment'. Some indexing and abstracting services do try to signal reviews. This is easier for the abstracting services, for they of course examine the articles in detail: *Chemical abstracts* uses R to mark reviews in its index; *Biological abstracts* augments the title with the word 'review'. *Science citation index* uses (R) to indicate a review article, but the basically clerical nature of the citation indexing method must mean that some slip through the net. With computerized data bases reviews can simply be tagged with a special code, once they have been identified: AGRICOLA and CAS both do this. Fortunately, the need for special tools is gradually being recognized and there are now available a handful of bibliographies confined to reviews: *Bibliography of medical reviews* is a cumulated listing based on the corresponding section in the monthly *Index medicus*; *Bibliography of reviews in chemistry* derived similarly from *Chemical abstracts*, but ceased publication after 1962 for lack of support, although it was revived in 1975 in the shape of *CA reviews index*, a computer produced KWIC index produced twice a year with some 20,000 review articles per issue. The annual compilation by D A Lewis *Index of reviews in organic chemistry* (Chemical Society), cumulated in 1971 and again in 1977 is a lone example of its type. N Kharasch and others *Index to reviews, symposia volumes and monographs in organic chemistry* (Pergamon, 1962-   ) goes back as far as 1940 in coverage in the three volumes so far published.

It is clear that bibliographically much remains to be done. And if what we read in the introduction to *Macromolecular reviews* for 1966 is true it must be done soon: '. . . the review article is becoming the primary [ie principal] source of information to a large majority of scientists'. One hopeful sign is the first general index to appear, the computer-produced *Index to scientific reviews* from the Institute for Scientific Information, publishers of the *Science citation index.* It obviously suffers from the problem referred to above, that of identifying unambiguously those articles that are indeed reviews, but it appears twice a year, covering some 2,600 journals as well as the various annual and other review publications. It lists over 20,000 entries a year.

The production of reviews

In 1969 the influential SATCOM report warned: 'the current production of reviews . . . is but a fraction of the present requirement, and the need is steadily growing. In addition, there is evidence that current reviews tend to cover the literature much less thoroughly and comprehensively than was formerly the case'. The Agriculture Information Review Committee concluded in 1979 that 'the information needs of many categories of agriculturalist might be better satisfied if more, appropriate reviews were available'. When this view was seen to be widely supported throughout the UK *Farmers' weekly* and the Commonwealth Agricultural Bureaux both agreed to consider including more review material in their publications and to look into the possibility of creating a new agricultural review journal.

A handicap of some gravity of which the student should be aware is the marked reluctance of many scientists and technologists to write these much needed reviews. Not only do they see it as a tedious, time-consuming, and exacting chore, but it does not in their eyes contribute to their professional reputation anything like so much as the production of a good research paper. Some even regard it as 'non-productive work'. The paradox here of course is that these same scientists and technologists *as readers* are those most vocal in their cries for more review articles and state-of-the-art papers. It has been found that when authors do agree to write a review they regularly underestimate the time necessary: up to one third of the reviews commissioned for *Reports on progress in physics* are lost because of 'excessive delay'. Even the offer of an honorarium, unheard of for the normal research paper, does not bring forth reviewers in the numbers needed.

High hopes were once entertained that the burgeoning information analysis centres would make a substantial contribution here, but the 1970 ASLIB study of such centres in Britain had to report that 'evaluating the collected information and producing critical and state-of-the-art reviews . . . in present practice . . . is seldom done, although it still should be borne in mind as a goal worthy of attainment'. The most sophisticated computerized systems in existence, or even in prospect, have little to contribute here. To produce a coherent, digested account of advances in a field of study and to relate it to what has gone before requires a human being.

The seriousness with which the scientific and technological community regards the present crisis was demonstrated by UNISIST (see chapter 10): one of the working parties marked as an area of crucial need the development of better mechanisms for review and critical evaluation of scientific effort. In 1976 the Centre for Agricultural Publishing and Documentation (PUDOC) in the Netherlands concluded a major study of review articles by not only recommending that scientists should be given the opportunity to write during working hours, but also suggesting that preparing reviews should be part of a scientist's job specification. Encouragement of the scientific community to produce more reviews was also stressed at the first international conference of scientific editors at Jerusalem in 1977. More recently both the Physics Information Review Committee and the Biological Information Review Committee have urged the same, while recognizing the problem of persuading suitable authors. The biologists 'would like to see such authors receive the same kind of recognition as do those who present the named lectures of our learned societies'. In his studies of this form of literature over a number of years Woodward has emphasized a point not often made about the value to the scientist personally: 'Collecting the literature, assimilating the data, and making coherent sense of the material must provide a comprehension of the subject far deeper than any subsequent reader can obtain'.

Further reading

'Critical reviews [a symposium]' *Journal of chemical documentation* 8 1968 231-45.

J A Virgo 'The review article: its characteristics and problems' *Library quarterly* 41 1971 275-91.

A A Manten 'Scientific review literature' *Scholarly publishing* 5 1973-4 75-89.

A M Woodward 'Review literature: characteristics, sources and extent in 1972' *ASLIB proceedings* 26 1974 367-76.

A M Woodward *Problems and possible investigations in the study of the role of reviews in information transfer in science* (ASLIB, 1975) [British Library R & D report no 5234].

A M Woodward 'The role of reviews in information transfer' *Journal of the American Society for Information Science* 28 1977 175-80.

Chapter 13

CONFERENCE PROCEEDINGS

As sources of information for scientists, formal meetings to hear of their colleagues' latest thoughts and discoveries have a long and respectable history. Indeed, it was the need first felt in the seventeenth century for some kind of framework for such meetings that formed a primary motive for the foundation of the first learned societies in science. Furthermore, as we have already seen (chapter 9), it was the desire for a literary vehicle in which to report the proceedings of such meetings that provided some of the impetus for the earliest scientific periodicals. Today such conferences range from small gatherings of the local branch or specialist section of a professional or learned society such as the Royal Society of Chemistry or the Plastics and Rubber Institute to the great international scientific congresses with thousands of delegates from all over the world. The majority of scientific conferences are still organized by the societies: 80% of those in physics, for example. Most commonly described as conferences or conventions they may also be known as symposia, seminars, sessions, assemblies, workshops, round tables, clinics, institutes, colloquies, study groups, summer schools, or teach-ins. Many are still announced simply as meetings. A 1965 United States estimate suggested that about an eighth of expenditure on information goes on conferences and meetings (about $50 million at that time), and a more recent estimate in the field of physics is that conferences cost as much as the expenditure on library and information services. Another survey showed that 94% of professional scientific societies in the US organize annual meetings at which papers on original research work are presented. If we confine our attention to international meetings, we discover that the number arranged each year amounts to thousands in science and technology alone, and the papers read at each quite frequently run into hundreds (as at the American Society of Mechanical Engineers or American Chemical Society meetings) and occasionally into thousands (2,100 at the Second International Conference on the Peaceful Uses of Atomic Energy). H Baum of the World Meetings Information Center reckoned that in 1972 more than 25,000 papers of interest to chemists were presented at meetings. Since then not only has the number of conferences
262

covered by *Chemical abstracts* more than doubled, but the number of papers presented at each conference has gone up by more than half. Indeed it is said that a chemist could spend a lifetime doing nothing but attend conferences.

Quite obviously they play an important role in scientific and technological communication. Many of the papers presented often report research work several months before publication in the periodicals: indeed a feature of conferences is the large number of interim reports on incomplete work which would not otherwise see the light of day until the final report is ready — often very much later. They are clearly 'an early and important outlet in the dissemination process'. Common too as conference papers are 'state-of-the-art' surveys, which as indicated in the previous chapter are among the most sought after of all types of sources in an increasingly turbulent sea of information.

As a medium of communication such conference papers have the great advantage of oral presentation, with questions from the audience, immediate informed criticism and comment, and follow-up contacts. Indeed, conferences can still furnish for the individual research worker many of the benefits of the person-to-person communication that was commonplace in the golden age of science when it was still possible for a scientist to know every other worker in the field. Quite truly it has been said by the investigators on the Project on Scientific Information Exchange in Psychology, which initiated the pioneering series of studies of the value of conferences to the scientist, that 'the convention offers, among all channels, the greatest range, both in degree and number, of opportunities for scientific communication. Considering but a single paper, an attender can choose to establish any degree of contact with its content or its authors, from merely glancing at the abstract in the programme to attending the session and approaching the authors to discuss specific questions or to pursue common scientific interests.'

And this still does not make explicit the most important function of such gatherings: in the words of a *Nature* editorial on the topic, 'the special value of conferences and symposia is that they bring people together in ways which permit informal exchanges of ideas and information'. The Physics Information Review Committee commented in 1978 that 'The role of different types of meeting, for different groups, is a neglected area of study, particularly in the UK. It is a rarely studied problem on which some light should be thrown'. They pointed out that a conference on the right *topic* might be the wrong *type* of conference for a particular subject field or a particular community of scientists. F W G Baker of the International Council of Scientific Unions has recently suggested that 'The conference as a method of information exchange will partly be replaced by the extension of teleconferences, into which people will be able to switch as they wish'. But he cautiously adds in a note that 'It seems unlikely, in the immediate future, that the face-to-face of teleconferences will replace the breath to breath of personal contact'. There is evidence to show that these unplanned 'corridor' conversations when delegate meets delegate are often for individuals quite as valuable scientifically as the formal papers when the speakers address the assembled conference. Such opportunities are particularly valuable for young scientists and technologists still in the process of building up that network of personal contacts regarded as a vital element in the scientific communication process.

Criticism of conferences

Of course conferences are not without their critics, who point out that much of what is heard from the speakers repeats what is elsewhere in the literature: for example, the contents of almost a third of the reports at the XVIIth International Congress of Psychology had been previously published in a scientific journal. Investigations by the Center for Research in Scientific Communication at Johns Hopkins University would support these views: of those attending conferences over a half were already familiar to some extent with the content of what was presented. Indeed about three quarters of the authors had made some report of their work before the meeting (in a thesis, research report, oral presentation, etc). Recent surveys of engineers and engineering managers in the United States have revealed a marked decline of interest in conferences, which they find time-consuming and inconvenient; this is balanced by an increase in the use of printed material, which can be read at home, or on a journey, or whenever they have a few minutes to spare. As an instance of the operation of the Principle of Least Effort (see chapter 1), this is particularly intriguing inasmuch as it tends towards *increased* use of the literature, rather than the other way round, as is more usual. There is a body of opinion unwilling to admit that conference papers (and research reports) have more than a temporary value. The editor of the *Journal of chemical documentation* (now *Journal of chemical information and computer sciences*) once said he was convinced that 'Certainly authors of reports and meeting papers are not so naive as to consider these two forms as equivalent to publications in good technical journals. In no way can we think of reports and meeting papers as part of the permanent literature in any discipline of science. They are suitable half-way communications to a limited audience. The author of a report or a meeting paper of technical value is professionally committed to writing a publication version and submitting it to a good technical journal for review by his peers for a place in the permanent literature.' Since for many conferences papers are not subject to any screening before presentation, the complaint of uneven quality is often heard. Complaints are made too that speakers are allowed, or even encouraged, to deliver virtually the same paper at conference after conference. Occasionally one hears the charge (particularly at technological conferences) that some papers are sales-oriented. Conversely, physicists have commented recently that the scientific 'temperature' of many conferences is too high: topics biased towards academic research, particularly those at the research-front, are inappropriate to many physicists, especially those in industry and in the schools.

Criticism of the practical organization of conferences is regularly heard. Conference schedules are frequently overcrowded, perhaps with no more than a few minutes allowed for each paper, following one another in rapid succession. One commonly adopted solution, parallel sessions, can frustrate as many participants as it satisfies. The Physics Information Review Committee, mostly practising physicists themselves, have told us that 'It is now commonplace to remark that the standard of presentation of papers is frequently, too frequently, abysmal. The commonest faults are illegible slides, slides too crowded with information, and a style consisting mainly of a head-down, monotonic reading from typescript'. But perhaps the most serious criticism is an implied criticism, not of the conferences themselves but of their published proceedings. Conference papers are very similar in content and form to papers published in periodicals, which are the most frequently used source in most scientific and technological disciplines. Yet despite the high value placed on conferences as such by most scientists and technologists, surveys have shown low use of published proceedings. And since this is a problem of direct and particular relevance for the librarian, its investigation will take most of the remainder of this chapter.

It would be as well to start with the warning by two previous investigators that conference proceedings are 'a source of unending bibliographical confusion', posing an 'intractable problem − the variety and complexity of the "conference" in all its maddening pre-publication and post-publication aspects'. We have already seen, for instance, that conferences can range in appeal from the merely local to the worldwide, and in attendance from a handful to several thousand. They can also be 'one-off' (eg Conference on Water Utilization and Effluent Treatment, Edinburgh, 1969), or annual (eg British Psychological Society annual conference, Southampton, 1970), or irregular (eg the International Congresses on Polarography: 1st, Prague, 1952; 2nd, Cambridge, 1959; 3rd, Southampton, 1964, etc). The term 'international' may mean no more than that there are delegates or speakers from more than one country. On the other hand, it may mean that the papers are in two or more languages. In some circles it is held that no meeting can be 'international' unless it is sponsored by an international organization. There may be one specialist topic for the conference, or it may be fairly general, or there may be no theme at all! It may be planned as highly scientific, for experts only, or may be deliberately designed to facilitate exchanges across disciplinary boundaries.

Pre-conference literature

If variety is found in the conferences themselves, even more can be observed in the documents they generate. Before the conference commences, in addition to the expected announcements, calls for papers, programmes, etc, it is increasingly common to find preprints of the actual conference papers, which may give the full text, or an abstract, or an otherwise abbreviated version. These may be duplicated (often reproduced from the authors' original typescript) or they may be printed; usually they are issued as separates, but could be in book form, eg 600 papers of *Electron microscopy: fifth international congress, Philadelphia, 1962* (Academic Press, 1962) in two volumes. A number of the major professional societies have major preprint programmes: the American Institute of Chemical Engineers, for example, issues over a thousand such meetings papers each year, often in microfiche format. They may appear in various sizes (for the same conference), at various times, and may be incomplete (ie, some papers but not all). It is not unknown to find them unnumbered and unpaged, and without the title, or date of the conference. Of course, in many cases they are not intended to have a permanent value: their aim is to act as a basis for discussion, allowing the speaker more freedom to comment. Conference-goers in all disciplines will confirm, however, that it is still not clearly agreed how a speaker should use his preprint. Should he assume that it has been read by the delegates, and take it from there? Should he read it word-for-word from beginning to end, even though it has been in delegates' hands for weeks and is probably before their eyes while he is speaking? Both approaches are to be found, even within the same conference. For the librarian, however, the real difficulty is that they are usually supplied only to registered participants, and are hardly ever available to librarians in the normal way. Unfortunately, this does not prevent them being cited in the literature and asked for by a library's readers!

Literature published during a conference

This does not bulk large, being confined to texts of opening and closing addresses, lists of participants, texts of resolutions (draft and final), etc, but can be very difficult of access. Indeed, it is often impossible

266

to lay hands on much of it without actually attending the conference.

Post-conference literature

These are the documents most generally understood by the term 'conference proceedings' and usually comprise the published texts of papers delivered (corrected where necessary), discussions arising, and sometimes minutes, resolutions, etc. They appear in a variety of forms:

a *As a book*: eg *The metal science of stainless steel: proceedings of a symposium, Denver, Colorado, 1978* (Warrendale, Pa, Metallurgical Society, 1979); *The pre-Cambrian and lower palaeozoic rocks of Wales: report of a symposium, Aberystwyth, 1967* (Cardiff, University of Wales Press, 1969) In the case of *Tunneling under difficult conditions: proceedings of the International Tunneling Symposium, Tokyo, 1978* (Oxford, Pergamon, 1978) the text comprises reproductions of the actual typescript conference papers, with diagrams, bound in book form. Not infrequently, they extend to more than one volume, eg *Conference of Commonwealth Survey officers, Cambridge, 1967: report of proceedings* (HMSO, 1968) in two volumes (though the thirteen chapters are also available separately); *Complete proceedings of the fourth International Congress of Biochemistry, Vienna* (Pergamon, 1960) in fifteen volumes. Some of the best-known examples are issued as separate volumes in a regular series such as the Society for General Microbiology Symposia, eg *Virus growth and variation: ninth symposium* (Cambridge UP, 1959) or the Institute of Biology Symposia, eg *The problems of birds as pests: proceedings of a symposium, 1967* (Academic Press, 1968). Other well-known series are the Ciba Foundation Symposia, the Cold Spring Harbour Symposia, the Society for Experimental Biology Symposia, the Institute of Physics Conference Series, the Royal Society of Chemistry Faraday Discussions and Faraday Symposia.

The examples so far quoted are issued by the normal range of publishers, whether university, governmental, commercial, or other. It is quite common, however, to find the publishing undertaken (and presumably underwritten) by the sponsors of the conference, eg *The biological effects of oil pollution on littoral communities: proceedings of a symposium, Pembroke, 1968* (Field Studies Council, 1968); *Conference on nucleonic instrumentation, Reading 1968* (Institution of Electrical Engineers, 1968); *Motorways in Britain today and tomorrow: proceedings of a conference, London, 1971* (Institution of Civil Engineers, 1971). One variation that can cause bibliographical confusion is the practice with regular international congresses, usually held in a different country on each occasion: commonly the proceedings are issued by a publisher from the country concerned. Since the last war the International Congress of Mathematicians has gathered at four-yearly intervals at Harvard, Amsterdam, Edinburgh, Stockholm, Moscow, Nice, Vancouver, and the proceedings have subsequently been issued by a different publisher each time.

The American Astronautical Society adopted an interesting method in publishing the 90 papers of its meeting on the commercial utilization of space. 30 of the papers were printed in full in conventional book form as volume 23 of *Advances in astronautical science*. Summaries of the remaining 60 were also printed in the volume, but the full text was given in microfiches inserted into pockets at the back of the book.

b *In a periodical*: There is room here for a wide range of bibliographical permutations. As has been mentioned, the reporting of meetings has been one of the major functions of the scientific periodical for three hundred years, and it is still common to find the papers of conferences so printed as part of one of the regular issues of a journal, particularly the organ of a learned or professional society. Indeed, if the value of the papers warrants it, the whole of a particular issue may be given over to the proceedings, making it a 'conference number'. Publication as an issue of a periodical in this way is the method recommended by the scientific community itself, speaking through the International Council of Scientific Unions: this should ensure that the papers are subjected to the normal scrutiny of referees. Some journals, in fact, are devoted largely if not exclusively to recording meetings, eg *Annals of the New York Academy of Sciences, Institution of Mechanical Engineers proceedings*. Volume 182 (1967/8), part 3H of the latter, for instance, comprises (at £15) *Thermodynamics and fluid mechanics convention, Bristol, 1968*. Frequently it happens, however, that the proceedings of a particular conference are too extensive to be included within a single issue of the journal, and they have to be spread over two or more issues, sometimes extending over months or even years. And it is very common for the individual papers from a conference to be published on the initiative of each author in a number of different journals, often quite haphazardly, and not always with a clear indication that they are papers that have been presented at a particular conference.

A frequent compromise giving great flexibility is for the periodical to issue the proceedings as a supplement or a special number, additional to its regular sequence of issues, eg *Chromosomes today: proceedings of the second Oxford chromosomes conference, 1967* (Edinburgh, Oliver and Boyd, 1969), which is a supplement to the journal *Heredity*, volume 29.

c *In both forms*: Dual publication is not common, so far as complete

268

proceedings are concerned, though it does occur, eg *Biological interfaces... proceedings of a symposium sponsored by the New York Heart Association* (Churchill, 1968), which was simultaneously published as a supplement to the *Journal of general physiology*, volume 52, number 1, part 2, 1968. What is more frequent is the publication of individual papers in journals as well as in conference proceedings. R W Frei *Recent advances in environmental analysis* (Gordon and Breach, 1979) comprises 20 of the 24 papers presented at the 8th Annual Symposium on the Analytical Chemistry of Pollutants, held in Geneva, though all of them were also published either in the *International journal of environmental analytical chemistry* or in *Toxicological and environmental chemistry reviews*.

d *In a report series*: A small proportion of conference proceedings appear in this form, commonly published by governmental bodies.

The librarian and conference proceedings
As if these and other vagaries of publication were not enough, the librarian has to contend with what is probably the greatest problem — conference papers that are not published at all! In a now classic study of 'lost' conference papers (itself read as a conference paper in 1958) it was shown that of 383 papers from a sample of four US conferences 48.5% were never published. A more recent study of over three hundred conferences of relevance to physicists held in the UK in 1973 suggested that 'few published proceedings are likely to appear — as one measure *Physics abstracts* is likely to have no more than 50 of these in the years from 1973 onwards'. In her excellent guide *Use of mathematical literature* (already mentioned in chapter 2) A R Dorling quotes from the quite unusually scrupulous account rendered by one scientific publisher of the discrepancies between the list of papers in the proceedings he is publishing and the papers actually read at the conference in 1968: the papers by one author are to be published later in extended form; the papers of another have already been issued by a rival house; one paper was withdrawn because what was presented turned out to be inconsistent; another was withdrawn because its results at the last minute were found to be included elsewhere; papers by two authors were not received in time; another paper had been accepted for publication elsewhere before the conference; one paper was not included because of its length, though it was later published in a volume to itself; another paper contained only part of the material presented at the conference, the rest being contained in a privately circulated manuscript.

Of course, as was pointed out above when discussing conference preprints, not all papers are intended to have archival permanence: by definition, for instance, interim reports are designed to be superseded by final reports. We have been told recently that physicists, for example, are 'less concerned that information will be "lost", than that the perhaps incomplete and unrefereed accounts of presentation and discussion will be treated as part of the primary archive, be cited and used by other researchers and therefore possibly degrade the standards of scholarship and science'. Indeed it has been recommended that unrefereed conference proceedings should carry a warning: 'Not to be cited in any research paper'. Some conferences deliberately discourage publication so that participants can report and discuss more freely work that is not sufficiently advanced for wider dissemination, eg the Gordon Research Conferences in the US where participants at the hundred or more week-long gatherings are advised: 'In order to protect individual rights and promote discussion no information presented . . . may be used without authorization from the individual making the contribution . . . Recording lectures by tapes and other means and photographing slides are prohibited'. For some workers, too, oral delivery to a select audience might be a much preferable alternative to the hit-or-miss method of publication in a journal. But, as with preprints, this does not prevent *references* to such papers appearing in the literature, and demands for copies being made in libraries. A related problem is publication of papers in a form different from the original presentation: according to one survey only 38.2% of the sample were published later in 'susbstantially the same form as in the congress preprints and transactions'. A substantial minority of papers only appear in print in abbreviated form as an abstract. It is common for discussion to follow the presentation of a paper at a conference; though much of what is said may not be worth preserving, this is not always the case. Indeed, the British Standard on conference proceedings recommends that abstracts of discussions should always be included. Yet two out of every three published proceedings ignore them: quite exceptional is the treatment afforded by the Institute of Fuel in the two volumes of the proceedings of its Incineration of Municipal and Industrial Waste Conference, Brighton, 1969, where the first volume prints the papers given and the second is devoted to the discussion thereon.

An oft-heard complaint is that proceedings often appear very late, quite commonly two years after the conference, and sometimes three, four, or even more years later. Of course, publication in book form is well-known to be slower than journal publication, and the conference editor is compelled to wait for *all* the contributions to come in before he can publish (unlike the journal editor who can always relegate long-delayed papers to the next issue). One authoritative estimate that six months should suffice is probably over-optimistic. A survey of four hundred conferences carried out at what was then the National Lending Library for Science and Technology showed that less than a quarter achieved this. The arrangement by date of conference adopted by *Directory of published proceedings* (see below) reveals in every issue several proceedings published a dozen or more years after the event.

With the variety of publication patterns outlined above, combined with the intrinsic problem of conferences deriving from the separation in time of the event (the conference) and the record (the proceedings), it would be surprising if there were not acquisition problems for the librarian. The discovery that wanted proceedings are out-of-print is a common experience, particularly where the papers have been published in periodicals (and not enough extra copies printed), or where they have been published by the sponsors of the conference (who without experience of the market often underestimate the demand). The difficulties would be grave enough if conferences which are regularly held adopted a consistent pattern of publication, but the librarian sometimes finds three or four different methods (or no method at all!) adopted in succeeding years. And even where the same method is adopted for each of the conferences in a series (publication in book form, for instance), it is frequent to find that publisher is different in each case, especially with the major international conferences, as has already been mentioned.

That the problem is not diminishing is witnessed by the very significant increase in the number of published conference proceedings over the last twenty years. That this is not only an absolute but a relative increase the student can quite easily gauge for himself by comparing, say, the proportion of such volumes of proceedings to total titles in classes 500 and 600 in the current *British national bibliography* volumes with similar figures from earlier years. There is no doubt that they are very frequently asked for in libraries: the British Library Lending Division is acquiring separately-published conference proceedings at a rate of over twelve thousand a year and dealing with requests for them amounting to 170,000 annually. They make up a substantial part of the literature of science and technology. Because of their form (relatively short treatment of highly specialized topics) published conference papers serve many of the same ends as periodical articles. As we have seen their mode of presentation makes them particularly well suited as 'state-of-the-art' reports, eg *Ideas in modern biology . . . proceedings of the XVIth International Congress of Zoology* (Garden City, NY, Natural History Press, 1965) is made up of 'reports from 19 of the world's foremost biologists in a detailed examination of the major ideas in modern biology'; *Plutonium, 1960: the proceedings of the second international conference on plutonium metallurgy, Grenoble* (Cleaver-Hume, 1961) aims to present 'a balanced picture of the state of knowledge of plutonium metallurgy in 1960'. This role is often recognized by the titles given to such works: the European Brewery

Convention proceedings which appear regularly every two years used to have *Progress in brewing science* as cover-title, and the third International Conference on Semiconductors, Rochester, 1958, was published as *Advances in semiconductor science* (New York, Pergamon, 1959). In new and developing fields the publication of the proceedings of a conference can mark a watershed, serving to make an important and often unifying contribution, with the volume often remaining a standard work for some years, eg *Proceedings of the first international conference on operational research* (EUP, 1957); *Biodeterioration of materials ... proceedings of the first international biodeterioration symposium, Southampton, 1968* (Elsevier, 1968).

The criticisms referred to earlier of papers read at conferences apply of course with even more force to published proceedings, particularly the accusation of poor quality material. This is a grave charge, and the answer lies with the conference organizers. So long as there is no editorial check corresponding to the 'refereeing' that most learned journals impose on potential contributions, and so long as sponsors for prestige reasons feel obliged to publish in full the proceedings of 'their' conference, without applying the same criteria as they would to a monograph or treatise, complaints of this kind will continue.

Bibliographical control
A further hazard in the path of the librarian seeking to help readers in pursuit of conference proceedings is the sheer difficulty of identifying what is required. Probably the largest collection of recent conference papers in science and technology is at the British Library Lending Division, and yet the library staff complain that requests 'present considerable difficulty. This is partly due to the fact that borrowers often find it impossible to obtain a complete and accurate reference to the conference report required.' The reason for this is technical, but quite simple: no one knows how best to describe a conference report. Even the 1967 *Anglo-American cataloguing rules* (Library Association, 1967), which devotes two pages to the topic, has not really cracked the problem. The opinion of one of the rules' most penetrating analysts is worth quoting: 'The appalling inconsistency and complexity of presentation of such proceedings is not fully explored in the new rules, and cataloguers will continue to find great difficulty in dealing with them. I have a feeling that this is the outermost limits of author-title cataloguing, the point at which the system becomes inapplicable.' Ten years later the Head of the Cataloguing Section in the British Library Bibliographical Services Division was still reporting that the cataloguing of conferences was one of the 'two most debatable areas of practice ...We hope that the new edition of *AACR* will help us with confer-

ences'. The second edition appeared in 1978 with the critic just quoted as the British editor. Its preface characterizes it as 'having the same principles and underlying objectives as the first edition, and being firmly based on the achievement of those who created the work'. Its effect remains to be judged, but it can do little for all the conferences referred to in the literature to date. In practice, we find conferences cited by one or more of the following: the official name of the meeting, the corporate body (frequently, bodies) sponsoring, the editor of the proceedings, the subject dealt with (which may not be mentioned in the official title), the place (eg Rome Conference), the authors of individual papers, etc. And if one decides to rely on the title-page for a description, as often as not one finds much of this vital data absent; and one discovers that it is a frequent habit of commercial publishers to substitute a more eye-catching title for good measure.

Even when the reader (or librarian) knows precisely what he is looking for, all is not plain sailing. To locate individual papers it is quite often necessary to turn to indexing and abstracting services. Although the position has improved since the ASLIB investigation published in 1961, which showed that of 386 papers in ten conference publications only 30% were abstracted and 26% listed in the English-language abstracting tools, there are still far too many conference proceedings ignored. In 1965 *Biological abstracts* announced that it would no longer abstract individual papers from conference proceedings. Not all the blame rests upon the abstracting services, however. Speaking on behalf of ICSU/AB member services, T M Aitchison complained in 1973 of 'the difficulty of tracing conferences and persuading the conference publishers to make the publications available'. INSPEC makes considerable efforts to obtain conference proceedings, covering about 20,000 items a year, over 15% of the total data base. The monthly *International aerospace abstracts* uses 'conferences' as a heading in its index. *Biological abstracts* too now indexes some 80,000 meetings papers a year in its sister journal *Biological abstracts/RRM* (ie *r*eports, *r*eviews, *m*eetings). Even less well covered are preprints of conference papers: the delay in the appearance of the formal published proceedings often means that the preprint may have to serve for many months (or even years) as the only version of a paper available. Not many services follow the practice of *Engineering index* which treats such documents as part of the regular literature. *Chemical abstracts* holds them for eighteen months to see if they appear in a regular journal; if not they are then abstracted.

The reader with the required volume in his hands could well be forgiven for assuming he was at the end of the trail, but some of these conference reports amount to several hundred pages and, as an earlier ASLIB survey showed, 59% are without a subject index, 66% without an author index, and half of them have *no index at all*, necessitating a page-by-page search. No improvement, and, if anything, a decline was evident in the NLL survey a dozen or so years later: only 22% of titles boasted a subject index and 26% an author index. There are those who are prepared to argue that indexes (particularly subject indexes) to some conference proceedings do not warrant the intellectual effort required. They take the line that such volumes do not make a coherent subject unit like a monograph or textbook. On the other hand, however, there are instances where it has been felt worthwhile to provide cumulated indexes, eg to the 1951-63 joint computer conferences of the American Federation of Information Processing Societies. It is significant that the recent British Standard on the presentation of conference proceedings says that 'An author and contributor index and a subject index shall be included'.

a *Advance information.* In an attempt to keep pace with the ever-advancing demand for conference proceedings there have been in recent years spectacular improvements in the bibliographical tools available. Nevertheless, because of the unique way conference papers are generated, the wisest counsel for the librarian wishing to keep abreast of this category of material is to identify *in advance* the appropriate meetings and then make contact at the earliest possible moment with the sponsors, or organizers, or publishers, in order to be sure of obtaining required publications.

Fortunately, there is a wide selection of lists to help locate future conferences. Apart from general lists with which the student will be familiar, such as the annual *International congress calendar* (Brussels, Union of International Associations), which looks some ten or twelve years ahead and is particularly useful for tracing Third World conferences, there are at least three devoted to scientific and technical meetings: *Forthcoming international scientific and technical conferences* (ASLIB) is British and appears four times a year; the others are American and appear quarterly, *Scientific meetings* (Poway, Cal, Scientific Meetings Publications), and *World meetings* (New York, Macmillan), the most elaborate, described as 'a two year registry of all important future medical, scientific, technical meetings', in two separate parts — 'United States and Canada' and 'Outside United States and Canada'. The British Library Lending Division sends out six hundred letters each year to conference organizers identified from this list. Advance conference information of this kind is one of the forms of news data particularly suitable for access by videotext via the ordinary TV screen (see chapter 11), and INSPEC has been involved in UK trials.

b *Published proceedings.* But locating a conference and locating its proceedings are two different problems: the compilers of these lists of conferences-to-come warn the user to expect a 5% error rate, because
274

'a few papers listed will not actually be presented and a few papers that are presented will not be listed'. Some do actually indicate plans for the publication of proceedings, but the librarian often places more reliance in the current bibliographies at his disposal, which do not list proceedings until they have actually appeared. Many such proceedings in book form are of course listed in the regular bibliographies of current books, but there are also current lists confined to conference proceedings. Again there are lists specially for science and technology, the most comprehensive of which is *Directory of published proceedings: series SEMT* [ie science, engineering, medicine, technology] (Harrison, NY, Inter-Dok) appearing monthly and cumulating annually, and *Proceedings in print* (Arlington, Mass) five times a year with an annual cumulated index appearing separately. In recent years coverage of the latter has been extended to all published proceedings regardless of subject. The only British-compiled list (although of course its coverage is international) is in point of fact a library accessions list: the monthly *Index of conference proceedings received by the BLLD*. As a matter of policy the library tries to acquire *all* conference proceedings and so this is by far the largest and does have the very positive advantage of locating an actual copy for consultation or loan, etc. A cumulation covering the years 1964 to 1973 indexes 46,500 conferences under 27,500 keyword headings. The 1974 to 1978 cumulation adds a further 53,000 in its two volumes, and annual cumulations have appeared regularly since. The file is held in machine-readable form and can be consulted on-line via BLAISE. Another library listing is the New York Public Library *Bibliographic guide to conference publications* which has appeared annually since 1975, covering works catalogued during the year in question together with additional entries from Library of Congress MARC tapes.

   c *Individual conference papers.* Most of these bibliographies include not only separately published proceedings but also those which appear complete as part of periodicals: until recently, however, there was no index (other than the general indexing and abstracting services) to individual conference papers. This gap has now been filled by the monthly *Conference papers index* (Louisville, Ky, Data Courier), formerly known as *Current programs*, which lists scientific, technical and medical conferences with details of the papers under 17 main headings, with subject keyword and author indexes, cumulating annually. It is important to note that it is an index to papers *presented*: 'Publication of a paper is not a prerequisite for inclusion'. Over 100,000 are listed annually; machine-readable tapes can also be leased and the data base is accessible on-line from a number of service suppliers. For published papers we have had since 1978 *Index to scientific and technical proceedings* (Philadelphia, Institute for Scientific Information) covering some 100,000 papers from 3,100 volumes of published proceedings a year. Complete coverage of what ISI estimates to be 10,000 scientific and technical meetings a year is not attempted and only the most important are chosen.

It should be remembered that many specific subject fields within science and technology have their own conference lists, eg the quarterly *World calendar of forthcoming meetings: metallurgical and related fields* (Metals Society) and *Meetings on atomic energy: a quarterly world-wide list* (International Atomic Energy Agency). Many journals too make a feature of lists of meetings: these are particularly useful for local and national meetings, many of which do not get into the lists mentioned earlier. *Chemistry and industry*, for instance, not only publishes a weekly schedule but issues a six-months calendar of meetings as a supplement twice a year. Each twice-monthly issue of *Physics abstracts* includes a separate conference index. *Confer-alert* is a monthly loose-leaf service from INSPEC providing data sheets on current and forthcoming conferences in its field up to five years ahead. Unfortunately rare are subject lists confined to published proceedings, such as 'Conference publications abstracted and indexed in the *Engineering index* database' which appears annually in the index volume.

d *Retrospective lists.* For retrospective bibliographical coverage these various lists are of limited value because most of them have only been in existence a few years. The only extensive compilation covering the whole field of science and technology is the computer-produced *Union catalogue of scientific libraries in the University of Cambridge: Scientific conference proceedings 1644-1972* (Mansell, 1975) in two volumes, indexing some six thousand titles under an average of four keywords each: the earliest is a 1644 conference of falconers in Rouen. Prior to its appearance the searcher had to rely on the appendix 'Periodic international congresses' in the *World list of scientific periodicals*, which is primarily a location list without bibliographical detail and excludes 'one-off' conferences. It is true there are special subject lists, eg *Bibliography of international congresses of medical sciences* (Oxford, Blackwell, 1958), and the scientist and technologist can also turn to the general bibliographies such as *International congresses and conferences, 1840-1937: a union list* (New York, Wilson, 1938); *International congresses, 1681 to 1899: full list* (Brussels, Union of International Associations, 1960), and its supplement covering 1900 to 1919 (1964); and for more recent years the *Yearbook of international congress proceedings: bibliography of reports arising out of meetings held by international organizations* (Brussels, Union of International Associations), which covers the eight years prior to its year of publication, with its predecessors covering the years 1957 to 1959, *Bibliography of*
276

proceedings of international meetings (1964). Retrospective indexes locating individual conference papers are rarely encountered: a computer-compiled example worth examining is W Duffy and J A Miller *Low temperature physics: a KWIC index to the conference literature* (California, University of Santa Clara Press, 1971) which lists nearly three thousand papers from conferences between 1958 and 1969.

Further reading

F Liebesny 'Lost information: unpublished conference papers' *Proceedings of the international conference on scientific information, Washington, 1958* (Washington, National Academy of Sciences − National Research Council, 1959) 475-9.

C W Hanson and M Janes 'Coverage by abstracting journals of conference papers' *Journal of documentation* 17 1961 143-9.

C W Hanson and M Janes 'Lack of indexes in reports of conferences' *Journal of documentation* 16 1960 65-70.

P Poindron 'Scientific conference papers and proceedings' *UNESCO bulletin for libraries* 16 1962 113-26, 165-76.

H Baum 'Scientific and technical meeting papers: transient value or lasting contribution' *Special libraries* 56 1965 651-3.

B E Compton 'A look at conventions and what they accomplish' *American psychologist* 21 1966 176-83.

'The flow of information from international scientific meetings' *UNESCO bulletin for libraries* 24 1970 88-97.

W D Garvey and others 'Research studies in patterns of scientific communication: II The role of the national meeting in scientific and technical communication' *Information storage and retrieval* 8 1972 159-69.

P A Haigh 'Conferences and their proceedings' *NLL review* 2 1972 7-10.

H Baum 'A current-awareness service based on meetings − the need, the coverage, the service' *Journal of chemical documentation* 13 1973 187-9.

P R Mills 'Characteristics of published conference proceedings' *Journal of documentation* 29 1973 36-50.

P J Short 'A librarian's tool-kit for tracing conference proceedings' *ISG news* no 8 (April) 1978 13-6.

S W McLintock 'Scientific and technical conferences and their proceedings' *ISG news* no 9 (August) 1978 7-9.

P Atkin and P Seed 'The acquisition of conference proceedings at the British Library Lending Division' *Interlending review* 7 1979 47-51.

Chapter 14

RESEARCH REPORTS

Research reports make up perhaps the most important category of what has come to be called the 'grey' literature, not readily available through normal book trade channels. Over twenty years ago a writer in *ASLIB proceedings* warned us that 'Report literature is fast superseding the scientific periodical and the journal of the learned society as the most important medium of scientific communication, particularly in those fields of science concerned with national defence'. As we have seen, periodicals are still with us today, though under seige, but the research report has consolidated and even extended its position as one of the most vital sources of primary information and a major rival of the journal article. According to D W King the number of scholarly scientific articles published in the US increased by 42% between 1960 and 1974, but the number of research reports in the National Technical Information Service collection (see below) grew more than four-fold, the sharpest rise of all the various forms of scientific and technical literature. In the last decade a number of surveys of literature use by scientists and technologists have shown that reports regularly comprise around 10% of documents used.

Of course, scientists and technologists have been writing reports on their work since the earliest days: what has changed over the years has been their method of communicating these reports. As has been described above, the informal letter and the society meeting only sufficed while workers were few enough to permit this personal communication; then for two-and-a-half centuries the scientific journal became the accepted method of reporting new work, supplemented by the circulation of reprints in an endeavour to retain some individual contact with fellow workers. The vast growth of science and technology has now emphasized the inadequacies of the journal (see chapter 9) and the research report issued as a separate document has emerged as the most successful alternative. Its advance over the last forty years, prompted in the first instance by the need to digest and exploit the tens of thousands of secret scientific and technical documents captured from the enemy at the end of World War Two, has gone hand in hand with the staggering increase in government spending on scientific and technical research (particularly defence expenditure), for the research report is particularly suited in a number of ways to serve as the major communication medium for 'big science' and for team research. For example, it is characteristic of government research (and research under government contract) that the scientists and technologists should be obliged to provide frequent accounts of the progress of their investigations for their masters. Many such reports get no further, of course, but even where it is intended that they should be available to the general public, the journals are not always the most suitable vehicles. In the first place, they are too slow; secondly, the report may have too limited an interest to warrant disseminating so widely; and thirdly, journals do prefer accounts of completed work rather than work in progress. In any case, much mission-oriented governmental research is in new or interdisciplinary fields, with no established journals to serve them. Another characteristic of government-funded research and development is its concentration on defence, nuclear energy, space — all topics cloaked in varying layers of secrecy. The results of such research, when communicated, must obviously have limited circulation, and the best medium so far tried is the research report. It is worth noting the high relative importance of the research report in the Soviet Union. Another advantage of the research report is that it can be as long and detailed as required, whereas an article in a journal is always subject to stringent space limitations, and sometimes to stylized restrictions of form. One survey by W Kuebler found that reports averaged fifty pages in length whereas published articles averaged only eight. The further point is sometimes made that in some areas of scientific and technical activity research workers have not the time or inclination to refine and polish a report to meet the usual acceptance standards of the learned journals.

As the last sentence implies, many reports are fairly rough and ready documents. The text is usually in a more primitive state than a periodical article or a book, not being refereed as a rule, and sometimes not even being edited. Typically, they are not printed, but consist of pages of duplicated typescript (with diagrams and, increasingly, computer printout) stapled into a cover. So protean are they, indeed, that one despairing commentator claims that only two generalizations can be made about them: they are usually written in English or some approximation thereof; and they are mostly reproduced on rectangular pieces of paper. The vast majority are not published in any normal way: common practice is to produce no more than fifty or a hundred copies for distribution, supplemented where necessary by secondary reproduction (photocopy or microform). They appear in vast numbers, although, not surprisingly, no one appears to be sure how many: according to the 1963 Weinberg report some 100,000 are issued each year in the USA alone, though estimates made since then have ranged as high as half-a-million. Described sometimes as scientific, technical, or laboratory reports, they mainly describe government-sponsored research, though there has been an increase in recent years in the issue of reports on current research by universities and colleges.

Company reports
Private industrial research reports play a particularly important role within individual companies, where they are often found with positive recommendations for action, and are highly reader-oriented. They are a major resource of the company library, indeed of the company itself. As R D Mannix of Unilever Research has told us, 'These documents hold the corpus of knowledge on which the company bases its commercial direction, its innovation and its efficiency. For many library users in industry these reports are much more important and much more frequently consulted than the published literature'.

By definition they are not generally accessible. Some may be secret for military or official reasons, like many governmental reports, but much more probably the information is simply commercially valuable, and if available to market competitors would harm the company's interests. If an industrial firm wishes to make information available outside its own walls it does not do so as a rule by issuing its own reports, but by taking out a patent or by publishing in the open literature.

Criticisms
It is regularly claimed that such reports form a vital part of the research literature: 'whole technologies such as aeronautics and applied atomic energy have been built up almost exclusively on the information contained in reports'. In some subject fields there are more research reports than periodical articles, eg ceramics, metals. But there is a difference of opinion among scientists and technologists: some feel that it is a waste of time trying to keep abreast of this 'shadow literature', since any information of lasting value will eventually find its way into the regular journals. Research reports are regarded in certain quarters as a notorious example of the redundancy which is widespread in the scientific communication system: in a study of over a thousand such reports produced by psychologists in 1962 it was discovered that the main content of a third of them had been published in a scientific journal by 1965, more than half of them with virtually no differences. Yet on the other hand some editors refuse to publish papers the substance of which has already appeared in a report. Others will not permit their authors to make reference to reports on the grounds that they are not generally

281

available. One disturbing consequence of this ambivalent status was highlighted by the editor of *Nature*: 'Those responsible for learned and scientific journals are increasingly aware of . . . a growing stream of complaints from the authors of technical reports that their priority for some new idea or experiment has been stolen by the author of an article in the more familiar scientific literature . . . there is no doubt that the publication of original results or ideas in technical reports is a professional hazard for many scientists'.

Complaints about quality too are heard: the Head of the Office of Science Information Services of the National Science Foundation said in 1975 that 'Today, critics of scientific and technical reporting argue that up to 50% of all reports contain faulty interpretation of data, inadequate documentation of methods and analyses, and poorly supported conclusions'. And as has been noted, many are progress reports only, destined to be superseded in the normal course of events. Some may not even be that, and may merely contain, in the words of the Director of Information of the American Institute of Physics, 'A lot of balderdash thrown in just to keep someone in the fund-raising end of the company happy'.

Research reports and the librarian
What there is no gainsaying is that reports are frequently cited in footnotes and bibliographies of all kinds and the librarian is regularly asked to supply them for his readers. Obviously, they raise many acquisition problems, haphazard and semi-published as they are, in limited editions, by a wide variety of bodies. He cannot rely on his usual source of supply, the book trade, and is often obliged to identify and approach the issuing organization himself. So easy is it to bring out a research report, however, that there are probably many thousands of issuing sources, ranging from two-man research teams to the largest companies and universities. The situation is further confused by the many instances where the originators of the report are not the issuing body.

A unique problem of reports is security classification: one definition of an unpublished report is 'one which, at the time of its issue, cannot be obtained by the public as of right'. Reports to which access is limited are known (rather confusingly for a librarian) as 'classified'. Since there are commonly various grades of security classification — in order of ascending sensitivity: restricted, confidential, secret, and top secret — as well as a 'declassified' category, the librarian often finds in practice that even classified reports may be available if he can show specific authorization or 'a need to know', eg from the Defence Research Information Centre of the Ministry of Defence, or its US counterpart the Defense Technical Information Center of the Department of Defense. Commonly too, individual *copies* are numbered so that they can be strictly accounted for, and the librarian working with such reports may be obliged to add to his burdens by the need to control and even to 'log' their movements. 'Unclassified' reports, of course, are normally available to all; on occasion, however, the librarian may find himself frustrated even here by further restrictions on access.

A feature of research reports is their issue in series, characterized by a letter-or number-code. Originally adopted for security reasons (so that a document could be specified without revealing either its author or its title), the practice is still common of referring to a report by its number alone, eg FRL-TN-39; PB 155056; AD 250778. The fact that each of these three numbers refers to the same document merely serves to highlight another complication. And as these codes are not controlled or co-ordinated in any way we may sometimes encounter the reverse of this problem: different reports from different organizations that by mischance have been allocated the identical number. Such numbers may be accession numbers, contract numbers, series numbers, etc. One thing they rarely are is subject classification numbers, and so they carry no clue as to subject. Indeed they may be deliberately contrived not to reveal but to conceal their origin or availability. In his book on reports literature C P Auger cited one announced in *Scientific and technical aerospace reports* (see below) with no fewer than thirty characters: N72-28275 (NLL-M-20984—(5828.4F):NEN-3005). Over 20,000 such codes are listed in L E Godfrey and H F Redman *Dictionary of report series codes* (New York, Special Libraries Association, second edition 1973); they describe the situation as 'chaotic', with an 'astounding number of codes blossoming each year'. Complementary to some extent is D Simonton *Directory of engineering scientific and management document sources* (Newport Beach, Cal, Global Engineering, 1974).

Report literature has been regarded by the regular bibliographical tools as 'a minefield in which only the wary venture'. Although there are signs of improvement, INSPEC, for instance, now covering several thousands a year, by many it is completely ignored, eg *Current technology index, Applied science and technology index*; by some it is covered partially and intermittently. Even INSPEC warns the user that its coverage 'is in no way comprehensive'; CAB claims only that 'So-called grey literature is processed whenever possible'. When they do attempt to include reports, as does *Chemical abstracts* and *Metals abstracts*, they often do it inadequately, apparently deriving the entries second-hand from the official abstracting sources (see below) and not always making clear, for example, that some have limited availability.

Bibliographical control

The inadequacies of the abstracting and indexing services, the unique peculiarities of the form, and pressure from librarians and others, have in recent years so stirred the consciences of a number of governments that they have made serious attempts at rationalization and control. After all, it is the governments that are ultimately responsible for the overwhelming majority of research reports, for many of those that they do not produce themselves are issued by non-governmental organizations under government contract. Quite rightly, governments feel that the results of this research, financed by the taxpayer, should be disseminated freely unless security is likely to be jeopardized. Some, taking their responsibilities a step further, have set up official documentation centres to collect all domestic reports and to act as issuing agency, and in the case of some of the more advanced centres, to provide bibliographical control through lists, indexes and abstracts. Increasingly too, such centres are acknowledging their responsibility not merely to collect and make available such reports, but actively to encourage their use by scientists and technologists and in particular to ensure their application to the task of technological innovation. Some centres are limited to specific subject areas, eg aerospace, nuclear energy, or to a specific clientele, eg military contractors. In some cases the larger issuing agencies have devised their own bibliographical systems.

US government research reports

The world's largest producer of research reports is the United States government and it is no more than fitting that its method of control should serve as a model. The keystone of several interlocking systems is the National Technical Information Service, the lineal descendant of the office responsible for the captured enemy documents after the War, established in 1970 and incorporating the former Clearinghouse for Federal Scientific and Technical Information at Springfield (Va). Described as the 'focal point for the collection, announcement, and dissemination of unclassified US government-sponsored research and development reports', each year it adds a further 80,000 titles to its collection of over a million. As pointed out in *The use of biological literature* by R T Bottle and H V Wyatt, 'The reports taken together represent one of the greatest collections in the world of non-confidential technical information'. The key to this storehouse is the twice-monthly computer-produced abstract journal *Government reports announcements and index*, with 2,500 items per issue. In addition to the usual access by author and subject, this computerized index permits reports to be traced also by corporate author, contract number, or accession/ report number — a feature of particular value for this type of literature.
284

All reports are available for purchase either in hardcopy or microfiche form, and NTIS thus distributes over four million copies a year. Rapid alerting services are provided by *Government reports topical announcements,* available semi-monthly in each of 46 separate subject categories, although these are gradually being replaced by a more frequent and more flexible series of abstract newsletters. These comprise summaries of new reports in the form of weekly newsletters photocomposed by computer and issued in 27 subject categories, eg *Energy, Physics.* The last issue of the year includes a subject index. A different approach to current awareness is provided by *Highlights digest,* comprising two- to four-page news sheets drawing attention to some five thousand selected reports a year in 57 subject areas. NTIS *Tech notes* are single-page leaflets monitoring the latest developments in applied technology issued every two weeks in eleven subject categories, eg computers, energy. SRIM (Selected research in microfiche) is a standing-order service automatically providing subscribers every two weeks with the full texts of reports in their chosen fields of interest, selected from over five hundred categories. Published literature searches amounting to over a thousand have been prepared by NTIS staff in anticipation of demand, but *ad hoc* searches are available to order, including on-line searches by NTIS analysts. Indeed, for those who wish the whole of the NTIS Bibliographic Data File on magnetic tape is available for lease. It is also available for on-line searching. A more recent addition to this array of bibliographical apparatus is the NTIS *Title index* to the whole document collection since 1964. Available on computer-output-microfiche (see chapter 21) this provides access by author, report number and subject keywords.

Rather confusingly, a number of the larger report-producing agencies have their own systems of bibliographical control as well. Two major examples are the Department of Defense, with its twice-monthly listing *Technical abstract bulletin* and its *Index* (although understandably many of the reports are secret), and the National Aeronautics and Space Administration. *Scientific and technical aerospace reports* (STAR) is a twice-monthly abstract journal produced by NASA, but it covers *world-wide* report literature in its field. The third major governmental producer used to be the US Atomic Energy Commission, but its functions were taken over by the new Energy Research and Development Agency. For a while ERDA continued the publication of the AEC world-wide report listing, *Nuclear science abstracts,* but finding in 1976 that its coverage was virtually identical with the improved *INIS atomindex,* it chose to cease publication. Coverage of the remainder of the reports was taken over by ERDA's own abstract service, under a variety of names, but that in its turn was superseded by *Energy research abstracts,* twice monthly from the US Department of Energy, which now lists not only reports but all other materials emanating from USDE or its agents.

To some this might suggest that there is now a danger of too much bibliographical control of reports, and there is certainly evidence of overlap and duplication. Indeed, so extensive has been bibliographical effort in this sphere over the last few years that there are fears on other grounds also, fears for the comparative freedom and informality of the medium itself: '...what has gone on in scientific communication in the last twenty years is the formalization of the technical report. I'm wondering whether this isn't having some form of effect on the value of those reports ...will not real, vital, and rapid communication leave those forms and go to some other form that is less well controlled?'

We should also remember that over 20,000 research reports a year are printed and published and made available in conventional fashion by the US Government Printing Office, the world's largest scientific publisher. Many of these appear in long-established and well-known series, eg US Geological Survey *Professional papers*, US Bureau of Mines information circulars, and all are listed in the *Monthly catalog of United States government publications*.

UK government research reports
Just as in the US, many such reports are officially published and made available through Her Majesty's Stationery Office, including one of the world's oldest report series, the Aeronautical Research Council *Reports and memoranda* started in 1909. These are all included in the *Monthly list of government publications*. As for the semi-published reports, until quite recently the only reliable way (outside the atomic energy field) was the do-it-yourself method of identifying from directories and other similar sources those bodies likely to issue reports in a particular field of interest and approaching them direct. Within the last decade access to these British reports, amounting perhaps to 3,500 a year, has been transformed by two bibliographies. Current awareness is catered for by the monthly listing from the British Library Lending Division, a brave attempt at comprehensive coverage of the grey literature, and since 1981 renamed *British reports, translations and theses*. This effort is backed by what was claimed in 1973 as the largest library collection of 'open' reports in the world and now amounts to over two million (including copies of all available NTIS reports), all, of course, available for loan. For abstracting and indexing the searcher has the twice-monthly *R & D abstracts* (formerly restricted but since 1968 available in an 'unclassified' version) and its very thorough indexes, which lists the science and technology reports available from the half-million-strong collection of the Department of Industry Technology
286

Reports Centre at Orpington – covering some 80% of all government-sponsored research and development. TRC also provides a selective fast announcement service, *Techlink*, in the shape of one-page summaries, in 52 subject areas. Enterprisingly, TRC acquire from NTIS the twice-monthly magnetic tapes from which, among others, *Government reports announcements and index* is prepared. This they process to produce their *Subject index to US Government reports announcements*, a list of titles (and document accession numbers) arranged in the order of the subject field to which NTIS has assigned each document. TECHSCAN is a twice-monthly SDI service compiled from a composite data base comprising *Government reports announcements and index* together with *R & D abstracts*. The librarian borrowing reports from either of these two major collections will find, in practice, that he is usually supplied at cost with a microfiche copy for retention rather than a paper copy on loan.

The largest single producer of reports (excluding the Ministry of Defence) has been the UK Atomic Energy Authority. With its statutory duty to communicate to British industry whatever information will enable the various applications of atomic energy to be exploited to the full, the UKAEA has an elaborate and sophisticated bibliographical organization. Hundreds of research papers are published annually in the open literature, but for the usual reasons mentioned above reports have been found more satisfactory in many cases. All have been listed since 1955 in the monthly *List of publications available to the public*. An excellent *Guide to UKAEA documents* by J Roland Smith (HMSO, fifth edition 1973) gives a full account of the system.

A significant attempt at bibliographical control on a European scale has recently been made following the seminar at York in 1978 sponsored by the Commission of the European Communities. It was decided to set up national grey literature 'authorities' in the UK, France and West Germany to co-operate in identifying, collecting, listing and making available research reports within Europe. In 1981 the CEC launched project SIGLE (*S*ystem for *i*nformation on *g*rey *l*iterature in *E*urope), to run on an experimental basis for two years. Belgium and Denmark joined the original consortium of three, and other countries expressed their interest. The British Library Lending Division is the 'authority' for the UK. An on-line data base is to be created for access via EURO-NET/DIANE (see chapter 11).

Reports of scientific expeditions
Making an interesting little bibliographical category of their own are expedition reports. It commonly happens that the biological, geological, and other specimens brought back to museums, universities, and

learned societies provide material for continuous study over many years. Indeed of the great expeditions of the last century much of the material still awaits research workers. The scientific results therefore may be published over a very long period as a series of research reports. Frequently the series is named after the ship, eg 'Challenger' reports, 'Discovery' reports. Technically they are not periodicals, since they will come to an end someday, and they are not really books. Nevertheless, they are often entered in bibliographies of periodicals, eg *World list of scientific periodicals*, in some bibliographies of books, eg *British national bibliography*, and in some abstracting and indexing services, eg *Biological abstracts*. A useful list of expeditions compiled mainly to help cataloguers in establishing correct forms of entry is E Terek *Scientific expeditions* (Jamaica, NY, Queens Borough Public Library, 1952).

Further reading

R C Wright 'Report literature' J Burkett and T S Morgan *Special materials in the library* (Library Association, revised edition 1964) 46-59.

J C Hartas 'Government scientific and technical reports and their problems' *Assistant librarian* 59 1966 54, 56-9.

K E Jermy 'Handling industrial (scientific and technical) confidential report material' *ASLIB proceedings* 18 1966 206-17.

J C Hartas 'Technical report literature — its nature and some problems' Bernard Houghton *Information work today* (Bingley, 1967) 77-87.

J P Chillag 'Problems with reports: particularly microfiche reports' *ASLIB proceedings* 22 1970 201-16.

B Houghton *Technical information sources: a guide to patent specifications, standards and technical reports literature* (Bingley, second edition 1972) 91-111.

J P Chillag 'Don't be afraid of reports' *BLL review* 1 1973 39-51.

C P Auger *Use of reports literature* (Butterworth, 1975).

J L Hall 'Technical report literature' W E Batten *Handbook of special librarianship and information work* (ASLIB, fourth edition 1975) 102-23.

A H Holloway and others *Information work with unpublished reports* (Deutsch, 1976)

J M Gibb and E Phillips 'A better fate for the grey, or non-conventional literature' *Journal of research communication studies* 1 1979 225-34.

Chapter 15

PATENTS AND TRADE MARKS

The torrent of research reports described in the previous chapter is perhaps the most conspicuous sign for the librarian of the involvement of the modern state in science and technology, but the whole of the literature bears witness to the fact that governments are now major scientific and technical publishers. Just as now there are few facets of our lives that are not subject to some degree of official control or scrutiny, so there is scarcely a subject about which there has not been some official publication, and no category of information or form of literature unrepresented in the bibliographies of, for example, Her Majesty's Stationery Office or the US Government Printing Office. It is important for the student to realize that a so-called 'official' or 'government' publication may not be overtly official in nature. While the *Highway code* obviously is, D M Rees *Mills, mines and furnaces* (HMSO, 1969), 'an introduction to industrial archaeology in Wales', is not, and could equally well have been a commercial publication. So far as science and technology is concerned, the conventional categorization of official publications as a class apart has far less practical application for the student of the literature than it very obviously has in the social sciences. What is significant about the following titles is not merely that they are all official publications but that one is a dictionary, one is a handbook, one a textbook, guide to the literature, directory, book of tables, monograph, bibliography, research journal, abstracting journal, review journal, etc: (US) Bureau of Mines *A dictionary of mining, mineral, and related terms* (Washington, USGPO, [1968]); United Nations *Fertilizer manual* (1968); *Admiralty manual of navigation* (HMSO, revised edition 1970-); J F Smith and W G Brombacher *Guide to instrumentation literature* (Washington, USGPO, 1965); *Corrosion prevention directory* (HMSO, second edition 1978); Meteorological Office *Tables of temperature, relative humidity and precipitation for the world* (HMSO, 1960); C H Gibbs-Smith *Aviation: an historical survey* (HMSO, 1970); Ministry of Technology *Solid-liquid separation: a review and a bibliography* (HMSO, 1966); *Tropical science; World fisheries abstracts; Agricultural science review.*

This wide embrace by governments of science and technology is a twentieth century phenomenon, but there is one area of activity that has been of critical concern to the state for several hundred years, and which has produced an extensive and unique form of scientific and technical literature. This of course is the field of inventions and the duty of the state to encourage the development of industrial techniques while at the same time ensuring that it does its utmost to provide for the protection of the rights of inventors. It is thought that the earliest law for the grant of patents of invention was a decree of the Venetian Senate in 1474, although there is evidence that Venice had been making patent grants as early as 1416, and some claim that they go back as far as 500 BC. It was under this legislation that Galileo was granted a patent in 1594 for an irrigation device, though the first known grant was for a barge with special lifting gear awarded by the city of Florence in 1421 to Filippo Brunelleschi, architect of the Pitti Palace and the dome of the Cathedral. The oldest system with a continuous history to the present day is the British (or English, as it was originally), which for practical purposes may be said to start from the Statute of Monopolies of 1623, though as in Venice patent grants had been made for many years before that. The first English grant was for a method of making glass, given in 1449 to Thomas Utynam, who supplied the stained glass for Eton College chapel. A particularly common practice was to use the grant of a patent as a bait to entice foreign craftsmen to come to England with their skills so that local apprentices might learn them. The requirement of disclosure was not to come till much later. Patents as a form of scientific and technological literature were born when the Patent Law Amendment Act of 1852 directed that all patents subsequently granted should be printed. Bennet Woodcroft, appointed 'Superintendent of the Specifications', grasped his opportunity and printed all the retrospective patents also, until that time existing in manuscript copies only, from Number 1 of 1617 (for 'Engraving and printing maps, plans, &c') to Number 14,359 of 1852. In their distinctive covers they were described as a 'Blue-coated encyclopaedia of human progress'.

Essentially, a patent is a bargain struck between the state and the inventor. The state guarantees to the inventor the sole right for a certain period of years to make, use, or sell his invention, in order that he may reap a fair reward for his labours and to encourage him to further efforts. History has shown that this stimulates capital investment in new manufacture, and many important new industries have grown up under such a state guarantee. In return for its protection, the state (ie the community) obtains the invention — first of all at the market price, and then, after the expiry of the guaranteed period, to manufacture for itself, if it so desires, quite freely. More importantly for our purposes, the community is let into the secret of the invention in the form of the patent specification – a detailed disclosure in written form. Published and on sale to anyone at a very modest cost each, 'this magnificent unsurpassed printed record of patents of invention', as it has been described, amounting to more than a million and a half since 1617 forms a basic and unique primary source of scientific and technical information, including over the years the first unveiling of Whittle's jet engine, Hansom's cab, Mackintosh's waterproof, Cockerell's hovercraft, Friese-Greene's cinecamera, Baird's television, Marconi's wireless, Dunlop's pneumatic tyre, etc. On a world-wide scale the total number approaches 30 million, including Bell's telephone, Diesel's engine, Nobel's dynamite.

The word 'patent' means open, and is an abbreviated way of referring to 'royal letters patent', ie open (not private) letters addressed by the sovereign to all subjects, announcing the grant of some privilege. In the case of patents of invention (there are of course other kinds of letters patent) this privilege is a temporary monopoly granted to the inventor. All major countries (with and without sovereigns!) have some such system, many of them based to some extent on the UK model.

UK patent procedure

The old British patent system has recently undergone a major transformation under the 1977 Patents Act. This marked the first fundamental rethinking since the 1852 Act: perhaps that was the reason its passage through the House of Lords set a new record with over eight hundred amendments being considered. What follows is a non-legal outline of the new procedure which came into effect in mid-1978, though until all the applications under the old procedure have cleared the system, both will be running side by side. In contrast to the old, the new procedure follows what is known as a deferred examination method.

To obtain a patent the inventor has first to describe his invention in the form of a specification, with drawings if necessary, and often drafted with the aid of a patent agent (see below). This he then submits as an application with the appropriate fee to the Patent Office in London. Today most inventors are employed by industrial and commercial companies and think up their inventions in the course of their daily work. The UK system permits companies to apply for patents, and most patents are thus company patents, although the inventor's name does appear on the documents. The application is assigned a running number on filing, eg 8212345, with the first two digits indicating the year, ie 1982.

Obviously to have any chance of acceptance the invention must be new, and the existence of a previous patent is not the only ground for rejection: any prior publication (even by the inventor himself) can disqualify. Before the formal examination to determine whether it is acceptable or not a Patent Office examiner makes a preliminary literature search for novelty. Within eighteen months of the priority date (which is the date of the first filing either in the UK or in certain foreign countries that have signed the appropriate International Convention) the application is assigned a seven-digit number (a new sequence commenced in 1979 with 2,000,001) and the specification is published together with the results of the examiner's literature search. In accordance with the internationally agreed code the letter 'A' is affixed to the number on the printed specification to signify *first* publication, eg 2,123,456A. On the front page appears an author abstract (more precisely, an abstract provided by the applicant) of not more than 150 words.

The literature search report in many cases dissuades the applicant from proceeding further, usually by showing him that his idea is not as novel as he thought. If he still wishes to proceed the next stage is the full formal scrutiny by the Patent Office examiners. An invention must be useful (although this is usually generously interpreted), not obvious to those 'skilled in the art', not illegal (eg, a man-trap), and not against natural laws (eg a perpetual-motion machine). In general terms it must be a 'manner of new manufacture': this would include products, processes, methods of controlling and testing, as well as improvements to existing 'manufactures'. At this stage the examiners also take into account any criticisms or objections from others who feel their rights are being affected. Frequently the examiners suggest modifications for discussion with the inventor, but eventually (in the case of successful applications) agreement is reached and the application is accepted.

The specification (with amendments, if any) is then published again with the same seven-digit number, followed by the letter 'B' to indicate *second* publication. From priority date to final acceptance the average lapse of time is three-and-a-half years. The student should note that under the old system the specification was not published at all until it had been examined and accepted. There is provision for a 'C' document in those few cases where the 'B' document has to be reprinted with considerable amendment following action for revocation.

One small historic point is worth noting, because it marks a significant break with the past. The legal grant no longer takes the form of royal letters patent. UK patents of invention are now granted by the State not the Crown, and the document the inventor receives is not a parchment deed stamped with the royal seal but a certificate from the Comptroller-General, the permanent official at the head of the Patent Office. The period of protection is twenty years from the date of filing: four years protection is automatically given, but for the remainder annual fees are payable on a sliding scale.

Although we are mainly concerned in this chapter with patents as a source of information, their primary purpose is to serve as weapons of competition: as such they are often brandished in legal disputes. In the first instance these are referred to the Comptroller, who in many cases is able to resolve the matter. If his arbitration fails, recourse is had to the courts; initially to the Patents Court (Appellate Section) and ultimately to the House of Lords. Accounts of leading cases appear in the regularly published *Reports of patent, design, trade mark and other cases*. Operating as a kind of 'technological solicitor', not only in litigation matters but at the application and examination stages, is the patent agent. A register of this specialized branch of the legal profession is maintained by the Chartered Institute of Patent Agents in accordance with the rules and regulations of the Department of Trade under the Patents Acts, and published each February.

British patent law and practice are matters of some intricacy, and in the course of a few paragraphs it has been impossible to give more than the sketchiest outline. Standard texts to consult are H Brett *The United Kingdom Patents Act 1977* (Oxford, ESC Publishing, second edition 1978) and T A Blanco White and others *Encyclopedia of UK and European patent law* (Sweet and Maxwell, 1977-), which is loose-leaf.

The value of patents

It cannot be denied that patents are probably the most neglected of the primary sources of scientific and technological information: 'the ugly duckling of the information world'. In the ACSP survey of the information needs of over three thousand physicists and chemists (see chapter 12) patents scored lowest of all as useful sources of information. Even mechanical engineers, as surveyed in 1967 (see chapter 12), make surprisingly little use of them: over two-thirds of the 2,702 respondents hardly ever used them, and less than 6% claimed to use them once a month or more frequently. A number of surveys have shown particularly low use of provincial depository collection of patents. Even the inventors themselves neglect patents as a source of information: we are told that one of the commonest reasons for so many applications having to be amended is 'ignorance of the state of the art on the inventor's part'. T A Blanco White, probably the leading legal authority on UK patents, is of the opinion that 'even in well-established industries it is common to find that important discoveries for which the industry seems to have been waiting for years could have been made by anyone who had taken the trouble to piece together the information contained in patent specifications'. And yet patents are perhaps the most up-to-date form of technical literature in existence, novel by definition, the prime record of the progress of industry. The general shift in recent years towards 'quick issue' patents in many countries, including the UK, with applications being published without examination a few months after filing, has enhanced this advantage that patents have over journal publication. Much of the information is not available elsewhere, and for many technologies, especially chemical, almost the whole of the very latest information is contained in patent specifications, A study of over 70,000 US patents issued in 1978 found that some 70% are a unique source of information; contrary to popular belief, patents are relevant, timely, and complete. The author warned: 'Ignore the literature and you will probably miss technology relevant to your work'. They are of value to academic scientists also, for many contain extensive discussion on the theoretical basis of the invention and concise accounts of the 'state-of-the-art'. The report of the literature search carried out by the patent examiner furnishes a valuable professionally compiled bibliography on the subject of the invention, though it is not necessarily exhaustive. These have been published with US patent specifications as a matter of course for several years, but are now standard features of all UK published applications (and of PCT and EPC applications, as we shall see). Mellon tells us that 'studies of chemical literature are not complete without a search of patents'. In numbers too they make up a substantial portion of the literature, with approaching half a million granted annually out of a million applications. Some it is true are 'equivalents', ie virtually identical versions published in a number of countries (for patent protection is not yet international, unlike copyright), but 60,000 are granted annually in the USA alone, and 30,000 in Britain.

Reasons adduced for their comparative neglect are that the information in patents is of little value, difficult to extract, and encumbered with jargon. It is true that, unlike papers in journals, patents do not set out to be informative. They are entirely functional, written to define a legal monopoly, and the inventor will normally divulge no more than the Patent Office insists upon, ie sufficient to enable one skilled in the particular art to carry out the process. They are frequently couched in 'patentese', and, in the words of the *New scientist*, 'as stimulating to read as the London telephone directory'. The indexes too, and the classification system, are aimed at searching the claims rather than tracing technical details which may well be in the body of the specification. Many scientists and technologists see them as stumbling blocks deliberately sited to prevent them pursuing a particular path: there is a distinct feeling that all that patents tell you is what you cannot do. As a writer in *The information scientist* warned, 'they are also items of industrial property, and as such are easily tainted with the subterfuges and strategies which many commercial organizations are tempted or forced to assume in the dog-eat-dog world of private enterprise'. The lengths to which companies may go can be instanced by the Polaroid camera, basically a single invention, but protected by 1,100 patents: it was on the basis of these that the company recently sued Kodak. Yet there is no doubt that they embody a vast amount of information: there are patents extending to no more than a page, but at the other extreme we have examples like British Patent 1,108,800 with 1,319 pages and 495 sheets of drawings (an IBM computer). In many cases the technical drawings and the descriptive details are sufficient for a skilled technician to construct a working embodiment of the invention. Patent enthusiasts like R P Veerasnij, Director of the USSR Central Patent Information and Technico-Economic Research Institute, even claim 'serious advantages of patent documentation over other sources of scientific and technical information. Each patent specification provides a specific solution, resulting from an invention, for a particular technological problem, whereas an article in a journal or a research report must be laboriously sifted for what is essential and not essential, trustworthy and exaggerated, before the reader can discover the author's main idea and apply it for his own purposes. This is unnecessary in a patent specification, since it has already been done by the examiner.'

Another widely held opinion is that patent office procedures take so long that the invention is already several years old when the specification is published. This is less true than it was, at least in the UK; but in any case, publication in the form of a patent specification is still the earliest disclosure in the vast majority of instances.

A more disturbing reason for this lack of attention to patents could be the not uncommon belief among scientists and technologists that any relevant information in patents will find its way into the regular literature in due course. An investigation by Felix Liebesny has shown that this is true of only a very small number of British patents (61 out of a random sample of 1,058). He concluded that 'If the patent literature is neglected by those seeking technical information, there is a considerable risk that a significant amount of important information may not be retrieved'.

So far as actually utilizing the information in someone else's patent is concerned, the first point to establish is whether the patent is still in force: a large number are not renewed and therefore lapse before the full term of twenty years. In any case much of the information they contain is not protected by the patent: this protection extends only to the claims. And of course, as many inventors have discovered to their cost, it is no infringement to manufacture in one country where the invention has only been patented in another: over half of US patented inventions are protected in the US only. General experimental work based on patented information is not normally objected to, although technically this is infringement. To go any further, to incorporate in a commercial process, for instance, it is necessary to approach the patentee for a licence, for which a mutually agreed fee is usually payable. The patentee can be obliged to grant a licence if he has not worked his invention on a commercial scale within three years of the grant: probably less than half of all patented inventions actually come onto the market. If the full term has elapsed, of course, the invention is no longer protected at all. There are many instances of ideas born before their time, and patented. When their true relevance emerges later the patent may well have expired and they are in the public domain for anyone to exploit. Neither should one overlook what is perhaps the most productive use of patent literature, the stimulus that even patented and protected literature can provide for new ideas and fruitful areas for research. In the classic example, quoted by Michael Hill, 'James Watt improved on Newcomen's engines by adding a separate condenser. Trevithick improved on James Watt's ideas by using high pressure steam. And so it goes on'. Important too in certain contexts is what may be called their 'intelligence' value; much can be gleaned about industrial competitors from a close study of their patents. And at the national macro-economic level in the US and the Soviet Union in particular, trends in patents are given particular attention as a tool for technological forecasting.

Bibliographical control

Other reasons for this disregard suggested in the survey of mechanical engineers are that 'not having access to the relevant guides to the literature, engineers do not know what exists in the patent publications', and that 'patent literature is not generally available at local level'. Possibly librarians and information officers could play a more positive role here, for physical layout, bibliographical coverage, indexing, both current and retrospective, and access to material is certainly better for patents than for a number of other primary sources of information like research reports, conference papers, theses. Indeed, Michael Hill, Director of the Science Reference Library, claims that they are 'arguably the most precisely managed and best organized store for retrieval that there is'.

To begin with, there is normally only one publishing outlet in each country, and its products are very strictly ordered: they are for instance issued in standard format and are serially numbered. As we have seen, the latest UK patent sequence started with 2,000,001 in 1979 (though there had been three different numbering systems for the million-and-a-half patents granted before that). Each week the new applications
296

filed, the applications published, and the newly-accepted patents are listed with name and subject indexes in the *Official journal (patents)*. Weekly on the day the applications are published also appear the abstracts (actually copies of the front pages of the applications), published in 25 separate pamphlet series grouped according to subject, eg metal-working, transport, organic chemistry, electric power. Annually all the pamphlets for each particular series are gathered into a volume with subject and name indexes. A cumulated *Index to names of applicants* is published annually.

Retrospective searching if the patentee's *name* is known consists in checking back through the name indexes as follows:

2,000,001-date: index published annually from 1979

1,500,001-1,537,580: single index (This numerical sequence will in fact continue until all the accepted applications under the pre-1977 Act procedure have been published, but the specifications will be indexed with the 1977 Act specifications in the annual indexes from 1979.)

1,000,001-1,500,000: index published every 25,000

340,001–1,000,000: index published every 20,000

1852-1930: index published annually

1617-1852: consolidated index.

Retrospective *subject* searches are best undertaken through the abstracts (under the old procedure written by the examiners and known as abridgements) and their indexes. During the search stage specifications are classified according to an elaborate system comprising 8 sections, 40 divisions (in 25 groups), 405 headings and 45,000 subdivisions, and the resulting class marks are printed on the published applications. The abstracts are arranged according to this scheme, and for the searcher the guides to it are the *Classification key* and the *Reference index*. Obviously to keep pace with the advance of science and technology any such scheme needs constant amendment: indeed the searcher who follows the trail back more than a few years will find himself involved in a variety of different schemes, eg

1963-date: current scheme (amended from time to time)

1931-1963: 40 (later 44) groups, 146 classes, 217 sub-classes

1909-1930: 146 classes, 271 sub-classes

1855-1908: 146 classes, alphabetically arranged

1617-1876: 103 classes, alphabetically arranged.

To help convert pre-1963 class marks into post-1963 and *vice versa* the searcher has available the *Forward concordance* and the *Backward concordance* respectively. Another useful aid to retrospective search is the *Fifty years subject index, 1861-1910*.

Patents specifications are available for purchase from the Patent Office Sale Branch at Orpington, Kent, but copies may be consulted free of charge at the Science Reference Library (situated in the Patent Office building in London). As well as all the published abstracts, indexes, keys, etc, further special facilities for the searcher are provided by the Science Reference Library: the Applications register, the only completely up-to-date record of applications pending, and the Stages of Progress Register, giving the present status of each patent, eg whether in force or lapsed, expired, licence granted, opposition to grant filed, etc. A futher special index is the Name Index to Specifications Published which is designed to cover the gap between publication of the application and of the printed *Index to names of applicants*. Available to all by post is the selected patent specification service whereby subscribers are supplied on publication with all patents within any one or more sub-divisions of the classification system, and the subject-matter tabulation service, under which lists of specification numbers are supplied instead of the specifications themselves. The Patent Office also provide subject-matter file lists for each code mark in the *Classification key*, showing all the patents issued since 1911 (or, as an alternative, since 1965) which have been indexed under that mark.

Extensive collections of indexes and abstracts are found at the provincial depository libraries, and frequently there are substantial holdings of foreign patents also. J E Wild *Patents: a brief guide to the patents collection* (Manchester Libraries Committee, second edition 1966) provides a very useful introduction to one of the largest, comprising over 5 million specifications and 30,000 bound volumes. A more general guide is Sheffield City Libraries *Patent holdings in British public libraries* (third edition 1973). An up-to-date two-page summary of provincial holdings appears from time to time in the *Official journal*: 'Patent holdings in British public libraries'. The old system has recently come under scrutiny and for the future the new Patent Information Network for the United Kingdom is to comprise 25 public libraries ranging from Aberdeen to Plymouth, together with the National Library of Wales at Aberystwyth. Six of these, to be known as Provincial Patent Libraries, will provide all UK and US specifications and all PCT and EPC published applications (see below), with all the associated journals, abstracts and indexes. In some cases they will also provide patent documents from other major countries. The remaining twenty libraries, to be known as Patent Information Centres, will contain only a selection of patent publications. All the libraries will have trained staff to help patent searchers and will provide a document supply service.

Other indexes
Supplementing the official Patent Office bibliographical tools are the regular indexing and abstracting services, many of which include

298

patents, eg *Science citation index, Photographic abstracts, World textile abstracts, Chemical abstracts*, where they amount to 30% of all documents cited. A number of services concentrate on patents only, eg *Footwear and leather abstracts, Polymer science and technology patents, Plastics abstracts.* It should be noted, however, that many abstracting services virtually ignore patents, eg *Engineering index, Metals abstracts.* INSPEC warns: 'Patent coverage is very selective'. It is true that abstracting and indexing services are much better known and more frequently consulted by scientists and technologists than the official patent abstracts and indexes but a 1972 survey of eight hundred consecutive British patents found that twelve months after publication only half of them had been covered by one or more of some forty major abstracting services. Fortunately, many primary and secondary journals include notes or abstracts of new patents as a regular feature, eg *Journal of applied chemistry, Production engineer, Modern plastics, Textile manufacturer, Electrical review, Metal finishing.* A feature of this field is the existence of specialist patent indexing companies, of which one of the most efficient and successful (and expensive) is Derwent Publications of London, providing an extraordinarily wide range of indexing and abstracting services covering over 600,000 patents from 24 countries per year. Subscribers can choose to be alerted on a country basis, eg *German patents abstracts, Soviet inventions illustrated, Netherlands patents report*, or slightly later on a subject basis, eg *Central patents index*, covering pure and applied chemistry in a dozen subject sections, such as *Plasdoc* (plastics, polymers, monomers), *Petroleum, Metallurgy*; or *World patents abstracts*, covering non-chemical patents in half-a-dozen weekly subject sections. The Derwent Patents Copy Service is a document delivery service often more rapid than the official suppliers. A variety of other aids are available, including abstracts on standard IBM cards for manual or machine searching, 16mm microfilm, and magnetic tape for in-house searching.

Patents from overseas

The stress laid on overseas patents by a commercial service such as this reminds us how essential it is even for advanced industrial countries to keep a close watch on inventions from abroad: the UK Patent Office collection of over 20 million patents is from over a hundred countries. As has been described above, patents are granted by individual states to their citizens as a device to stimulate national economic growth. It is therefore no surprise to find that because of the lack of agreement on international protection for inventions, which has already been referred to, an inventor is obliged to take out separate patents in as many countries as he feels monopoly rights will be useful. A major chemical firm like Dupont, for example, may file in as many as a hundred different countries in a dozen or more languages, and often at great inconvenience and expense. Indeed, in most countries many more patent applications are filed from abroad than by nationals; 60% of the specifications filed at the UK Patent Office are from overseas; contrariwise, to protect the 7,000 native British inventions a year about 40,000 patent applications are filed abroad. This does mean that the really important inventions get patented in all the major industrial countries — but one can never be sure! There is for instance a considerable overlap between British and US patents, but not sufficient to permit a worker in one country to ignore the patents of the other. Although starting much later than the British, with the Act of 1790, US patents now far exceed them in current output and total numbers (now well over the four-and-a-quarter million mark). Originally modelled like so many others on the old British system, the American patent system still operates on broadly similar lines, with bibliographical control provided by the weekly *Official gazette* with its annual *Index of patents: part I, List of patentees* and *part II, Index to subjects of invention.* Unlike the new UK system, the US still retains the method of examination before publication, with its consequent delay: thus the US publishes only granted patents, not applications.

A very practical reason for close observation of overseas patents is the practice in a number of countries, notably Belgium and the Scandinavian countries, of laying applications open to public inspection without examination for novelty about three months after filing. Many of these OPI documents are the equivalents of applications made in Britain or the US, which are not disclosed to the public until after the novelty search, or even, in the case of the US, examination and acceptance, perhaps two or three years later. And of course where the US Patent Office rejects an application it is not published at all: 30% of applications are lost this way. The advance information thus provided has obvious commercial as well as scientific value. In a country like Belgium, the numbers involved are relatively small, perhaps 250 a week, but recently the burden of examining an increasing flood of applications, many of them duplicate foreign filings, has persuaded a number of other countries to amend their practice in one way or another. Germany, for example, with a backlog in 1968 or over a quarter of a million applications, was one of the first to change to a deferred examination system. This is a half-way house between the Belgian practice of rapid disclosure and the US method of not publishing until after examination. Germany now publishes all applications after eighteen months, and does not examine at all unless specially requested. Second specifications are published after examination but this may be several years later, and is really only for legal purposes. It is well worth noting that this reform greatly increased the use of the information contained in patent specifications. France, Japan and, as we have seen, now the UK have also adopted a system of deferred examination. These are major industrial powers, responsible on their own account for large numbers of patents, but more to the point, they are countries where many American inventors desire protection and therefore file patent applications. In future many US patents will no longer be primary sources when published.

International co-operation

A more constructive reason for concern with overseas patents is the developing interest in co-operation between countries on patent matters. The tremendous growth in international trade since the second world war has compelled governments to face up to the quite unnecessary duplication and complexity that is caused by the continued existence of separate national patent systems. It has become increasingly unacceptable to have highly qualified examiners in a dozen or more countries searching simultaneously to see whether the identical invention should be granted a patent. Indeed, in more than one industrial country the whole patent system has virtually collapsed under the strain.

International co-operation on patents is not new: in 1883 in Paris was signed the International Convention for the Protection of Industrial Property, establishing the Industrial Property Union comprising eleven member states. Reformed several times, it now embraces some ninety countries throughout the world (the 'Paris Union'), who have all contracted to grant patents without discrimination on grounds of nationality, and to allow a period of grace, now one year, for an applicant in one country to file further applications abroad, while retaining the benefit of the home filing date for priority of his rights in the other countries. This gives him the chance to gauge the worth of his invention before committing further time and money on foreign applications.

Less enthusiastically received than this 'convention' patent has been the other main manifestation of co-operation, first mooted as early as 1904, the *International Patent Classification* (third edition 1979), initiated by the Council of Europe in 1954, and designed to replace gradually all the national patent classifications. Comprising 8 sections, 115 classes, 607 sub-classes, and more than 46,000 groups, it is now fully operational, and regularly revised every five years. Responsibility lies with the World Intellectual Property Organization (WIPO) at Geneva, one of the specialized agencies of the United Nations. WIPO also serves as the bureau of the Paris Union and has produced *Official catchword index to the third edition (1979) of the International Patent Classification* (Munich, Heymans, 1979). Virtually every country now enters the international classification mark on its specifications, as the UK has done since 1957, but few use the scheme to classify them, retaining their own scheme alongside, eg UK, US, Japan, Soviet Union. This of course defeats the objects of the scheme which are economy and simplification by means of an international agreement on work-sharing. PCT and EPC patents (see below) use the scheme exclusively, as do France and Germany, who have both abandoned their national schemes.

Plans for a world patent were laid as far back as 1909 in Germany, and we have also had an abortive British Commonwealth patent, for even in the high noon of empire an inventor had to take out thirty-five separate patents if he wished to protect his brainchild throughout the Queen's possessions. There are those who warn that an international patent is 'not feasible because of the different concepts... of the essential ingredients of a patentable idea', but the considered view of one experienced Assistant Comptroller at the Patent Office is that 'The main obstacles in the way of simplification arise more from human nature than basic philosophical differences.' What there can be no disagreement on is the fact that international collaboration in this field is an extraordinarily complex and difficult matter, not merely for scientific and technological reasons, but on account of legal, economic, linguistic, political, historical and geographical obstacles, to name but a few.

In Europe in the last ten years or so we have seen more action on this front than over the previous hundred; and world-wide there have been some moves made towards the objective of the international harmonization of patent regulations. It is fitting that the most encouraging advances should have been made among the great industrial nations of Europe, for that is where the pressures are at their most acute. The twenty-one participating countries of Europe have pinned their hopes on a concentric series of three linked agreements, somewhat complex in their effect.

a *Patent Co-operation Treaty (PCT)*. This was drawn up after a detailed study by WIPO and signed in Washington in 1970 by twenty of the eighty Paris Union countries, and later by fifteen more. Designed for eventual world-wide coverage, it takes the form of a *work-sharing* agreement to save time, effort and expense for the national patent offices and for the applicant. He files one international application at his own national or regional patent office in English, French, German, Japanese or Russian, but designates the countries where he wants a grant. This filing has the same effect as if national applications had been filed separately in each of the designated states.

The invention is then searched for novelty by a recognized International Searching Authority, eg the European Patent Office (see below)

302

or, outside Europe, the patent offices of the United States, Japan or the Soviet Union. It is then published by WIPO with the search report (in English and the language of the application, if different) after eighteen months. As with UK patent applications author abstracts appear on the front page: they also are in English and the language of the application, if different. It may also be examined for inventiveness by a recognized International Preliminary Examining Authority.

If the applicant still thinks it worth his while to proceed, his application, translated where necessary, is forwarded to the countries he has designated, where it becomes in effect a national application in each. If appropriate it may also be forwarded in the same way to the European Patent Office. The intention is that these countries will accept the validity of the international search, and of the examination where this has also been carried out in advance. The application will of course already have been published and it is hoped that each country will make the grant with the minimum of extra processing. In the UK this has meant republication, and translation into English where necessary, with an application number in the normal UK sequence (see above). *PCT gazette*, published by WIPO, appears every second week and includes abstracts of the published PCT applications, together with indexes.

The implementation of this agreement represents a major advance in international patent co-operation; not the least of its attractions is that it does not interfere with national laws or criteria.

b *European Patent Convention (EPC)*. Discussion began in the 1950s among the six original countries of the European Economic Community and the Convention was signed in Munich in 1973 by sixteen of the twenty-one participating European states. It goes beyond a work-sharing arrangement inasmuch as it involves *harmonization of law* and a considerable degree of centralization. Whereas the PCT stops short of the actual grant of a patent, leaving that to the national patent offices, in 1978 the EPC countries established a European Patent Office in Munich which actually grants a European patent.

A single application only is required, in English, French or German, made through the applicant's national patent office to the EPO. A search for novelty (world-wide and including non-patent literature) is made by the Searching Division at The Hague, formerly known as the Institut International des Brevets, founded in 1947 and a significant early example of practical international patent co-operation. The application and search report (in the language of the application) is published within eighteen months. As with UK patent applications author abstracts appear on the front page: they are in the language of the application only.

If required by the applicant a rigorous examination then follows.

Once granted, the specification is republished by the EPO in the original language, with translations of the claims in the other two languages. It is then forwarded to each of the designated countries, who do not republish it or assign to it a national serial number, but accept it as it stands. The European patent runs for twenty years and in each of the designated countries has the same force as one of its national patents. *European patent bulletin*, published by the EPO in English, French and German, appears twice a month and lists both the published applications and the granted patents, with indexes. Abstracts are published in a series of pamphlets each week as *Classified abstracts of the European Patent Office*; each pamphlet covers one sub-class of the International Patent Classification.

It is perhaps too soon to judge the long-term effects of the EPC, but clearly national applications in Europe could fall substantially, perhaps to a third of previous levels according to one estimate. But this of course is the object of the exercise, as the Deputy Comptroller of the UK Patent Office made clear: 'It is hoped that the new European system will be so good as to encourage all countries to take out European patents and therefore taking out patents by the national route, as is the practice today, will gradually decline'. But the deciding factor for many may be cost. On the very day that the PCT and the EPC came into effect in the UK the *New scientist* had this to say: 'Together these are intended to make foreign patenting easier but it will remain very, very expensive. British inventors with limited funds will have to continue patenting in the UK only'.

c *Community Patent Convention (CPC)*. Just as the EPC (known as the First Convention) is regarded as a regional patent agreement within the world-wide PCT, so within the EPC itself we now have a further agreement (known as the Second Convention) among the nine (now ten) states making up the European Economic Community. Signed in 1975 in Luxembourg this takes two important steps towards further harmonization of law by providing that firstly, an applicant via the EPC route can ask for his patent to have validity in all EEC countries; and secondly, after 1985 if a European patent grant includes any one EEC country it will automatically extend to the others. In effect, these provisions, especially the second all-or-nothing stipulation, have created a unitary patent covering all ten EEC states, a Community patent in other words, as opposed to the bundle of national patents that results from the PCT or EPC routes. It is not however a separate document, but simply one of the patents in the normal EPO sequence.

There are as yet no proposals to phase out the national systems within the EEC: possible a two-tier system might emerge, with important inventions being patented internationally, and the others only at national level. It may well be, however, that national laws affecting

304

national patents will be brought more into harmony.

To sum up the present position, patent documents valid in the UK are now of four kinds:

a UK grants made upon application direct to the UK Patent Office in the usual way. Both the *applications* and the *granted patents* are published in printed format by the UK Patent Office.

b UK grants made by the UK Patent Office upon applications forwarded via the WIPO route under the PCT provisions. The *applications* are published in typescript format by WIPO, and if accepted are republished in printed format by the UK Patent Office as *granted patents*.

c European patents granted by the European Patent Office which designate the UK as one of the countries where protection is desired. *Applications* are published in English, French or German in typescript format by the EPO, unless they come via the WIPO route and have already been published in one of these three languages. *Granted patents* are published in printed format by the EPO.

d Community patents granted by the EPO. Though they have a special status the *applications* and *granted patents* form part of the normal EPO sequence.

c and d are published in only one of the three official EPO languages, English, French or German, and are not as a rule republished by the UK Patent Office, though there is provision for an English translation to be made under certain conditions.

For the UK searcher the effect of these changes is that he has to search both UK Patent Office documents as well as European Patent Office documents, and many of the latter (currently over half) will be in French or German only, though in granted patents the claims will be in all three languages. He may also need to search the PCT published applications in cases where they do not result in a grant; these may be in any of the five PCT languages though there will always be an abstract in English.

Mechanization

A fruitful field for co-operation lies in the application of machines to patent searching, and much investigation has already taken place within the Committee for International Co-operation in Information Retrieval among Examining Patent Offices (ICIREPAT) founded in 1962 but since expanded in membership and reconstituted in 1968 as a committee of the Paris Union with its secretariat at WIPO. The International Patent Documentation Centre (INPADOC) was established at Vienna in 1972 by the Austrian government in association with WIPO 'to provide quick and reliable access to newly issued patent documents', including, for example, a service for the identification of 'families of patents', ie virtually identical patents published in several countries, a particularly knotty problem for patent searchers. INPADOC receives weekly magnetic tapes covering all patents issued from 46 countries, and the 7-million-item data base and the microfiche services based on it are both available to subscribers.

National initiative has produced the punched card searching system at the British Patent Office, for instance, and the microfilm of US patents, and the aperture card sets of German patents. On the commercial front Derwent Publications began to offer in 1976 an on-line searching service covering patents from 24 countries; also available on-line is the *Chemical abstracts* patent concordance; the IFI-Plenum data base on US patents can also be searched on-line; and there are others.

Trade marks

This is probably the most convenient place to discuss trade names, for they often carry the status of trade *marks*, another type of 'industrial property' protected by legislation in a fashion similar to patents. A trade mark is a mark used by a manufacturer or trader to distinguish his goods from similar wares of other firms. It should be remembered, however, that the nature of the property protected is quite different. A patent protects the inventor's *product*; a trade mark only protects a label. A trade mark registration as such cannot prevent any trader selling goods exactly similar to those of the trade mark owner, provided that he does not put the mark on them. In the words of the *Times*, 'A trademark is not merely a horse of a different colour, it is a different animal entirely'.

The simplest definition of a trade name is a name by which an article is known in commerce, such as Epsom salts, Coca-Cola, Portland cement, blue vitriol, plaster of Paris, Band-aid, Leica, klaxon, etc. Of course, not all such names are proprietary: many are common or generic, and for practical purposes can be regarded in the same light as, for instance, common names for plants, that is to say, as a particular kind of synonym. As such they are commonly included in encyclopedias and dictionaries of science and technology, eg *The condensed chemical dictionary* (New York, Van Nostrand, ninth edition 1977), *Kingzett's chemical encyclopaedia* (Bailliere, ninth edition 1966), and *Fairchild's dictionary of textiles* (New York, 1959). There are also dictionaries, particularly in chemistry, which make a feature of their coverage of trade names, eg *Gardner's Chemical synonyms and trade names* (Oxford, Technical Press, eighth edition 1978), and Williams Haynes *Chemical trade names*
306

and commercial synonyms (Van Nostrand, second edition, 1955). Translating dictionaries too will be found that make a considerable effort to include trade names, eg M Moureau and G Brace *Dictionary of petroleum technology: English/French, French/English* (New York, Scientific and Medical Publications of France, 1979). Brand name is a colloquial term to mean the same thing as trade name, usually in connection with food or domestic articles. A trade name should not be confused with a 'business name', which is simply the name under which a business trades, other than a personal name.

For hundreds of years, however, some specific trade names have been used in the way defined above, ie as a means to distinguish one manufacturer or trader's wares from similar goods of another: indeed in these days of mass production and intensive advertising it is probable in the consumer field at least that most goods reach the market under such trade names. We do not simply buy sherry, we ask for Bristol Cream; we are urged not to be satisfied with just petrol but to demand Shell; very few smokers are willing to purchase cigarettes irrespective of brand, for they prefer Camels or Gauloises, for instance. None of these names gives any clue as to the nature or use of the product, and yet it is difficult to see how we could do without them. It has been claimed that every individual in the civilized world is confronted by at least one trade name every single day. They are a convenience to the consumer, and an even greater boon to the manufacturer or merchant, who, in the words of a definition of over a hundred years ago, uses them 'in order to designate the goods that he manufactures or sells, and distinguish them from those manufactured or sold by another; to the end that they may be known in the market as his and thus enable him to secure such profits as result from a reputation for superior skill, industry, or enterprise'.

Trade names of this kind are proprietary, ie they belong to a particular manufacturer (or trader, etc) and in most industrial countries can be registered as trade marks to obtain the protection of the law. Of course, trade names that are registered in this way make up only one of several kinds of trade mark. Because of their form they are known as 'word marks' and amount to about 80% of the total throughout the world, but of course a trade mark may also take the form of a letter or a numeral, or a symbol, signature, design, picture, or other graphic device, or indeed a combination of any of these. Hall marks on gold and silver, pottery marks, proof marks on firearms, cattle brands, printers' devices, watermarks on paper, for example, have been known for many hundreds of years. Indeed, winemakers were in the habit of marking bottle tops with their own special symbols well before the time of Christ, and builders' marks have been found used in Ancient Greece. A more recent example, carried by Concorde, is the British Airways

'Speedbird', inherited from the British Overseas Airways Corporation, and generally admired as the simplest and perhaps the best airline symbol in the world. Attempts to register smells and sounds have so far been unsuccessful. A good trade mark, indeed, can be a prized business asset, realizable for hard cash. It is rumoured that one well-known trade mark changed hands for a quarter of a million pounds. The Coca-Cola Company is said to have placed a value on its marks of over $1 billion. Companies will go to extraordinary lengths to protect a trade mark: the 17-year struggle of the discoverer of aspirin, the Bayer chemical and pharmaceutical firm, to regain the right to its world-famous 'Bayer cross' trade mark has been described as 'one of the longest, most fiercely fought and most expensive commercial battles ever waged'. ICI have a full-time staff of three to protect the company's trade marks. Its famous roundel is now guarded by 2,500 separate registrations in the UK and overseas. And this is only one of ICI's 3,000 registered trade marks. Each year Rolls-Royce suffers four hundred trade mark infringements world-wide. Bass, the brewers, and proud owners of the world's oldest registered trade mark, a simple red triangle, have albums containing over two thousand examples of forgeries uttered in the last hundred years in the hope of cashing in on the renown of their famous beers.

The earliest (1876) and one of the most elaborate systems of formal registration is that in Britain, with over a million marks registered in the first hundred years, and so a brief description will serve as indicative of the rest. Indeed a feature of trade marks throughout the world is that so many of them display what an international authority has called 'a characteristic British accent'. This is so annoying to the French, for example, that in 1975 the French government took steps to reverse the trend. The Patent Office maintains the Trade Marks Registry, and marks are registered by a procedure similar to patenting an invention. An authoritative account may be found in T A Blanco White and R Jacob *Kerly's law of trade marks and trade names* (Sweet and Maxwell, tenth edition 1972). As with patents (see above) assistance is available from a specialized profession, the Institute of Trade Mark Agents, who issue a *Register of ordinary members* (1979).

Applications are first scrutinized to see if they are indeed original, not merely descriptive, distinctive, and not deceptive, and the marks are then advertised (with the assigned number) each Wednesday in the weekly *Trade marks journal.* Applications are currently being made at a rate of about 18,000 a year, of which around a third are rejected. If not opposed within the month they are registered and remain in force for seven years, though they can be renewed for further periods of fourteen years indefinitely, provided the appropriate fees are paid. Bass's red triangle, just mentioned, has been retained on the register in this
308

way since 1876. Marks are grouped into 34 subject classes, and registration is effective only within that class (although the same mark can be registered in more than one class if appropriate). The register is maintained in two parts: A, which gives the fullest protection and where there is no doubt that the mark is completely distinctive, and B, with a lower degree of distinctiveness, and therefore affording a less comprehensive protection. Currently there are about a quarter-of-a-million in force.

Each year an annual name index of applicants (including registered proprietors and users) is published, but no index of the marks themselves. However, since 1958 it has been possible to subscribe to the Patent Office service of weekly index slips of trade *names*. Current plans envisage computerization with microfiche replacing the slips.

There are two points worthy of mention here: the existence of a scheme for registration does not imply that unregistered marks are without legal protection. Many are quite obviously proprietary (and indeed a number date from pre-registration days): the courts will uphold the rights under common law of owners of such marks, but the onus of proof of ownership is obviously heavier where there is no formal register to refer to. The idea of an official list of unregistered trade marks was turned down by the Department of Trade enquiry which resulted in *British trade mark law and practice: report of the Committee* (HMSO, 1974). It did go on to suggest, however, that 'the Science Reference Library should be encouraged in their action of collecting together unofficial lists of trade marks and making them readily available'. Secondly, even registration is not 100% watertight if the trade name is so widely used as to become generic. Proprietors of trade names need to be constantly on the alert to ensure that their name is used for their own product and that alone. Formica spends over £300,000 a year to remind the public that the decorative laminate bearing that name originates only from them. Classic examples of trade names which have been lost in this way, because they have passed into the language (known in the fraternity as 'genericide'), are gramophone, linoleum, escalator, pianola, tarmac, sellotape, zipper, aspirin, and bakelite. Others that have teetered on the brink for some time are Hoover, Biro, Thermos, Xerox, Cellophane, and Photostat. Indeed, one sign that a trade name is drifting into the public domain is this dropping of the initial capital letter: newspaper and magazine editors regularly receive complaints about such lapses from lynx-eyed company lawyers anxious to preserve their clients' rights. But hints from such sources of legal action are often empty threats, under English law at least (though the Commission of the European Communities is hatching plans to change that). The protection that the law gives covers only the use of these names *in commerce*. For infringement to lie, trade in the

309

goods must occur. As a letter to the *Times* in 1978 put it, 'The Trade Marks Act is about trade, not freedom of speech'. But this does not prevent dissension: another letter to the *Times*, from the other camp three years later, expressed the view that 'If a writer has insufficient command of the language to need to resort to using a trade mark generically to describe the goods or services, then scholarship has come to a poor pass'. The practice of J Markus *Electronics dictionary* (McGraw-Hill, fourth edition 1978) is probably sound: he capitalizes all trade names, advising 'The correct use of trademarks will avoid unpleasant correspondence with corporate lawyers'.

Successful action may be taken, however: in 1980 the proprietor of Intralipid obliged *Index medicus* and MEDLARS to delete the term from their computer file of subject headings, MeSH. Certain names, often derived from places, such as sherry, port, champagne, mosel, which have indeed become generic have been hauled back under protection by special legal provisions such as the *appellation contrôlée* laws in France and the similar provisions in the 1971 German wine law.

There are many hundreds of thousands of trade names currently in use. It has been reliably claimed that there are now more trade names in the English-speaking world than there are words in the English language. As names, many of them convey no product information, and, as we have seen, bear no relation at all to what they represent, eg a Mercedes is a car, and Stork is a margarine. But that is not to say that such names are chosen at random: it is by no means unusual for a name to be carefully selected in the hope that the consumer will subconsciously associate the estimable characteristics of the name with the article in question, eg Sunlight, for soap, as early as 1884; Senior Service for cigarettes; Princess for cars. For his new car William Lyons had a list drawn up of more than five hundred fast-moving animals, birds and fish before he finally plumped for Jaguar. It can be a paralysingly delicate problem for a company about to launch a new product to think up a name that is not only original and acceptable for registration, but is also linguistically, psychologically, culturally, and sociologically appropriate to the goods in question. Indeed, the marketing vice-president of the National Biscuit Company said that to get names is the hardest thing in the world. Competitions are a common solution to this problem: Thermos was the winner of just such a public contest organized in Germany in the 1900s; Hovis was the winner in a competition open to the customers of the flour millers; Metro came out top in a ballot of British Leyland workers. They often have no meaning at all, being simply concocted words, chosen to be distinctive, eg Kodak. Some are even computer-generated, eg Exxon, the new brand name for Standard Oil in the United States; out of ten thousand names produced, 234 remained after a prolonged opinion poll and eventually these were reduced to the six finalists. Linguistic studies were undertaken in over a hundred languages, and the successful contender was deliberately selected because 'it says nothing and it means nothing': the double-x is found only in the Maltese language. As a help in thinking up such names there is *Hague's trademark thesaurus* (Chicago, Mortons Press, 1964), which comes complete with a set of calibrated word-formation dials. With its aid 'you can create synthetic names or trademarks of up to six letters that will be smooth-sounding, easy to pronounce and easy to remember'. According to one American estimate a hundred more are added to the vocabulary every month within the industrial field alone.

Enquiries about trade names are very common in scientific and technological libraries, usually in the form of a request for the manufacturer of a particular branded product, or for details as to its composition and other physical properties. For the latter, the searcher has to rely largely on the encyclopedias, dictionaries, handbooks and other similar sources of such data, eg the *Merck index of chemicals and drugs* (Rahway, NJ, ninth edition 1976), N E Woldman and R C Gibbons *Engineering alloys* (American Society of Metals, sixth edition 1979), G S Brady *Materials handbook* (McGraw-Hill, eleventh edition 1977), the five-volume *Colour index* (Society of Dyers and Colourists, third edition 1971). Historical enquiries too are common, usually about some mysterious-looking industrial artefact with a name marked on it: as a rule enquirers wish to know who made it, but sometimes (and this is more difficult) what it was used for. Where the enquirer is less interested in technical information and is seeking the name and address of the manufacturer (or owner, supplier, etc), the obvious sources are the trade directories, many of which list trade names, as the index to I G Anderson *Current British directories* (Beckenham, CBD Research, ninth edition 1979) clearly indicates, eg *Machinery buyers' guide* with ten thousand, *Wire industry yearbook* with two thousand. One of the most extensive listings of US names is in volume 10 of *Thomas' register of American manufacturers*: it is worth noting that many of the names listed, perhaps a majority, are not registered. Many trade journals make a regular feature of listing new trade names in their field, eg *Soaps, perfumery and cosmetics, Mechanical handling*.

As noted above, the official published indexes are inadequate for searches under the trade names (as opposed to searches under the names of applicants, proprietors, users, etc), but such is their importance that there are a number of commercially published search tools devoted solely to trade names. The most extensive British list is the Kompass *UK trade names*, appearing every two years with over 60,000 names in alphabetical order. Lapsed names are excluded, but it should be noted that lapsing is a matter of law, and does not mean that such a name is no longer in use. Neither of course does the inclusion of a name in such a list imply that it has legal protection or is officially registered. An American list concentrating on consumer products is E T Crowley *Trade names dictionary* (Detroit, Gale, second edition 1979), with over 150,000 names in its two volumes and annual supplements, *New trade names*, but it makes no attempt to distinguish between registered and unregistered names. Its companion volume is *Trade names directory: company index*, listing brands under companies and giving company addresses. An outstanding example of a multi-purpose American list which not only gives a short description of each material and its uses as well as the manufacturer's name, but also distinguishes between registered, unregistered, and generic or common names, is O T Zimmerman and Irvin Lavine *Handbook of material trade names* (Dover, NH, Industrial Research Service, second edition 1953) and its *Supplements* (1956-65). These are general works, of course: there are also lists for particular industries: eg *Benn's hardware directory* with over 13,000 names. We find such lists published by a wide variety of organizations, such as an industrial company, eg Pilkington Bros *Directory of glass names* (1971); a learned society, eg R H S Robertson *A glossary of clay trade names* (Glasgow, Mineralogical Society of Great Britain and Ireland, 1955); a trade journal, eg *Binsted's directory of food trade marks and brand names* (Food trade review, fifth edition 1975); a research association, eg Rubber and Plastics Research Association *New trade names in the rubber and plastics industries*, which is an annual compilation. An unusual and indeed unique list is J H Whitham and A Sykes *A register of trade marks of the Cutlers' Company of Sheffield* (second edition 1953) confined to the world-renowned industry of a single city.

Professional trade mark searchers were disappointed that the official 1974 enquiry mentioned above did no more than recommend pressing on with mechanization in the Trade Mark Registry, and for other marks could only draw attention to the 170 directories incorporating trade mark lists that had been identified by the Science Reference Library: since then we have had published B M Rimmer *Trade marks: a guide to the literature and directory of lists of trade names* (Science Reference Library, 1976) with some 450 titles. The Commercial Library and Information Service of the Manchester Public Libraries has also published a guide to over 150 such lists as one of its excellent Brief Guides to Business Sources: *Trade names lists* (1975).

It has to be remembered that the official registration of trade marks as practised now in over 150 countries is usually on a national basis, and therefore most trade names have a national currency only; the Benelux countries make up one of a very small number of exceptions. Of course, international companies can and do register the *identical*

312

trade mark in a number of countries, but legally these are still separate national trade marks. Where the same product is known by *different* names in a number of countries it sometimes becomes necessary to translate: Perspex is the British trade name for the transparent plastic polymethyl methacrylate, but in the United States it is known as Plexiglas.

As with patents and other forms of 'industrial property' there have beem attempts at international agreements for the mutual protection of trade marks in the signatory countries, eg the Madrid Agreement of 1891, most recently revised at Stockholm in 1967, which provides for the registration of marks at the International Bureau of WIPO in Geneva. Such marks are listed monthly in the periodical *Les marques internationales.* Only 23 states are party to this agreement, and for sound reasons of their own Britain, the United States, Japan and a number of the other major trading countries have never adhered to it. More recently, in 1972, at Geneva, a 34-nation meeting led to the Trademark Registration Treaty of 1973, providing for a single application to WIPO which would secure registration in contracting states designated by the applicant. This was signed at Vienna by only eight of the 49 participating countries and still awaits ratification. Within the EEC a European trade mark has been under active discussion since 1964, with a European trade mark office sited in London or perhaps Strasbourg.

Meanwhile, as an index to such internationally-registered marks there is the long-established A W Metz *Intermark-index* (Zurich, 1925-). This work has an added interest for the student for it makes a brave attempt to tackle the problem of indexing non-word marks, eg symbols, devices, etc, which have no obvious place in an alphabetical list. A further difficulty encountered by trade name searchers is eased by the associated computer-compiled publication V Gevers and A W Metz *Alphabetical and phonetical directory of international trademarks* (Zurich, 1970), covering the period 1949-69, which lists the names not only in their normal alphabetical sequence but also according to their phonetic equivalent (where this is different), eg Cellophane appears under 's' as well as under 'c'. In this way one can locate a name even though its spelling is not known. This can be of crucial importance if a searcher is trying to discover whether a particular name is already in use: in order to be registered new names have to be phonetically and visually distinct as well as lexically original.

Further reading

L J Haylor 'Scientific information and patents' *ASLIB proceedings* 14 1962 342-9.

D Caplan and G Stewart *British trade marks and symbols: a short history and contemporary selection* (Owen, 1967).

F Newby 'Patents as a source of technical information' B Houghton *Information work today* (Bingley, 1967) 63-75.

V Tarnovsky 'Patent information services' *ASLIB proceedings* 19 1967 332-41.

R M Duchesne 'Patent library service in the United Kingdom' *Journal of librarianship* 2 1970 196-204.

B Houghton *Technical information sources: a guide to patent specifications, standards and technical reports literature* (Bingley, second edition 1972) 9-63.

F Liebesny *Mainly on patents: the use of industrial property and its literature* (Butterworths, 1972).

K M Saunderson 'Patents as a source of technical information' *ASLIB proceedings* 24 1972 244-54.

F Liebesny and others 'The scientific and technical information contained in patent specifications – the extent and time factor of its publication in other forms of literature' *Information scientist* 8 1974 165-77.

Patent Office *Applying for a trade mark* (new edition 1975).

R Baker *New and improved . . . inventors and inventions that have changed the modern world* (British Museum Publications 1976).

Trade Marks Registry *A century of trade marks: a commentary on the work and history of the Trade Marks Registry* (HMSO, 1976).

M W Hill 'Patent documents: anticipated changes which will affect libraries' *Journal of librarianship* 10 1978 97-118, 146.

Patent Office *Applying for a patent* (revised edition 1978)

British Library *Access to patent documents and information* (1979)

V S Dodd 'Developments in patent documentation' *ASLIB proceedings* 31 1979 180-90.

European Patent Office *How to get a European patent: guide for applicants* (Munich, third edition 1979).

M Hill *Patent documentation* (Sweet and Maxwell, 1979)

Patent Office *Patents: a source of technical information* (revised edition 1979)

British Library *Access to patent documents and information: a summary of evidence* (1980).

M Stiling *Famous brand names, emblems and trade marks* (David and Charles, 1980).

Chapter 16

STANDARDS

Standards are more than a form of scientific and technical literature; without them day-to-day life as we know it would be impossible. We buy, for example, replacement electric light bulbs today without having to worry about whether they will fit our sockets. This was not always so, not until the manufacturers agreed to standardize, and British bulbs will still not fit American sockets. We can buy film for our cameras, however, in Britain or the United States, and be sure that it will fit, even though the camera is Japanese, or German, or Russian. What makes this possible is standards, national and international.

Simply and basically, standards (or standard specifications) are rules as to the quality or size or shape of industrial products. This definition can be extended to include processes, methods, terminologies, etc. The actual statement of the requirements is the *specification*; once it has been accepted by a recognized authority it becomes a *standard*. In practice, however, the terms are used so loosely and interchangeably (even by official bodies) that the student can regard them as synonyms. In a modern technological society they are essential, embracing almost every kind of scientific and technical activity. In Britain standards are reckoned to have an influence on more than half the gross national product. If a spanner will not fit a nut, if two cans of the same blue paint vary widely in shade, time is lost and frustration mounts. J E Holmstrom tells us: 'It has been estimated . . . that differences in the design of screw threads as between British and American practice added not less than £100,000,000 to the cost of the second world war'. And yet as long ago as 1841 in a paper to the Institution of Civil Engineers Joseph Whitworth was urging acceptance of a uniform system of screw threads. Germany was more far-sighted: in 1938 Hitler banned the use of the Whitworth standard thread in favour of metric sizes During the *first* world war too, standardization 'among the allied powers now fighting the Germans' had been proclaimed by Winston Churchill, then Minister of Munitions, as 'based on principles so obvious that they really do not at this time of day require even to be emphasized'.

315

Types of standards
A typical standard specification is a pamphlet, no more than a few pages in length, setting out measurements, methods, definitions, properties, often with tables or diagrams. For purposes of study they can conveniently be classified according to their purpose (although there are many examples of mixed types):

a *Dimensional standards*: these represent a deliberate effort to make things fit. They try to ensure interchangeability, so that the same products wherever and whenever they are made are identical in size, eg screwdrivers and screwdriver bits for recessed head screws, crown bottle openers, WC seats (plastics).

b *Performance or quality standards*: the aim here is to ensure that a product is adequate for its purpose, that it really will do what it is supposed to do, eg nylon mountaineering ropes, road danger lamps, sparkguards for solid fuel fires, children's high chairs, leather-covered cricket balls, pavement marking paints.

c *Standard test methods*: these help to ensure that materials and components do match up to performance or quality standards. They enable comparisons to be made on a scientific basis, eg methods for the chemical analysis of ice cream, determination of stiffness of cloth, sampling and testing of paper for moisture content, method for the measurement of noise emitted by motor vehicles.

d *Standard terminology*: by standard glossaries of terms in a particular field communication can be made more precise and accurate, eg glossary of packaging terms, nomenclature of commercial timbers. Symbols too are a mode of communication, and standardization has a role to play here, eg symbols for use in data processing flow charts.

e *Codes of practice*: these try to ensure correct installation, operation and maintenance, eg guarding of machinery, street lighting, electrical fire alarms, frost precautions for water services.

f *Physical and scientific standards*: these have a function different from the technical standards so far described, insofar as they deal with the physical quantities which form the basis of measurement in industry and commerce, eg length, mass, time, temperature, etc.

Standards simplify production and distribution for the manufacturer, ensure uniformity and reliability for the consumer, and save the time of both (and of everyone in between) by eliminating unnecessary and wasteful variety, such as the classic instance unearthed some years ago by a committee of the British Standards Institution of 96 different types of garden spade in production.

The British Standards Institution
The compilation of standards (and the achieving of agreement as to their implementation) is the task of industry, but it is common to find

316

in many industrial countries a central standards organization, often with government support, responsible for co-ordinating effort and issuing the standards. Although there are a number of bodies in Britain which devise their own standards, eg the Institute of Petroleum, the Cement and Concrete Association, most work through the main organization, the British Standards Institution, an independent body with its own Royal Charter, originally founded in 1901, the first in the world by several years. Responsible for over 9,000 standards, it has its own technical staff of over 200, but does its work through a series of voluntary committees representative of every part of industry. Its work is supported by government grant, although this was reduced in the 1970s. Great care is taken to consult interested bodies and individuals during the compilation of standards, and draft copies are widely circulated for comment. All the standards mentioned in the previous section of this chapter are British Standards and are to be found listed (with a brief description) with all the others in the *British Standards yearbook.* New standards (issued at the rate of over 700 a year) and revised standards are listed in the monthly *BSI news* and every two months in the cumulative *Sales bulletin.* Some forty subject lists (called 'Sectional lists') are available free of charge, eg Shipbuilding, Hospital equipment, Iron and steel. Particularly useful are the British Standards loose-leaf *Handbooks* which collect together or summarize several related standards, eg *Metric standards for engineering, Methods of test for textiles, Building materials and components.*

BSI operates a system of certification marking, under which by licence manufacturers are permitted to mark their goods with the BSI monogram (popularly known from its shape as the 'kite' mark). The mark is to be seen on a wide variety of products from dustbins to clothing, and it provides an independent assurance to the purchaser that these products are produced *and tested* in accordance with the requirements of the relevant British Standard. There is of course no compulsion on manufacturers to apply British or any other standards to their products; the whole system is based on 'voluntary consensus' and the standards have the status of recommendations. It is often convenient, however, for the law-makers to refer to British Standards in their legislation, normally on matters concerning safety or health. In such cases the adoption of a certain standard may become mandatory, eg on car seat belts. This is an increasing trend.

At BSI headquarters in London a useful library and information service on standardization is maintained, together with a complete set of British Standards for reference. There is also a collection of over a quarter-of-a-million international standards and standards from overseas available for loan at a modest charge. Complete reference sets of British Standards are also maintained at over 250 libraries and academic institutions throughout the UK and at over 200 institutions overseas.

United States standards

The American standards scene offers an interesting contrast, with a very large number of bodies preparing standards, eg government agencies, professional organizations, technical societies. The *Directory of US standardization activities* (Washington, National Bureau of Standards, 1975) describes the work of nearly 600 organizations. The central organization, The American National Standards Institute, formerly the United States of America Standards Institute, is different from the BSI. It functions as a clearing-house but does not itself play the major role in preparing standards (although it may publish them). Its central function derives from its power to approve standards prepared by one of the many other bodies as 'American National Standards'. Annually appears its *Catalog* listing some 6,000 titles, which is supplemented by the *Listing of new and revised American national standards* each alternate month. A large and comprehensive library of American and foreign standards from some 50 countries is maintained; the *Magazine of standards* is published quarterly and *ANSI reporter* twice per month.

Typical of the larger standard-preparing bodies are the American Petroleum Institute (over 100), National Electrical Manufacturers Association (over 200), Society of Motion Picture and Television Engineers (over 150), Underwriters' Laboratories (200), but by far the largest is the American Society for Testing and Materials, described as 'the world's largest source of voluntary consensus standards', responsible for over 6,000, mainly in engineering. All of these bodies publish their own standards, whether or not they receive approval and separate publication as ANSI standards. One particularly well-known example is the Society of Automotive Engineers whose annual *SAE handbook* in two volumes is in fact a compendium of almost a thousand standards and recommended practices. The current *Book of ASTM standards* fills 48 volumes, now issued annually with a separately published *Index* in the last volume. New and revised standards are listed in the monthly *ASTM standardization news*. ASTM is obviously a major research organization: its results are disseminated through the bi-monthly *Journal of testing and evaluation*, and over 600 other publications in the annual *List of publications*. A *Fifty-year index* covers all the ASTM technical papers and reports from 1898 to 1950, and a series of *Five-year indexes* extends the coverage to 1975.

The nation's central measurement standards laboratory is the National Bureau of Standards, the custodian of the basic standards of physical measurement, eg mass, length, time, temperature. (In the UK a number
318

of similar functions are the responsibility of the National Physical Laboratory.) While the bulk of industrial and other standards are prepared elsewhere, much of the basic research and testing in physics, chemistry, engineering, etc, is done at the NBS and the results published in the *Journal of research, Circulars, Handbooks, Research papers,* etc. *Publications of the NBS, 1901-47* (and supplements 1947-57, 1957-60, 1960-66, 1966-67, 1968-69, etc) is a comprehensive guide. Current publications, amounting to over 1,200 a year, are listed with abstracts in the *NBS publications newsletter* every two months and in the annual *Catalog.* Since 1969 the NBS Office of Engineering Standards Services has offered a special information service on the 16,000 published engineering standards collected from over 350 US trade, professional, and technical societies. And as a government agency (part of the US Department of Commerce) the NBS has a particular responsibility for helping to prepare standards for federal purchasing.

Government specifications
Modern governments are major purchasers of many kinds of industrial products, and therefore have a particular interest in standard specifications. In many cases BSI or similar specifications seem to suffice, but commonly, the length of their purse enables them to dictate special standards of quality or performance to their suppliers; in other words they write their own specifications. These are put together with some care, often with the benefit of the best advice, and it sometimes happens that they find a wider application throughout industry. A number of them (particularly military specifications) are confidential, others are distributed in the form of unpublished documents only to those directly interested, but two major British series are available through Her Majesty's Stationery Office; the DEF-STAN (Directorate of Standardization) specifications, and the DTD (Directorate of Research Materials, formerly Directorate of Technical Development) specifications, both series now issued by the Ministry of Defence. Lists appear at intervals, eg *Index of 'Defence' publications: Part 1, Defence specifications* (ninth edition 1976); *Index of DTD specifications* (1974) and its supplements.

The US government is the world's largest bulk purchaser, and as one might expect produces a large number of specifications, with over twenty thousand currently in use in the military field alone. In the USA there is a much sharper division between government and industrial specifications. A useful 27-page pamphlet on this complicated pattern was produced by the US General Services Administration *Guide to specifications and standards of the federal government* (Washington, 1969). Annually published by GSA is the *Index of federal specifications*

319

and standards, with a conthly cumulative supplement. There is also for military specifications the annual Department of Defense *Index of specifications and standards*, with cumulative supplements every two months. Of course it is not only the federal government that finds it needs standards: a guide to state standards is L L Grossnickle *An index of state specifications and standards: covering those standards and specifications issued by the state purchasing offices of the United States* (Washington, National Bureau of Standards, 1973).

Company standards
Many firms have their own standards engineer or department to enable them to keep abreast of developments and to act as a link with the main standardizing body, whether national, or trade, or professional. It is often the case, however, particularly in the United States, that a company is obliged to prepare its own standards, particularly when no appropriate published standard exists. Sometimes these take the form of suitably modified versions of published standards, and occasionally, the positions are reversed when a private company standard forms the basis of an industry-wide or even national standard, once the need is seen to be more widespread. For the librarian these 'in-house' standards can raise problems: understandably, many of them are confidential and available only to contractors, though this does not prevent them being referred to and asked for. BSI publish a useful pamphlet on *The operation of a company standards department* (revised edition 1979) as well as a *Guide to the preparation of a company standards manual* (1980).

International standards
All major industrial countries and many others, over eighty all told, have a national system of standards, deliberately modelled on BSI in many cases. Primarily of interest to industrial users within a country's own borders, they assume greater significance as international trade grows. Manufacturers of imported goods often find it good tactics (or may even be obliged) to observe national standard specifications: conversely, exported manufactured goods may help to familiarize foreign customers with national standrads. The influential position of Germany in photographic and electronic exports has made many other countries aware of DIN (Deutsche Industrie Normen) standards (for film speeds, for instance, and audio plugs and sockets). Issued by the Deutsches Institut für Normung, they are widely consulted outside Germany, and the 18,000 in force are listed in the *DIN catalog/index* with its bilingual subject index. Over 2,200 of them are published in

320

an English version also, listed in the annual *DIN English translations of German Standards*. A valuable monthly guide to new standards from all over the world is the BSI *Worldwide list of published standards*.

There have for many years been attempts at international standardization, with varying degrees of success. The International Organization for Standardization, closely associated with the United Nations Organization, has issued over 4,200 ISO standards for adoption (if they wish) by its national members, now numbering over eighty, eg BSI, ANSI. They are listed in English and French in the *ISO catalogue*, an annual list updated quarterly by cumulative supplements. It has been claimed as the largest international system of industrial collaboration in the world. ISO does not concern itself with the electrotechnical field because this has been an area of particular international activity with both the International Electrotechnical Commission (by far the oldest of the international bodies) and within Europe the International Commission on Rules for the Approval of Electrical Equipment issuing IEC and CEE specifications respectively for international consideration. The recent change in colour coding for the cores of three-wire flexible cables and cords affecting millions of European households is on the basis of a CEE specification, the cause of a fierce struggle between Britain and Germany as to whether the live lead or the earth should be coloured red. An even fiercer struggle is in progress over the proposed IEC world-wide plug and socket outlet. CECC standards are issued by the Electronic Components Committee of CENELEC, the European Committee for Electrotechnical Standardization. In 1971 the decision to publish European Standards was taken by the thirteen member states of CEN (the European Committee for Standardization), but only sixty or so have appeared so far, eg EN 14-1974 *Dimensions of bed blankets*, which is identical with BS 5129. In fact, all of these European Standards are mandatorily adopted as national standards by the member countries (now numbering sixteen), but the relationship between national standards and international standards is normally more complicated. For many national standards there do exist corresponding international standards, though the extent of agreement may vary: they may be identical, they be technically equivalent, or they may merely be related. This is a complex matter, but it can be seen demonstrated in the *British Standards yearbook* where every appropriate entry contains a reference to the corresponding international standard with a symbol to indicate the extent of the agreement. Each year, in fact, BSI implements without change over three hundred international or European standards. BSI provides a useful listing of international standards at the end of the *Yearbook*, and there is a monthly listing at the end of each *BSI news*. The ANSI *Catalog* also lists international standards. S J Chumas *Index of international standards* (Washington,

National Bureau of Standards, 1974) is a KWIC index to some 2,700 ISO, IEC, CEE and other standards.

At its Hemel Hemstead centre in Hertfordshire BSI has offered since 1966 its Technical Help to Exporters service, providing information and advice on every kind of foreign standard or technical or legal requirement that might be relevant to exporters. Information can be supplied about the particular demands of over 160 countries. Translations are supplied on commission and over six thousand translated documents are available for purchase, including many key standards, eg DIN 1045 *Concrete and reinforced concrete structures.* Some are gathered into compendia covering particular subjects, eg *Powered lawn and garden equipment.* Since 1977 it has also published *THE: technical export news.*

Tracing standards
Like research reports, standard specifications all have a code number, eg BS 3012, CP 327.404, AU 145, 2G 143 (all British Standard specifications). Since it is very common for them to be cited (particularly orally) solely by their number, librarians and information officers soon learn to recognize some of the main categories by their prefixes, eg NF (France), UNI (Italy), GOST (USSR), but there is as yet no one basic guide that they can turn to. The search to identify a particular code number commonly leads through the various indexes already mentioned in this chapter, bibliographies such as E J Struglia *Standards and specifications: information sources* (Detroit, Gale, 1965), dictionaries of abbreviations, and indeed any other likely source. Perhaps the most comprehensive is the work already mentioned (see chapter 14) as a guide to report series codes: D Simonton *Directory of engineering scientific and management document sources* (Newport Beach, Cal, Global Engineering, 1974).

Commonly sought are the British or American equivalents of foreign standards (or *vice versa*). Exact correspondence may not be found, and indeed may not be required; what is usually wanted is standards that are interchangeable, having the same result. Unfortunately there are only a limited number of bibliographical tools which are of use in relating the standards of one country to those of another. One example is the *World standards mutual speedy finder* (Tokyo, International Technical Information Institute of Japan, 1976) in six volumes with comparative tables for the UK, US, France, West Germany and Japan, as well as relevant international standards.

322

Further reading

A S Tayal 'Standard specifications in libraries' *UNESCO bulletin for libraries* 15 1961 203-5.

R Binney *British standard* (Newman Neame, 1966).

B Houghton *Technical information sources: a guide to patent specifications, standards and technical reports literature* (Bingley, second edition, 1972) 64-90.

C D Woodward *BSI: the story of standards* (BSI, 1972).

A M Allott 'The availability of standards in the United Kingdom: part 1. British Standards Institution publications' *ASLIB proceedings* 27 1975 227-37.

P Ricci 'Product and engineering standards' *RQ* 18 1979 351-4.

Chapter 17

TRANSLATIONS

Since the decline of Latin as the international language of scholars and scientists, communication has been increasingly impeded by the language barrier. At first, when the great bulk of work was reported in one or other of the major Western European languages (German, English, French — in that order of importance), most scientists were native speakers of one, and sufficiently familiar with the others to follow the more important discoveries. In 1900, for example, 93.5% of the chemical literature was in one of these three languages. Nowadays not only is research reported in a much wider range of languages (*Chemical abstracts* monitors literature in 56 languages), but second only to English among the top four in terms of the amount of scientific and technical literature published is a language with which the majority of the world's scientists and technologists are quite unfamiliar, namely Russian. Analyses made for the EEC suggest that language alone is responsible for a degradation in the transfer of knowledge from the world's authors to European users of between 40 and 50%.

English speakers in this field, as in so many others, are fortunate inasmuch as their language is more widely used than any other. In one sense this favoured position has handicapped them, and the general impression that the British are not very good at languages is supported by a number of surveys. As for the Americans, in 1979 a Presidential Commission warned that their 'scandalous incompetence in foreign languages' is threatening national security and economic viability. But whether they are indeed linguistic incompetents has never really been put to the test, because they have never been obliged, as say Dutch or Finnish or Czech scientists and technologists have, to learn a foreign language simply to be able to read the important literature in their own subject. A 1959 survey of Scandinavian scientists found that the *whole* of the sample could read Swedish, Norwegian, Danish, German and English, and three-quarters could also read French. A 1966 survey of British engineers found that only one in two hundred ever looked at publications that were not in English. Even so, 54% of the periodicals at the British Library Lending Division, when it was the National Lending Library for Science and Technology, 54% of the citations in

324

Index medicus and 57.5% of the source materials used by *Chemical abstracts* are in languages other than English. These *proportions*, it is true, continue to decline steadily each year, as English increases its dominance, but the total *number* of foreign items has grown because of the overall increase in the items indexed and abstracted. About two-thirds of this non-English material is in Russian, German, and French (in that order of importance), yet an NLL survey revealed that only one scientist in ten felt he could cope with Russian, two out of three with German, and nine out of ten with French. The numbers claiming fluency were very much smaller. Over three-quarters admitted coming across a paper in the previous twelve months that they would have liked to read but could not because of the language difficulty, and almost half of them had found themselves up against this language barrier *within the previous month.* A later survey at the University of Sheffield of the academic staff, research workers and research students showed that well over half feared that they might be missing important work done outside the English-speaking countries. A repeat by BLLD in 1979 of the NLL survey evoked 3265 responses and found that language skills had declined rather than improved. Not the least disturbing of these 'depressing' findings was that among younger scientists the situation was worse still.

That there is a problem, then, needs no stressing. The undoubted domination of English has led some scientists to urge its universal adoption as the language of science; their view is that the foreign language barrier should be overcome by the foreigner. But this would be to ignore the marked political and national pressure on many foreign workers to publish in their own vernacular, which is often more than sufficient to overcome the economic and professional pressure to publish in English. J M Ziman's more humane view is that 'it is for those of us who speak the various versions of English to respect this scientifically irrelevant factor, and to open our ears to foreign tongues and our minds to foreign thoughts'. With D J Urquhart many would agree that the ultimate solution lies with the working scientists themselves, and the aim should be to have within each research group one who can read French, one German, one Spanish, one Russian, one Chinese and one Japanese. Experiments are under way, for example in the courses in Japanese at Sheffield University and German at Nottingham University, to teach scientists to read scientific literature after a minimum period of instruction and without a deep understanding of the normal language. Others have warned, with Pope, that 'a little learning is a dangerous thing'. Attempts have been made from time to time to encourage scientists to write in artificial universal languages such as Esperanto or Interlingua, but without success. In 1979 a proposal was made to introduce Europeen, a mixture of the languages of the European Communities.

325

In the meantime, of course, the palliative is translation. In fact, scientific and technological translation is a substantial industry, processing an estimated 4,000 million words a year in the United States alone. In the UK Patricia Millard *Directory of technical and scientific translators and services* (Crosby Lockwood, 1968) lists 300 individual approved translators, expert in 50 languages and a range of subjects from aerodynamics to zoology. A similar list for the US is F E Kaiser *Translators and translations: services and sources in science and technology* (New York, Special Libraries Association, second edition 1965). More up-to-date though general in subject coverage are lists such as the *Index of members of the Translators' Guild* and the *Professional services directory of the American Translators Association*. An extensive international listing is S Congrat-Butlar *Translation and translators: an international directory and guide* (New York, Bowker, 1979). Unpublished lists of translators living in the area are maintained by many of the large city and regional reference libraries. Some of them have also been published, eg *A register of translators in Lancashire* (Preston, Lancashire Library, 1975). In addition to freelance and part time translators, and of course those working for translation agencies, there are many in full time employment as translators in industry and commerce, government service, etc. Some government departments and other large organizations in the UK will have as many as twenty staff translators. In our subject field translating effort is concentrated for obvious reasons on primary sources, and in particular on periodical articles. Such translations may be undertaken privately by individuals; many of them are produced on a regular basis by institutions (firms, universities, learned societies, research associations, government departments, etc) for their own members and others, eg the schemes operated by the UK Atomic Energy Authority, the Royal Aircraft Establishment, the Central Electricity Generating Board, the Metals Society, and the largest scheme of all, the US Joint Publications Research Service, a centralized service for government offices, agencies, and departments which currently produces a thousand pages of translations every working day. A number of specialist firms produce and issue translations commercially, eg Technicopy.

But translations are expensive, sometimes as much as a hundred times the cost of acquiring the original. This is because the production of a good scientific or technical translation calls for two separate skills: thorough familiarity with the *source* language, out of which the text is to be translated, and subject knowledge. It is also important to have a good literary style in the *target* language, into which the text is to be translated; this is normally the translator's mother-tongue. This combination is comparatively rare, and can command an appropriate remuneration, ranging from perhaps £15 a thousand words for simple matter in the better known languages like French or Spanish to £40 or more a thousand for turning a complex text into, say, Chinese. It is also a very time-consuming process: in the European Communities a rate per translator of five or six pages per day is regarded as the norm. It is a mistake to assume that such translation is merely a matter of word-matching: exact interlingual synonyms are rare, and even where they do exist it often happens that their connotations do not match. A distinction is sometimes drawn between 'literary' and 'non-literary' translations, on the grounds that the specialized vocabulary of non-literary subjects makes word-matching more feasible. Many experienced translators feel that this distinction is irrational and misleading. Even technical translating 'necessarily entails paraphrasing, attention being paid primarily not to what the original says but to its meaning and the way this is expressed'. A good translation, in fact, can even be an improvement on the original insofar as the content may be more clearly expressed and more easily grasped by the reader. Nevertheless it remains true that it is normally regarded as wasteful to translate the same scientific paper twice, yet literary works are translated time and time again.

Some way short of translation but nevertheless exceedingly useful is 'linguistic aid', which is the term the Science Reference Library uses to describe its free service of single sessions of up to thirty minutes per person, where a member of staff competent in the language will attempt an *oral* rendering of the gist of a particular document. Some ten or a dozen languages are encompassed by this service, though the SRL staff are frank enough to admit that so far as subject knowledge is concerned they rely very largely on the enquirer.

Machine translation
Using the vast manipulative power of the computer for machine translation was hailed by many a generation ago as the twentieth century panacea for the woes of Babel. Recently W J Hutchins has written of 'the disastrous and grossly expensive mistakes of the early work on MT. There is perhaps no other scientific enterprise in which so much money has been spent for so little return'. The early protagonists tried to run before they could walk, attempting to program the whole translation process from start to finish. They later confessed that 'it dawned on us that we were not at all sure how this instrument called language, used by us daily since childhood, actually worked'.

In recent years there has been a distinct revival of interest, though sights are now set lower and talk is more of machine-*aided* translation. In 1978 J C Sager of the University of Manchester Institute of Science and Technology told a well-attended ASLIB seminar on Translating and the Computer that 'There is no likelihood or danger of replacing human beings by machines for a very long time to come'. In 1980 the view of the International Translations Centre (see below) was that it is not the case that the computer can do all the work, certainly not in the decades to come'. Nevertheless some promising advances have been made. At the Chinese University of Hong Kong, for example, an automatic system has been used to translate Chinese into English: considerable pre-editing of the Chinese text for the computer is needed, as are a number of other high level interventions of human intelligence, but there is a product of the system on the market in the form of a cover-to-cover translation (see below) of the Chinese journal *Acta mathematic sinica*.

The political dimension too has become much more obvious in recent years: language is an increasingly sensitive issue, and governments faced with translation problems have turned to the computer for help. In countries with a deliberate bilingual policy, eg Canada, vast quantities of material have to be translated as a matter of routine. Even more obviously, international organizations encounter the language barrier, particularly if they have adopted an official multilingual policy. Within the EEC, for instance, translation takes half the budget of the European Parliament and one third of its staff. The Commission is pressing ahead vigorously with a development programme for machine translation, though Eric Gaskell, its Librarian, has expressed scepticism about a solution using automatic methods. There is some progress to report: EURODICAUTOM is an on-line terminology data bank used directly by the translating staff and now publicly accessible via EURO-NET/DIANE. Since 1976 the Commission has been developing SYSTRAN, an automatic translation system from the US where it was used by NASA during the Apollo-Soyuz space project to translate from Russian into English. It does supply rough and ready translations of a kind but considerable post-editing is required by the human translators. The Commission is pinning its hopes on EUROTRA, a new multilingual system covering all the languages of the European Communities and expected to be operational by the mid-1980s.

There are now an encouraging number of completely automatic systems that are fully operational, producing quite usable text without the need for any editing, but these are usually found only in very specific subject fields, eg weather forecasting, where the problems are more limited in extent and therefore more manageable. The production without human intervention of high quality translations in a broad subject field is still beset by formidable linguistic problems: it still seems to many that the more we find out about semantics and syntax the more we realize there is to discover. General opinion at the moment is that fully automatic translation will probably always be incompatible with high quality and a compromise must be accepted.

328

Availability of unpublished translations

Many of the translations described above are published, or at least made available to the general public, but many more are not. Yet there is rarely anything confidential about a translation as such if the original is already available in the open literature, and many of the institutions responsible for commissioning these *ad hoc* translations have no objections to allowing others to consult them. It would be hard to think of a more fruitful field for co-operative effort, and for spreading the high cost of translations over as many users as possible. The first question to be asked, therefore, by the scientist or technologist faced with a paper he wishes to read in a language unfamiliar to him is not 'Who can translate this for me?' but 'Has this been translated before?' It is to answer precisely such a question that in 1951 ASLIB established with government aid at its headquarters in London the Commonwealth (as it was then called) Index of Unpublished Translations (with copies in other Commonwealth capitals) and originally available free to all. With nearly half-a-million entries on cards, growing at a rate of 12,000 per year, this index is the obvious starting point in the search for a translation, and can be consulted by post, phone, or telex, though non-members now pay a charge since the withdrawal of government funds in 1975. It is purely a location index of available translations: ASLIB does not hold copies of the translations, but will put enquirers in touch with the organizations which do. Over 400 sources in the UK and elsewhere co-operate by notifying new translations, mainly from Russian and German into English with the emphasis on science and technology. The success rate for consultations, currently running at about 12,500 a year, is about 7%. A record is kept of each unsuccessful request, and if a further enquiry is made both requesters are informed in case they should wish to co-operate in having the item translated. As a further contribution to the elimination of duplicated effort the index also lists translations in preparation or under consideration. It is still regrettably the case that many organizations do not inform ASLIB of their translations. In some instances, it may be that industry is reluctant to reveal to its competitors its interest in a particular subject; in others the reason may be that the 'translations' are no more than rough-and-ready versions in need of substantial editing before release. A study some years ago showed that the estimated savings in avoidance of duplication exceeded the cost of maintaining the index by over 70%. ASLIB also maintains for its members a Register of Specialist Translators, recording over 220 approved names with both subject and linguistic qualifications.

Translation pools
One step further than the location index is the collection of translations available for consultation, photocopying or loan. Some libraries like the Science Museum Library and the Science Reference Library have built up extensive holdings by a positive policy of acquisition by purchase where possible, but also by encouraging the organizations or individuals responsible for commissioning translations to deposit copies in the library. Such libraries of course maintain indexes to their own collections, and some helpfully mark the original article to indicate that a translation is available. In Britain one of the largest collections, with half-a-million translations, growing at a rate of 18,000 a year, is at the British Library Lending Division. It is claimed that the required translation is available for about a quarter of all requests. Most used to be from Russian, and investigations here and elsewhere indicate that the demand for translations from Russian is probably equal to the total demand from all other languages, but since 1967 it has been broadened in scope to cover translations from all languages into English. Additions to the collection donated by other organizations are listed in the monthly *British reports, translations and theses*. A feature unique to the BLLD is its responsibility for the UK government sponsored translation programme. One aspect of this is the Article and Book Translating Service under which recent articles and books in all languages are translated at a reduced price, in return for a minimal share of editing the draft by the requester. Confined to Russian when it was originally set up in 1957, the service now finds that German, Russian and Japanese are the languages most in demand. About 500 such translations a year are currently produced: they are placed on sale and announced monthly in *British reports, translations and theses*, and in due course are added to the loan collection. A recent development is the collection of a number of these translations into suitable 'packages' for sale: half-a-dozen or so, with a common theme, are chosen from those that are more than two years old, eg *Collected translations on ornithology* (1974). As will be seen, a commendable feature of translation pools and indexes is their close co-operation: this is instanced by the practice of ASLIB and the BLLD of consulting each other before returning a negative reply to an inquiry. A further example is the inclusion by both institutions of the translations of other major translation pools, in particular the major US pools and the International Translations Centre.

In the US responsibility is shared between the National Technical Information Service and the National Translations Center. The former, a government agency (see chapter 14), functions as a collection and distribution centre for all governmental (including government-sponsored) translations. Like the research reports it also handles, these translations

330

are announced, indexed and abstracted in *Government reports announcements and index* twice each month. The National Translations Center at the John Crerar Library in Chicago, formerly operated from 1953 by the Special Libraries Association with government support, serves as the collection centre for all non-governmental translations. Over 200 organizations co-operate by furnishing copies of their translations and the collection is now well over 200,000. These NTC translations are also available from BLLD (and *vice-versa*) under an exchange agreement. A *Consolidated index to translations into English* was published in 1969. New accessions appear in the monthly *Translations register-index*: the translations from GRA and elsewhere (including BLLD) and commercially available translations are also included in the index portion, which cumulates twice a year. More recently it has assumed a role as a referral centre as well as a translation pool.

The International (formerly European) Translations Centre in the Library of the Technological University of Delft in the Netherlands is to some extent a pool but primarily an index. Founded in 1960 under the auspices of what is now OECD (Organization for Economic Co-operation and Development), it is a truly international exchange for a dozen co-operating national translation centres throughout Europe. Of course, a translation perfectly acceptable to a polyglot Swiss scientist or Dutch technologist may not suit the British or American worker. Clearly this limits to some extent the value of this kind of international co-operation across linguistic frontiers.

ITC are the joint publishers (with the Commission of the European Communities and the Centre Nationale de la Recherche Scientifique at Paris) of *World transindex*, a monthly listing of scientific and technical translations, both published and unpublished, of periodical articles and patents mainly from Slavonic and Asiatic languages into Western languages, primarily English, French and German, amounting to about 25,000 a year. Translations *from* Western languages into French have been included since 1978 and into Spanish since 1979, but other 'inter-Western' translations are excluded. It has been estimated that there are about 16,000 such translations each year. *World transindex* is also maintained in machine-readable form at CNRS and is accessible on-line via EURONET/DIANE.

Many major countries have their own national pools, eg CNRS at Paris; the Technische Informationsbibliothek at Hanover. There are also a number of international *subject* pools, eg Transatom, a central information clearing-house established at Brussels in 1960 to collect and share translations of nuclear literature. At this point too could be mentioned the US Joint Publications Research Service translations referred to above: these are listed in the monthly *Transdex index* with its annual cumulations in microform.

331

It is sometimes forgotten that the seeker after scientific and technical information may encounter a double language barrier; not only does he discover that some of the papers he desires to read are in an unfamiliar language, but his preliminary literature search to identify the papers he wants may also be impeded by indexes or bibliographies or data bases in foreign languages. As an aid to scaling this preliminary barrier we now have a number of multilingual thesauri (see chapter 4), multilingual indexes to classification schemes, common command languages for interrogating data bases, and the increasing possibility of machine translation of titles and descriptors.

Cover-to-cover translations
The stock-in-trade of the translation pools are unpublished translations of individual periodical articles (or conference papers, or research reports, or patents, etc), but since 1949 we have seen blossom over five hundred examples of a new kind of journal, translations from cover-to-cover of their foreign counterparts. Well over 90% of extant titles are translations from Russian, but there are examples from half-a-dozen other languages, eg *German chemical engineering, Electrical engineering in Japan, Acta physica sinica*, the physics journal of the Chinese Academy of Sciences. Some are commercially produced by general scientific and technical publishing houses, eg *Radio engineering and electronic physics* (Scripta) or by firms specializing in translations, but most are sponsored by learned or professional societies, eg *Soviet physics* (several series) by the American Institute of Physics, or by governmental agencies, eg *Welding production* and about a dozen others, commissioned by the BLLD from a variety of learned societies, research associations and commercial publishers.

The advantage of such translation is that it eliminates the hit-or-miss selection of articles, and does ensure the availability (if not necessarily the use) of important foreign scientific literature. The aim, of course, is to confine cover-to-cover translation to really worthwhile journals, leaving the translation pools to cope with single articles. It is open to the criticism of wastefulness, for even the best journals must occasionally publish articles of limited value, and there are instances of articles in Russian journals that are themselves translations from the English. But results from a National Science Foundation study 'show that even if only one paper in forty is of general interest it is cheaper to translate all forty than to determine which one to translate'. And of course it greatly facilitates bibliographical location and handling, though confusion is often caused in the frequent cases where the English title is not a direct translation of the original title. Naturally enough they are expensive; often several times the cost of the originals. A more
332

severe disability is the long and perhaps inevitable delay between the appearance of the original and the translated issue: six months is common, and twelve months or more is not unknown. It has also been observed that they suffer a very high death rate, though some are resuscitated after an interval, eg *Applied electrical phenomena* perished in 1970, only to reappear in 1973 as *Electrochemistry in industrial processing and biology*. A regularly updated list is produced jointly by BLLD and ITC: *Journals in translation* has close to a thousand titles, current and defunct, with a KWIC index to the English titles, and a useful guide is C J Himmelsbach and G E Brociner *A guide to scientific and technical journals in translation* (New York, Special Libraries Association, second edition 1972). In its Aids to readers series the Science Reference Library has published *Holdings of journals in translation (cover-to-cover translations and translations of selected articles)* (revised edition 1980), listing almost four hundred.

The titles chosen for cover-to-cover translation are almost invariably primary research journals, but there are a handful of secondary sources also, eg *Cybernetics abstracts, Corrosion control abstracts* both of which are versions of the corresponding sections of *Referativnyi zhurnal*; and *Russian chemical reviews*. Occasionally the journals are not translated from cover-to-cover entirely, but only selected articles are chosen, often with synopses of the articles that have not been translated, eg *Petroleum chemistry USSR*. Not cover-to-cover translations at all, strictly speaking, but conveniently mentioned at this point are those journals containing selected translated articles gathered from a number of originals, eg *Review of Czechoslovak medicine, Steel in the USSR, Geochemistry international, International chemical engineering*; and those comprising the translated tables of contents of a predetermined selection of journals, eg *Science periodicals from mainland China*. As we have already seen (chapter 10), there are also several examples of abstracting journals from overseas which specialize in providing not full translations but English-language abstracts of scientific and technical literature in their own language, eg *Hungarian technical abstracts, Bibliography of agricultural sciences in Japan, China science and technology abstracts*.

Translations of books
By their very nature books contain mainly secondary material and relatively few get translated: UNESCO's annual *Index translationum* is the best guide, and its arrangement (under countries) by UDC allows an approach by subject. It does of course include translations into many languages, but the great majority are into English. Its two great drawbacks are that it is very late in appearing and it is far from complete.

333

Of great assistance in its use is *Index translationum: cumulative index to English translations* (Boston, G K Hall, 1973). A feature of Russian books, however, is that they do frequently contain reports of original work: the BLLD has a limited programme amounting to perhaps a dozen items a year for translating significant titles that fill a definite gap in Western knowledge and making them available in typescript/photocopy/offset editions, eg Y I Zaitsev *From sputnik to space station* (1974), V A Vinokurov *Welding stresses and distortion* (1977). Some 120 of these are listed in *Translated books available from the BLLD* (1976) and its 1978 supplement. The US Joint Publication Research Service have published over a thousand books in translation since 1965, and the Israel Program for Scientific Translations has translated over 800 Russian books since 1960. The American Mathematical Society has a series 'Translations of Mathematical Monographs' with some fifty titles to date, mainly from Russian but with the occasional Chinese example. There are a number of other similar series.

Translations from Japanese

One of the countries that have grown to world stature in science and particularly technology over the last generation is Japan, and this is reflected in the literature, with the number of periodicals in the field trebling in ten years. The translation problem that this could cause, however, is diminished to some extent by the practice of publishing many Japanese journals in English, eg *Japanese journal of microbiology, Journal of biochemistry* (Tokyo). About 40% of original papers in the nuclear field are in English. In science and technology as a whole well over 10% appear in this way, and many of the others have English summaries or English contents pages, eg *Seikagaku* (Biochemistry), *Konchu* (Insects). A list which indicates those journals providing such summaries is P C R Mason *A classified directory of Japanese periodicals: engineering and industrial chemistry* (ASLIB, 1972). The need for an extensive programme of translation has not yet been felt, although the situation is being kept under review, for there are signs that this English-language content is decreasing. Japanese translations are included in the indexes and pools, and there are a handful of cover-to-cover translations, eg *Japanese journal of applied physics in English*.

Translations of patents

Patents can sometimes be located in translation indexes and pools (eg in *World transindex* and *Translations register-index*), but the practice noted earlier (chapter 15) of simultaneous application for a patent in more than one country opens up to the searcher the chance of an

334

English-language equivalent to his foreign patent. South African patents have a particular value here, since they are laid open to public inspection quickly. It should be added, however, that they are relatively few in number (about 130 a week), and available only as photocopies from the South African Patent Office. The existence of an equivalent patent can be traced in, for instance, the concordances published by Derwent Publications, or *Chemical abstracts*, or in the indexes of applicants maintained at the Science Reference Library, or through the INPADOC Patent Family Service. And of course there are several examples of journals providing English-language *abstracts* of foreign patents, eg *French patents abstracts.*

Information about translations
A major problem revealed by the NLL survey of scientists and technologists referred to above is inadequate dissemination of information about the availability of translations. This chapter will have demonstrated that bibliographical control is fairly adequate: what still remains to be done is to ensure that the potential users are made aware of this. In the University of Sheffield survey mentioned earlier 91% of those interviewed had never used any of the indexes to existing translations. Only 17% of the NLL sample knew of *Technical translations* (the predecessor of *Translations register-index*) and only 7% used it; only 21% were aware of the ASLIB index and only 6% used it. And these, it should be remembered, were from a group of workers 76% of whom admitted coming up against the language barrier in the previous year. An even more remarkable revelation is provided by the list of 92 journals that the respondents suggested should be translated cover-to-cover: no fewer than 20 of them *were already available* (and listed in a variety of different sources). A recent survey by ITC found that 58% of the requests received were for translations that in fact had already been *published*, and were therefore readily traceable. The repeat survey in 1979 by BLLD found that only 17% had tried to locate an existing translation, only 11% were aware of *World index of scientific translations* (the predecessor of *World transindex*), and 29% (a much higher proportion than in 1965) had taken no action at all to check the existence of a translation. That ignorance of this kind is not confined to scientists and technologists can be seen from the existence side by side at least for a time of two cover-to-cover translations of the same Russian original, eg *Moscow University physics bulletin* (from the rival publishers Allerton Press, New York, and the Aztec School of Languages, West Acton, Mass); and *Cybernetics* (from Plenum Publishing Corporation, and JPRS).

Further reading

C W Hanson *The foreign language barrier in science and technology* (ASLIB, 1962).

D N Wood 'The foreign-language problem facing scientists and technologists in the United Kingdom — report of a recent survey' *Journal of documentation* 23 1967 117-30.

V K Rangra 'A study of cover to cover English translations of Russian scientific and technical journals' *Annals of library science and documentation* 15 1968 7-23.

J M Lufkin 'What everybody should know about translation' *Special libraries* 60 1969 74-81.

C F Foo-kune 'Japanese scientific and technical periodicals: an analysis of their European-language content' *Journal of documentation* 26 1970 111-9.

H Tybulewicz 'Cover-to-cover translations of Soviet scientific journals' *ASLIB proceedings* 22 1970 55-62.

W J Hutchins and others 'University research and the language barrier' *Journal of librarianship* 3 1971 1-25.

J B Sykes *Technical translator's manual* (ASLIB, 1971).

C M A Knul 'Towards a follow-up care program for *ad hoc* translations' *ASLIB proceedings* 25 1973 220-6.

'Translations in the UK' *ASLIB proceedings* 25 1973 264-7.

F Kertesz 'How to cope with the foreign-language problem: experience gained at a multidisciplinary laboratory' *Journal of the American Society for Information Science* 25 1974 86-104.

G K L Chan 'The foreign language barrier in science and technology' *International library review* 8 1976 317-25.

Commission of the European Communities *Overcoming the language barrier: third European congress on information systems and networks, Luxembourg, 1977* (Munich, Verlag Documentation, 1977).

I Pinchuck *Scientific and technical translation* (Deutsch, 1977).

ASLIB *Translating and the computer: proceedings of a seminar, London 1978* (Amsterdam, North-Holland, 1979).

B J Birch 'Tracking down translations' *ASLIB proceedings* 31 1979 500-11.

S R Ellen 'Survey of foreign language problems facing the research worker' *Interlending review* 7 1979 31-41.

W Glover 'Services provided by Aslib relating to translations' *ASLIB proceedings* 31 1979 525-9.

B Snell 'Electronic translation?' *ASLIB proceedings* 32 1980 179-86.

D van Bergeijk 'Developments in the bibliographical control of translations with particular reference to the activities of the International Translations Centre' *Interlending review* 8 1980 128-31.

D van Bergeijk and M Risseeuw 'The International Translations Centre: the language barrier in the dissemination of scientific information and the role of the ITC' *Journal of information science* 2 1980 37-42.
'Machine aids for translators' *ASLIB proceedings* 33 1981 265-323.

Chapter 18

TRADE LITERATURE

Of all the primary sources of scientific and technical information, probably the most neglected by librarians is the trade literature produced by industrial and commercial companies. Issued in a tremendous variety of forms, ranging from single sheets to multi-volumed sets, by manufacturers or dealers to describe and illustrate their goods or services, of course such 'product data' as it is often called is basically advertising, and its aim is to sell a manufacturer's wares or enhance his prestige. One authoritative estimate is that in the UK alone there are over 30,000 companies producing trade literature. But like the advertisements in technical journals, much trade literature is far removed from the general consumer advertising familiar to us all, and is directed at a specialized audience of some sophistication. Commonly it is very technical: in the case of chemicals, for instance, as Crane points out, it will 'frequently summarize the chemistry of the compounds, give extensive information on physical properties, tell how to use them in various ways, and give references to the literature'. In very many cases it would not be untrue to say that the aim is as much to inform potential customers, users, and others, eg students, teachers, research workers, about commercially available materials, equipment, and processes, as to stimulate sales as such. Increasingly too in recent years manufacturers have been stepping up the informational content of the literature to encourage users themselves to find new applications and new markets for particular products. Minute attention is often given to the production of the various leaflets and folders and brochures, not only to the actual writing (which frequently has to appeal at several technical levels at the same time), but to physical layout and production. In some companies as much as half the advertising budget is devoted to trade literature. It has been estimated that in the field of electronic engineering, for instance, suppliers use 200,000 pages of trade literature to describe their wares. Since 1946 we have had a British Standard Specification *Sizes of manufacturers' trade and technical literature (including recommendations for contents of catalogues)* (BS 1311: 1955), and since 1966 an *American standard for trade catalogues* ASA Z39.6 – 1966).
338

Some of these publications are merely trade *catalogues*, ie basically little more than enumerations of available goods, with brief details and sometimes supplementary indexes or keys. Prices are usually omitted, although separate price lists are sometimes available on request. Often of course this information is of commercial rather than strictly scientific or technical value. Even so they serve a vital function for the scientist and technologist: the chemist who needs a substance with certain characteristics, or the engineer looking for a piece of equipment to perform a specific task finds such catalogues invaluable, for without their aid he may not be able to ascertain easily whether they are available commercially and may thus be obliged to synthesize or build for himself. For industrial designers and architects in particular they are indispensable. But what raises manufacturers' publications to the level of a primary source of scientific and technical information is the continuous flow of sheets, folders, pamphlets, bound and loose-leaf volumes, on new products and processes, theory and applications, containing original data that has not yet appeared in the regular literature. Much of it may never be incorporated into the journals or books, eg quantitative data on the physical, chemical and other properties of industrial (and particularly proprietary) products, and therefore some trade literature remains as a valuable supplement to the reference books and textbooks. As promotional material it is commonly more lavishly presented and illustrated than the average professional society paper, for example, especially with graphs, flow-charts, tables of data, use of colour, etc. This can give it a value irrespective of the language of the text. And as *industrial* literature, it is invariably firmly rooted in practice, thus forming a particularly appropriate supplement to the theory and principles of the research paper, the monograph, and the textbook.

Martin J Thomson of the Science Reference Library has pointed to the interesting link between trade literature and patents: for commercial reasons, as the student will remember from chapter 15, 'applicants for patents generally . . . disclose only the bare minimum necessary to secure the patent monopoly rights. Once the patent has been granted and the invention has been put into production, the patentees can afford to be more liberal with their information. Product brochures or company periodicals are excellent vehicles for this information. Accordingly such publications may contain more details and data than the original patent specification . . . Furthermore, if a company develops an invention or an idea which it does not wish to patent, it can protect its future rights to exploit the invention by publishing details in the company's literature so that they form part of the prior art'.

It would be a mistake to assume that trade literature has only ephemeral importance. It is being increasingly realized how valuable

339

are retrospective collections of trade literature for studies such as industrial archeology, business history, and the history of science and technology, and perhaps even social history and local history also. It has frequently been discovered that contemporary manufacturers' brochures are often the only source of information on various museum objects or industrial relics, particularly of the nineteenth and early twentieth centuries. Where such catalogues and leaflets are illustrated (as they often are) this hitherto neglected source is proving even more useful. They are now appearing with increasing frequency in the lists of antiquarian booksellers: recently on offer was a 24-page 1881 manufacturer's catalogue of manual fire engines, fire escapes, and fire extinguishing appliances. British Leyland have a very interesting collection of early trade literature on the car industry, and the collection at the Science Reference Library goes back to the 1850s. In May 1978 the Manchester Polytechnic Library mounted an exhibition of historical trade catalogues from its collection and published a brief illustrated guide. Unique so far is L B Romaine *A guide to American trade catalogs, 1744-1900* (New York, Bowker, 1960). An interesting parallel from the field of social history is the interest in reprinting illustrated mail order catalogues from the same period, eg *Sears, Roebuck catalog for 1897* (New York, Chelsea House, 1968) and *The Army and Navy Stores catalogue, 1907* (David and Charles, 1969). And when it comes to studying the history of a firm the value of a file of their trade publications needs no pointing out. Many firms of course have had their history written, either by an independent historian (with or without the firm's blessing), eg W Manchester *The house of Krupp* (Michael Joseph, 1969), or by someone specially commissioned for the task, eg C H Wilson *The history of Unilever* (Cassell, 1954), and published through the normal book-trade channels. Such works are obviously not part of trade literature. There are, however, thousands of examples of company histories, ranging from pamphlets to multi-volumed compilations, that are not issued in this way, but are published by the companies themselves for prestige or other advertising purposes. Many are admirable, scholarly publications, of great value to the economic historian in particular, but all are perhaps best regarded as a kind of trade literature, eg W & T Avery Ltd *The Avery business, 1730-1918* (1949). A pointer to the size of this field is the 480-page bibliography by J M Bellamy *Yorkshire business histories* (Crosby Lockwood, 1970). A more recent listing covering some three thousand firms in the North of England is D J Rowe *Northern business histories: a bibliography* (Library Association Reference, Special and Information Section, 1979).

It should be mentioned that not all trade literature is published by individual firms: trade associations sometimes issue catalogues listing
340

their members' products, eg *British chemicals and their manufacturers* (Association of British Chemical Manufacturers). Works of this kind are very similar in layout and use to the conventional trade directories (see chapter 6). In fact trade directories are often regarded as a form of trade literature, though their publication in most cases through normal book-trade channels would seem to indicate that they are best looked on (as in this work) as a specialized form of directory. As has been mentioned, however, some trade directories do make a feature of including what is undeniably trade literature, eg the manufacturers' data sheets and catalogue pages describing products and services in *Concrete year book.*

Forms of trade literature
The typical piece of trade literature is a folder or pamphlet, glossily produced but commonly of an awkward non-standard size, and there is no doubt that thousands of such pieces are distributed by manufacturers daily. What distinguishes such publications from general advertising is the wealth of technical detail and the very solid body of information conveyed. Substantial pamphlets with dozens of pages of well-written text and diagrams are common, eg Shell Chemicals Ltd *Building with plastics* (1965), Ferodo Ltd *Friction materials for engineers* (1961), Vinyl Products Ltd *Emulsion paint problems – causes and cures* [1966], and booklets of a hundred or more pages are frequent, eg NV Philips *Audio amplifier systems* (1970), ICI Ltd Dyestuffs Division *Rubber chemicals for footwear* (Manchester, 1961). They may be no more than a single page, eg the data sheets on British Oxygen Chemicals Ltd range of products issued as *BOC information.* Loose-leaf format has obvious advantages for keeping such works always up-to-date, eg *Colt ventilation and heating manual.* One manufacturer whose literature fills fifteen loose-leaf binders even goes so far as to provide a set of bookshelves to house them. Some catalogues appear in serial form, eg *Elastomers notebook* (Du Pont), though they are usually too irregular to rank as journals.

Titles of this kind really do demonstrate how these works try to inform as well as persuade, and even more substantial examples can be found, some of them almost the equivalent of a standard work in their field, eg C E A Shanahan *Chemical analysis of flat rolled steel products* (Richard Thomas and Baldwins Ltd, 1961). ICI Ltd Dyestuffs Division *The dyeing of nylon textiles* (Manchester, 1962). In some cases, they are indistinguishable from regular textbooks or monographs, save for the fact that they are issued by an industrial firm and not a publishing house, eg Sir Joseph Lockwood *Flour milling* (Stockport, Henry Simon Ltd, fourth edition 1960) is the basic text on the subject. A number have attained the status of recognized reference books in their fields, eg Yorkshire Engineering Supplies Ltd *Bronze: a reference book* (Leeds, [1962]), *Alcoa aluminium handbook* (Pittsburgh, Aluminium Company of America, 1962). The number of trade publications in the form of bibliographies is a further indication of the sophisticated approach to the user, eg ICI Fibres Ltd *Select bibliography on nylon* (Pontypool, 1966), AEI (Manchester) Ltd *Bibliography on mass spectrometry, 1938-1957 inclusive* (1961), and *Radiographic abstracts* (Ilford Ltd). The amount of literature produced by a number of the major companies is so great that some have felt it necessary to produce bibliographies of their own publications, eg ICI Ltd Dyestuffs Division *Technical publications subject index up to June 1963* (Manchester, seventh edition 1964): some indication of the range of material is given by the list of series covered – sales circulars, chemicals pamphlets, technical information series, technical circulars, pattern cards, swatches, manuals.

A special form of this literature is the customer's handbook, or maintenance manual, service manual, or user's guide, as they are variously called. These are basically textbooks and/or reference books prepared by the manufacturer for his customers on how to install or operate or maintain or repair his particular equipment. Perhaps the best known examples of this type are the workshop manuals for the various makes of cars, but there are similar compilations for most kinds of scientific and technical hardware, such as electron microscopes, furnaces, lathes, centrifuges, etc. Some are necessarily very elaborate, eg the series of volumes known as the IBM Systems Reference Library, covering the hardware and software of all IBM computers and peripherals.

House journals

One of the most distinctive forms of trade literature is the periodical published by a particular industrial or commercial firm or public corporation, eg *Southern electricity* (Southern Electricity Board), *Shell in agriculture, Atom news* (UKAEA), *Dupont magazine, Ciba-Geigy review, Welder* (British Oxygen Co). Known as house journals or house magazines (also as house organs in USA), like the other forms of trade literature they are basically advertising publications, but in some instances they also have great information value. In the United Kingdom the total number probably approaches two thousand, and in the United States perhaps five times that number, although a large proportion are designed for internal consumption, ie by the companies' own employees or shareholders, and may indeed be restricted to them, eg *Vickers news, The lamp* (Exxon Corporation), *Monsanto magazine, Nobel times, Shell UK oil news, British Aerospace news.*

These serve the function of newspapers within a firm, and contain information about, for instance, personnel changes, suggestion schemes, expansion plans, contracts won, although a number do have roles other than communication and morale-building. The importance attached to them by the companies and the care with which they are directed at their particular audiences can be seen in the fact that the Esso Petroleum Co Ltd have a whole range of such journals: *Esso oilways international, Esso newsline, Esso magazine, Esso air world,* and *Esso farmer.* The international character of many modern companies can be seen reflected in the separate language editions of a number of their journals, eg *Philips technical review* appears in Dutch, English, French, and German. Of course many of them have little scientific or technical interest, eg bank reviews.

The journals of most concern to the student are those which circulate outside the companies and these fall into three main categories:

a *Prestige*: usually aimed at the non-technical reader, and often lavishly produced, these are not necessarily concerned with a company's own products, but more with creating goodwill and preserving a favourable public image, eg *Ciba-Geigy journal, Aramco world magazine, Oil lifestream of progress* (Caltex Petroleum Corporation), and one of the best of all, the now sadly defunct *Far and wide* (Guest Keen and Nettlefolds).

b *Scientific/technical*: these are clearly aimed at a knowledgeable audience and qualitatively may be the equal of some of the research and technical journals described in chapter 9, eg *IBM journal of research and development, Sulzer technical review, X-ray focus* (Ilford Ltd), *Point to point communication* (Marconi), *Platinum metals review* (Johnson Matthey), *Steel research* (British Steel Corporation), *GEC journal of science and technology, The Bell System technical journal,* and one of the best known and widely respected of all, *Endeavour* (ICI Ltd), published quarterly in five separate language editions. Sponsored by the National Science Foundation, a joint research team from the Graduate Library Schools of Drexel University in the US and Antwerp in Belgium located 113 titles in the US that they judged 'scientifically or technically valuable', 115 in the UK, and 38 in France. They also found that about 4 or 5% of the articles are republished in the regular journals.

c *Popular*: these are similar in appeal to the commercially produced popular subject periodicals described in chapter 9, eg *Decorating review* (Wall Paper Manufacturers Ltd), *Air BP* (British Petroleum Co Ltd), *Aerial* (Marconi Co Ltd). Motoring journals are particularly well represented, eg *Vauxhall motorist, Ford news, Austin-Morris express, Specialist car* (British Leyland).

And yet when the joint research team reported on their study of

343

house journals they called their article 'The hidden literature'. They had concluded that 'House journals are generally regarded as, and proved to be, an extremely elusive form of literature'. Only a small proportion can be found in standard bibliographies such as the *World list of scientific periodicals*, the *British union-catalogue of periodicals*, or the *Union list of serials*.

As might be expected, many house journals are fly-by-night publications, but sound, respectable and long-established titles are also found, eg *Ford times* (1908-), *The lamp* (1918-), *Marconi review* (1928-), *Shell aviation news* (1931-). Many are equipped with annual indexes, eg *Electrical components and applications* (formerly *Mullard technical communications*), and a cumulated index is occasionally found, eg *Endeavour*, 1942-61. They manifest all the bibliographical idiosyncrasies that other periodicals show, but to an extreme degree. They change their names without notice and without comment; they use the same title for different publications; issues appear without a number or even a date; many are irregular and come and go spasmodically; and so on.

Some eight hundred titles are listed in the section 'House journals' in the UK volume of the annual *Benn's press directory*, and they are included on a selective basis in *Ulrich's International periodicals directory* — indicated 'house organ'. The New York Chapter of the Special Libraries Association prepared a useful listing in 1971, *Technical house organs*, but by far the best list is the directory of titles that takes up most of I J Haberer 'House journals' in *Progress in library science* 1 1967 1-96. Separately published directories are British Association of Industrial Editors *British house journals* (second edition 1962); *Working press of the nation: volume 5 Internal publications directory* (Burlington, Iowa, National Research Bureau), an annual which has replaced the *Gebbie house magazine directory*. Of particular value inasmuch as it provides a location is D King and M Thomson *House journals held by the Science Reference Library* (Science Reference Library, 1978), listing over seven hundred titles. A handful of the more important titles are included in indexing and abstracting services, but coverage is often selective and very patchy. An object lesson on the bibliographical difficulties in this field is provided by the section 'House organs' in the *Standard periodical directory* of 1973: titles and addresses cannot be relied on, several of the listed companies can no longer be traced, and some of the journals listed ceased publication over forty years ago.

Problems of trade literature
The reasons for the widespread neglect of trade literature in all save a few libraries are not far to seek, for it abounds with problems —
344

of acquisition, arrangement, retrieval, and use.

Since virtually all such literature (including house journals) is available free of charge from the manufacturer, simply for the asking, the student might well wonder whence comes the acquisition problem. In point of fact, it is this very availability which causes one of the major difficulties: like research reports trade literature is outside the usual source of literature supply, the book trade. Booksellers are naturally reluctant to deal on a large scale with producers of literature other than regular publishers, and even more disinclined to handle free material. This means that librarians are obliged to employ direct or do-it-yourself acquisition procedures, by first identifying appropriate manufacturers from trade directories, advertisements, and other sources, and then writing either for particular items or with a request to be placed on the mailing list. Apart from the lists of house journals just mentioned, bibliographical control does not exist, for trade literature is either ignored or deliberately excluded from most current bibliographical lists (including abstracting and indexing services). The quarterly *COPNIP list* published by the Committee on Pharmaceutical Nonserial Industrial Publications of the Special Libraries Association is probably a unique example of a current list devoted to trade literature. The best sources of information on new trade publications are the scientific and technical periodicals, a number of which make a feature of noticing or at least listing new titles, eg *Engineering, Metallurgia, R & D, Chemical week*. In some cases the notice includes a photograph of the front cover of the trade publication, eg *Engineering materials and design*. As part of a recent detailed study of the use of trade literature the Science Reference Library produced a list of such periodicals. Advertisements also commonly indicate the availability of further information on request, and a number of periodicals provide pre-paid tear-out postcards for their readers to use in sending for such literature. Commonly each advertisement is numbered so that all that is required is for the reader to encircle the appropriate number and post the card, eg *Chemical and engineering news, Modern plastics, Textile month*. Scarcely ever is trade literature reviewed. Librarians in industry are better placed, as a rule, since they can often rely on exchange arrangements, and contacts with visiting salesmen.

But acquisition is a simple task compared to the organization of a collection. Despite the British and American standards mentioned earlier the variety of sizes and shapes encountered is immense, no doubt because to a manufacturer whose products have to compete with rivals a publication in a striking or unusual non-standard format has a head-start, and for a collection composed mainly of folders and pamphlets even the simple question of storage needs careful thought. As to arrangement and indexing, opinions differ, and this is not the place

345

for discussion: it will suffice to indicate the nature of the problem. Ideally, any system should provide for access by name of manufacturer, name of product, trade name, and subject; yet one trade catalogue may describe hundreds, even thousands of different products. And perhaps more than any other form of scientific and technical literature the information content, and therefore its value, varies unpredictably. A particularly acute problem is maintenance: it is true that the frequent loose-leaf volumes are very efficient for keeping up to date, but only at the expense of a deal of time spent in filing; and the reluctance of manufacturers to date their literature makes the task of weeding even more difficult. The provision of accurate and up-to-date prices — understandably regarded as essential by users making comparisons — is a particular headache. Characteristically too, those who consult trade literature will not tolerate delay. And because access to the information in trade literature is denied by the indexing and abstracting services, adequate arrangement, indexing and maintenance is more than usually crucial: without it a collection is virtually unusable. Trade journals are obviously simpler to handle: the most common method is to treat them in the same way as other periodicals, incorporating them in the general sequence. This does have the effect, however, of separating them from the rest of the trade literature collection.

Encouragingly, libraries are paying more attention to trade literature: there are perhaps a hundred collections in the UK at present. A number of countries have established national trade literature collections, eg Czechoslovakia, Poland, Rumania. In Britain, as has been mentioned, the Science Reference Library has been giving particular thought to trade literature in the last few years, and has completed a survey of 65 trade literature collections. It has also greatly strengthened and reorganized its own substantial collection, which now contains literature from over 8,000 companies, mainly British, with a further 16,000 British and foreign companies whose publications are available in microform. There still lingers what is perhaps an understandable suspicion among some librarians about what is basically advertising matter, but they would do well to remember that in those libraries with trade literature collections a number of surveys have indicated that users find trade catalogues more useful than patents.

Commercially available trade literature services
A partial solution to the librarian's problems is to subscribe to one of the 'package libraries' or 'catalogue services' which are now increasingly available. Known also as product information services, for an annual fee they will provide within a particular subject field an indexed collection of trade literature in standard format: the newer services will
346

also guarantee to maintain the collection, usually on a monthly basis. A back-up enquiry service is usually available where the required item is not already in the packaged collection. The idea of such services is far from new, but the entrepreneurial drive and efficiency of some of the recently founded examples have certainly persuaded a number of libraries of the value of trade catalogues, university libraries in particular, where previously trade catalogues had been largely ignored.

The conception of assembling in standard format the catalogues of several manufacturers goes back at least fifty years, as can be seen in the publications of organizations like the Standard Catalogue Co Ltd of London; eg the four volumes of the *Architects standard catalogues* for 1981 comprise over three thousand pages on building materials, components, and services, one third of which are the manufacturers' *own* leaflets and brochures. In the United States, the Reinhold Publishing Corporation have for many years been providing a similar consolidated bound set of manufacturers' literature with the title *Chemical engineering catalog*. Best known of all is probably the massive series of volumes from McGraw-Hill, *Sweet's catalog file*, covering mainly engineering and building.

Perhaps because the traditional building trade is less a manufacturing industry than an industry assembling already manufactured articles (door handles, double glazing, drain pipes), its trade catalogues have always been of great importance. The 1977 survey by the Science Reference Library of UK trade literature collections found that a third were devoted to the construction industry. This accounts in part for the commercial success of perhaps the best known of the British trade catalogue services, Barbour Index Ltd, with over 150,000 pages of manufacturers' literature from 4,500 companies in 92 binders. More dynamic in approach than the older services, the 'package libraries', such as this and those covering 16,000 companies in the engineering field issued by Technical Indexes Ltd, are more flexible, more extensive, and better indexed. Added to this is the frequent servicing and a regular schedule for complete replacement. The price to the subscriber is kept artificially low by persuading many manufacturers not only to standardize the format of their literature and provide it free of charge but actually to pay for having it distributed.

In 1963 the Microcard Corporation started a service to supply in the form of 6in by 4in microfiches some 14,000 catalogues of the companies listed in *Thomas' register of American manufacturers* (see chapter 6). Use has demonstrated the aptness of microforms for rapidly changing collections of this kind, particularly now that the availability of self-threading cassette microfilm has extended the range. There are now on the market a number of product information services using microforms, and some former 'hard-copy' services have added

347

microfiche or microfilm also, and some have switched almost entirely to microforms, eg Technical Indexes Ltd. Further developments in microform technology led to systems using 6in by 4in ultrafiches with a 1:200 reduction ratio, giving as much as three thousand pages per fiche, though these have not proved as successful as was hoped. Representing a halfway house between full text and microform is the *Master library of water well equipment and maintenance data* (Trevose, Pa, Micro-Graphix) which reproduces over 10,000 pages of manufacturers' literature, reduced so that each page of the volume contains nine mini pages, to be read with the aid of the magnifying glass supplied.

These services are not without their critics: most of them are selective, some have proved to be very short-lived, and those which aim for complete coverage rarely achieve it: trade literature in serial form, for instance (see above), is usually ignored. Not all manufacturers are willing both to supply literature and pay for the privilege (although some systems make no charge to manufacturers). One criticism in particular levelled at microform is its lack of colour, often so important in trade literature.

Further reading

E M Baer and Herman Skolnik 'House organs and trade publications as information sources' American Chemical Society *Searching the chemical literature* (Washington, 1961) 127-35.

J Goodwin 'The trade literature collection of the Smithsonian Library' *Special libraries* 57 1966 581-3.

E B Smith 'Trade literature: its value, organization and exploitation' W L Saunders *The provision and use of library and documentation services* (Pergamon, 1966) 29-54.

R A Wall 'Trade literature problems' *Engineer* 225 1968 453-4, 489-91.

J L Bailey 'Trade literature, its nature, significance, and treatment' *Business archives* December 1969 19-23.

D Kennington 'Product information services – some comparisons' *ASLIB proceedings* 21 1969 312-6.

R Barber 'A retrieval system for product data' *Information scientist* 4 1970 3-10.

N Kelbrick 'Trade literature as a library material' *Library Association record* 73 1971 65-7.

M Ford 'The Technical Indexes system for the control of trade literature' *ASLIB proceedings* 24 1972 284-92.

M C Drott and others 'The hidden literature: the scientific journals of industry' *ASLIB proceedings* 27 1975 376-84.

M Wilden-Hart 'The acquisition of trade literature' *ARLIS newsletter* no 26 (March) 1976 3-7.

M J Thomson *Trade literature: a review and a survey* (Science Reference Library, 1977).

I L Travis 'Trade literature at the National Museum of History and Technology' *Special libraries* 70 1979 272-80.

Chapter 19

THESES AND RESEARCH IN PROGRESS

It is common knowledge that universities normally require a candidate for a research degree to write a thesis (sometimes called a dissertation) under the supervision of the academic staff, and that before the degree can be awarded the thesis has to be examined and approved by a recognized authority in the field. When it is recalled that in the UK two thirds of all theses are in science and technology, that in the US the number of doctoral dissertations written each year more than tripled between 1960 and 1980, and that doctoral theses are usually required to show evidence of original work, it will be seen that they could make a substantial contribution to the primary literature. It is true that their main function is to allow the candidate to demonstrate his grasp both of his subject and of research method, but perhaps half of them do appear later (often condensed or amended) as articles in learned journals or conference papers or even monographs, which suggests they do have a value beyond the walls of the university. A thesis is not the same thing as a monograph or a research report of course; it is part of the university examination process. But C J Boyer, an obvious protagonist of the doctoral dissertation as an information source, reminded us in his own doctoral dissertation on that very topic that 'Each dissertation represents a refereed paper, supervised by an adviser whose competence in the field is acknowledged by the position he holds within the university and subject to the criticism and guidance of two to six other similarly distinguished individuals. The research completed under such stringent conditions is surely of no less value than that completed in workshops and laboratories outside the halls of academe'.

What is not sufficiently realized is that the original theses themselves are available. Britain and America do not follow the practice of universities on the continent of Europe which often require the candidate to have his thesis printed (sometimes as many as several hundred copies, which are then found very useful for exchange purposes), and typescript copies only are preserved, but these are usually available on interlibrary loan, or for microfilming or photocopying, or at the very least for consultation in the university library. Even in those cases where the research results have been published in the literature, consul-
350

tation of the thesis will often bring to light background information and experimental detail, and of course the thesis is at the disposal of investigators well in advance of publication.

There is much evidence to show neglect of thesis literature by libraries, and by scientists and technologists themselves. Boyer complained of the 'almost total void of knowledge about the dissertation as a vehicle for the dissemination of research results'. Collections of theses in research libraries in the UK and US are negligible, with most universities acquiring only their own. Citations of dissertations make up no more than a tiny percentage of those in the scientific literature. Boyer contrasted the 15,000 US doctoral dissertations in the pure and applied sciences produced in 1970 with the number of scholarly book titles published in the same year, perhaps 5,000 all told.

All this is not to say that every such thesis deserves study. Many, perhaps most, deservedly never see the light of day again once the examiners have given their approval and the degree is awarded. Indeed the thesis as a medium of scholarship has been the object of severe criticism by some. Robert P Armstrong, for example, a professor of anthropology, has catalogued 'the dissertations' six deadly sins', accusing them of 'amateurism, redundancy, trivialization, specializationism, reductionism and arrogance'.

Still less can it be argued that every thesis deserves publication: historically, of course, the thesis was written to be defended orally, not published. The scientific or technical monograph that is merely a 'warmed-over dissertation' rouses particular ire. Writing of university presses, which are often expected to publish dissertations, Henri Peyre tells us 'There is discontent in the staff of many presses with the unreadability, turgidness, unwieldy length and occasional poverty of substance in many of the manuscripts proposed to them by professors and their PhDs'. In such cases perhaps we ought to be grateful for what the budding authors criticize as the publication 'barrier'. This Armstrong sees as 'an institutionalized inhibition upon the erratic and chaotic development of literary forms' persisting not through conspiracy but 'by virtue of a kind of social contract among publishers, writers, and readers'.

Nevertheless, great strides have been made in recent years not only in making more accessible (usually in microform) for those who want them the vast majority of theses that remain unpublished, but, equally important, in bringing their existence to the notice of potential users. Nevertheless, each university is still very much a law unto itself. Some publish regular lists, either separately, eg the annual *Titles of dissertations approved for . . . degrees in the University of Cambridge,* Massachusetts Institute of Technology *Publications and theses: list for the academic year*; or in an annual report or calendar. Others used to publish lists that are now discontinued, but many publish no

351

list at all. In a few subject fields investigators are lucky enough to have a listing of their own, eg the annual *List of theses in preparation and of theses completed in Departments of Geography in universities in the British Isles.* Some abstracting services do attempt to include them, eg *Chemical abstracts, Computing reviews*; and in some fields the specialist journals may list important theses of interest, eg in the January issue of *Chemical engineering progress.* For the most part, however, the basic current bibliographies are those compiled on a national basis. France and Germany have had comprehensive official listings for nearly a hundred years, but the only nation-wide bibliography in Britain is the twice-yearly ASLIB *Index to theses accepted for higher degrees by the universities of Great Britain and Ireland and the Council for National Academic Awards* which started in 1950 and now contains over ten thousand entries a year. Since 1971 some British doctoral theses have been listed by the British Library Lending Division in the monthly now entitled *British reports, translations and theses.* It is interesting to note, however, that when the Agriculture Information Review Committee admitted in 1979 that it was 'not sure that these indexes are familiar and accessible to potential agriculturalist users', it was probably understating the truth.

There is a particular difficulty experienced by users of these lists and indexes inasmuch as they provide no more than the thesis title as a guide to its content. There are grounds for believing that the titles chosen by the authors are less informative than they could be, for while they are all serious (and not eye-catching or fanciful, like the titles of some periodical articles), many are ambiguous, or misleading, or simply non-committal. The special problem this raises, of course, is that the potential user cannot just glance at the original in his library, as he can with a conference paper or journal article. With a thesis he has no alternative but to set in motion the quite expensive special machinery for obtaining access to a copy. This is why the *abstracts* of their theses that some universities publish are so particularly valuable, eg the University of Glasgow's annual *Summaries of theses approved for higher degrees in the faculties of science and engineering.* Since it is almost an invariable rule that candidates should provide an abstract when submitting their thesis, the suggestion has been mooted by ASLIB that these abstracts should be collected and published, at least for science and technology, as a quarterly journal. Though this has never materialized, if participating universities provide ASLIB with abstracts they are made available on microfiche simultaneously with each issue of the *Index to theses.*

In this respect the US has shown the way with the remarkable *Dissertation abstracts international*, a monthly compilation (the separately published section B covers the sciences and engineering) of author

352

abstracts of doctoral dissertations from over four hundred co-operating universities, mainly in the US and Canada. Each issue is arranged by subject, but contains an author index and a keyword title index; these are cumulated annually. A retrospective index covering the previous 29 years was separately published in nine volumes in 1970. Developed by University Microfilms, the service will provide on demand copies (Xerox or microform) of the dissertations thus abstracted. Not all US and Canadian doctoral theses are included (the University of Chicago and Massachusetts Institute of Technology have only recently begun to co-operate), but it is claimed that they are all listed in the annual *American doctoral dissertations.* The whole system is now computerized and searches of the data base can be made on-line or can be requested direct from the publisher.

Attempts were made with varying success to persuade universities in Europe to participate, but eventually in 1976 section C began to appear quarterly as *European abstracts,* now including some six thousand doctoral and post-doctoral dissertations a year. The company's earlier ambitious plans for a copy service matching its US service had to be scaled down in the face of the extraordinarily complicated and traditional practices of Western European universities, in some cases reaching back to the Middle Ages. The decision was taken to aim for bibliographical control first, a major task in itself with over three hundred institutions in over a dozen countries granting some 40,000 degrees each year. And, as was noted above, there is already a widespread international exchange system for conventionally printed editions of dissertations. Currently something over a hundred institutions are co-operating, with participation varying from virtually total in the case of Austrian universities to feeble in the case of the UK, with no more than half-a-dozen universities and a similar number of polytechnics in the scheme. British universities adopt a much more proprietorial stance over theses than either the US or the continental European countries, and now that the BLLD acquires all the *Dissertation abstracts international* theses it is paradoxically far easier for British libraries to borrow an American thesis than a British one. In the face of opposition from many universities the BLLD has persevered valiantly in its attempts to make British doctoral theses more widely accessible by offering to microfilm them and make copies available through their normal loan and photocopy services. The scheme has achieved increasing success, with about forty universities and the Council for National Academic Awards now co-operating, and currently about two-thirds of the UK doctoral theses are being covered. BLLD now holds a total of well over 350,000 theses.

Retrospective access to theses has also made great strides. The yawning bibliographical gap prior to the start of the ASLIB *Index* in

1950 has now been closed by the *Retrospective index to theses of Great Britain and Ireland, 1716-1950* (Oxford, European Bibliographical Centre, 1975-7). Over 40,000 theses are covered by the five volumes, four of which are concerned with science and technology. Coverage is somewhat patchy, however, dependent as it was on the returns submitted by some twenty universities, and the indexing is based on the titles alone rather than examination of the theses themselves. Some individual universities have published retrospective lists of their own, eg Sheffield University *Index to theses, 1920-1970* (1974). The BLLD has negotiated an agreement with the Center for Research Libraries, Chicago, for access to a collection of 600,000 foreign dissertations of the last eighty or ninety years, mostly from Western Europe. Bibliographic access to the corresponding American theses has been transformed by the publishers of *Dissertation abstracts international*, whose *Comprehensive dissertation index, 1861-1972* and its *Supplements, 1973-* attempts to cover every doctoral dissertation from degree-giving institutions in the United States. Available either on microfiche or in now more than fifty bound volumes, it lists the over half a million theses from 390 institutions alphabetically under title keywords in subject volumes, of which perhaps half are scientific or technical, eg volumes 1 to 4 cover chemistry. Most of the other retrospective bibliographies of American theses are subject lists, eg M L Marckworth and others *Dissertations in physics: an indexed bibliography of all doctoral theses accepted by American universities, 1861-1959* (Stanford UP, 1961), J and H Chronic *Bibliography of theses written for advanced degrees in geology and related sciences at universities and colleges in the United States and Canada through 1957* (Boulder, Pruett, 1958), with its continuations for 1958-63, 1964, 1965-6, and 1967-70. Australia has *Union list of higher degree theses in Australian university libraries* (Hobart, University of Tasmania Library, 1967) and New Zealand has D L Jenkins *Union list of theses of the university of New Zealand, 1910-1954* (Wellington, New Zealand Library Association, 1956) and its supplements. To track down such listings we now have M M Reynolds *Guides to theses and dissertations: an annotated bibliography of bibliographies* (Detroit, Gale, 1975), which identifies and describes over 2,000.

Master's theses present something of a problem: to begin with, they are not always required at the master's level, and in many continental European universities master's degrees, as such, are not granted. In research value they cannot compare with the doctoral thesis and are excluded from most bibliographies (the ASLIB *Index* is an exception). Yet they do exist, and access to them is often required. A highly selective listing of mainly US titles appears in University Microfilms twice-yearly *Masters abstracts*, comprising some two thousand titles a

354

year that have been recommended for microform publication by the candidate's head of department or graduate school dean, but there is an attempt at a more comprehensive bibliography with some 10,000 titles a year from 250 universities in the annual *Master's theses in the pure and applied sciences, accepted by colleges and universities of the United States and Canada*, compiled by Purdue University.

There are of course, limited numbers of non-university theses. Certain professional qualifications are awarded on the basis of a thesis submission. In the United Kingdom too in recent years we have seen an increasing number of higher degrees awarded by the Council for National Academic Awards on the basis of a thesis, usually submitted by a candidate studying at a polytechnic. These are listed, as has been noted, in the ASLIB *Index* and to a a limited extent in *European abstracts*.

One disturbing restriction of recent years, with the growth of university research sponsored by outside bodies, particularly industrial firms, has been the successful insistence by some of these sponsors that the results of the research are to be kept secret. This of course is a negation of all a university stands for, and in response some universities have altered their higher degree regulations so as to preclude such restrictions.

Research in progress

The requirement that a doctoral thesis must incorporate original work places on the candidate an obligation to see that his research does not duplicate what has already been done. This, of course, implies a literature search, but it also means that the candidate must try to ensure he does not choose an area that someone else has already chosen, even though he may not yet have published his results. In other words, he needs to know about research *in progress*. Of course, not only aspirants to a higher degree feel this need. Such is the pace of scientific and technical advance that this kind of research intelligence is important for all workers at what has been described as the 'cutting edge of science'. This is not merely to avoid duplication. A knowledge of research in progress in a field allied to one's own can lead to fruitful collaboration and mutual assistance. It may often assist in identifying trends and forecasting future developments, particularly important for scientists and technologists in industry. It can sometimes indicate possible sources of support for research in a particular field. It locates experts and centres of expertise in specific subjects. It may involve no more than an exchange of visits or of letters, but at the very least it is a point of contact and source of information. Yet in a modern industrial state the pattern of research can be very complicated. Current research and development expenditure in Britain is running at over £4,000 millions per year, and in the United States it exceeds $50,000 millions. Though the proportion has been declining in recent years, in both countries something close to half is supported from public funds, and, if one excepts defence research, is not too difficult to find out about. The various bodies awarding research grants usually publish lists on a regular basis, eg US Public Health Service annual *Research grant index*, the Science and Engineering Research Council annual *List of research grants*. Those arms of state with research programmes of their own almost always publish reports, eg Agricultural Research Council *Index of agricultural and food research*; Health and Safety Executive *Health and safety research*. An example of an official overall national guide is the annual Cabinet Office *Government research and development: a guide to sources of information*.

Subject oriented lists are occasionally encountered, eg American Chemical Society *Directory of graduate research*, published biennially, which covers research in chemistry at US universities; a similar British guide, Institute of Physics *Research fields in physics at United Kingdom universities and polytechnics* (fifth edition 1978); the series of annual guides produced by the Department of the Environment, such as *Register of research: Part 1 Building and construction*, which cover ongoing research in universities, government organizations, and private companies.

This last paragraph reminds us of the importance of the university sector in scientific and technological research. As a guide we have the three-volume *Research in British universities, polytechnics and colleges*, compiled by a special unit within BLLD, and the successor, though not without a deal of travail and a three-year gap, to *Scientific research in British universities* which sank under severe fire in 1975. Planned to appear annually, the first two volumes cover the physical sciences and the biological sciences respectively. It remains to be seen whether it will be an improvement on its unfortunate predecessor, which used to contain only sketchy information on each project, many of which were in fact little more than pious intentions. The British Library's original plan to include all UK research has been abandoned with general regret in the face of the problems, despite the optimistic recommendations in the feasibility study that preceded its publication.

Considerable help is furnished by directories of scientific and technical organizations, particularly those in the form of a research guide (see chapter 6). Access to information about research in progress in industry is less easy to come by. Such are the pressures of commercial competition that tight security measures are not unusual. Guides like *Industrial research in Britain* (Guernsey, Hodgson, ninth edition 1980) do indicate areas of research interest of industrial companies, and

356

approaches by genuine enquirers from outside the company are sometimes entertained.

A dream for some years has been a central index (possibly computer-based) containing details of all current scientific and technical research projects within a particular country. From this index enquiries could be answered and lists produced, as and when needed. Proponents of such a scheme argue that its value will extend beyond individual scientists anxious to keep abreast of current research, and will provide information, so far lacking in convenient compass, for assessing a nation's research and development programme. UNISIST has shown particular interest in this concept, and guidelines have been prepared on the conduct of a national inventory of current research and development projects.

In this field the US has shown the way with the Science Information Exchange, established at the Smithsonian Instituion with federal aid in 1949 originally for medical research projects only. Biology was added in 1953 and in 1960 coverage was extended to the physical sciences. It functions as a clearing-house for information on research planned or in progress but *not* published. Details of over 200,000 projects are stored in a computerized data base which is used to provide a variety of search services ranging from a simple custom search to 'research information packages', as well as current awareness services for individuals on a monthly, quarterly or other basis, right up to on-line interrogation of the data base. According to Roger Christian it 'has been of incalculable value in reducing duplication and overlapping R & D efforts in the US'. SSIE has recently compiled in co-operation with UNESCO a directory of research information systems and services: *Information services on research in progress: a worldwide inventory* (Budapest, OMKDK, 1978), covering 179 centres in 53 countries. Over half of the systems were found to be less than six years old.

On-line access is particularly valuable for such registers: one of the first data bases accessible via EURONET/DIANE was the Permanent Inventory of Agricultural Research Projects in the European Community (AGREP), a rare example of an international register of research in progress.

A register confined to books is operated by the National Book League in London. Entitled 'Books in progress: register of literary and technical research' it was established in 1978 and covers all subject fields.

Further reading

I R Stephens 'Searching for theses, dissertations, and unpublished data' American Chemical Society *Searching the chemical literature* (Washington, 1961) 110-20.

J Plotkin 'Dissertations and interlibrary loan' *RQ* 4 (January) 1965 5-9.

'Science Information Exchange (SIE): a national registry of research in progress' *Scientific information notes* 1 1969 43-6.

D J Urquhart 'Doctoral theses' *NLL review* 1 1971 8-9.

F J Kreysa 'SSIE — an information center which stores foresight' *Journal of chemical documentation* 12 1972 19-22.

C J Boyer *The doctoral dissertation as an information source: a study of scientific information flow* (Metuchen, NJ, Scarecrow, 1973).

P M Chettle *Studies for a British register of current research in the sciences and technology* (Southampton University Library, 1976) [BL R&D report no 5328].

E Harmon and I Montagnes *The thesis and the book* (Toronto University Press, 1976).

D Davinson *Theses and dissertations as information sources* (Bingley, 1977).

D F Hersey 'Information systems for research in progress' *Annual review of information science and technology* 13 1978 263-95.

C Urquhart *Provision and use of thesis information* (British Library Research and Development Department, 1979) [Report no 5513].

Chapter 20

NON-PRINT MEDIA

Scientists and technologists are obliged to investigate many phenomena that words alone are not adequate to describe, such as colour spectra, the songs of birds, the textures of fabrics. To read about these is not sufficient: they must be seen, or heard, or touched. The literature, it would seem, is no help, if by literature we mean textual sources – the written word. Over the years, however, the writers of literature have devised a number of parallel, auxiliary sources in an attempt to convey information that mere text cannot. The best known of these devices is pictorial representation (which of course actually preceded writing chronologically) used to illustrate (literally, to light up) textual matter. Such illustrations have amplified the information content of printed books in science and technology since the 82 engineering wood-cuts in Robertus Valturius *De re militari* ([Verona,] Johannes Nicolai, 1472); today, advances in printing technology and publishers' distribution methods over the last generation have permitted a larger proportion of current output than ever before to be illustrated, often in colour. Everyone is familiar with the role of book illustrations: there is no reason to dwell on them further. Many of them, if not most, are helpful and convenient and agreeable rather than absolutely essential to convey the sense of the text they illustrate. Over six thousand such illustrations from 178 books are cited in J W Thompson *Index to illustrations of the natural world* (Syracuse, NY, Gaylord, 1977). The visual material to be described in this chapter, on the other hand, is indispensable: the information it carries cannot be communicated by textual means.

Like scientists and technologists, librarians also have their problems of nomenclature, and the best term to describe these non-textual sources is a matter of minor dispute. The title chosen for this chapter is accurate, but because of its negative form it is not completely satisfactory. A suggested alternative, 'non-typographic documents', is similarly negative, with the added disadvantage that the meaning of 'document' could not yet be said to extend to models or specimens, even if it were agreed that it includes gramophone (or phonograph) records and videocassettes. 'Metabooks', a specific coining, may yet catch on, but the most widely used term is probably 'non-book materials'. But even the proponents of this term, the National Council for Educational Technology, make haste to excuse themselves: 'We apologize for the unattractive term "non-book materials" — even worse, the abbreviation NBM ... One day, perhaps, the word "document" will be commonly accepted as connoting simply an embodiment of evidence, whether it be in print or pictures or whatever, and we shall take for granted that arrangements for the handling of documents should make provision as a matter of course for all media. In the meantime, the case must not be overlooked; and as an expression "non-book material" is probably no more offensive and no less apt than its several rivals. "Non-print document" is perhaps more accurate but is not yet in common use'. The British Standard *Glossary of documentation terms* (BS 5408: 1976) has chosen 'non-book media' which it defines as 'media transmitting information or instruction by other than typographic means, eg by sound, pictorial representation, projected images, 3-dimensional representation etc.' While this definition itself is quite clear, the term chosen is still ambiguous: newspapers and magazines are 'non-books', and they are also commonly referred to as 'media', but they are clearly not intended to be covered by the definition of 'non-book media' given here. 'Non-print media', the term used in this chapter, is listed by the *Glossary* as a 'non-preferred synonym', though it is not a 'deprecated' term. It is retained here because it quite clearly excludes newspapers and magazines.

Maps and atlases
An obvious instance is a map: to convey the information content of a street plan of London or a contour map of the Cambrian Mountains is virtually impossible by any other normal means. Maps (and atlases, which are no more than volumes of maps) are widely used outside the purely geographical field. The science of geology is largely dependent on maps: the *Geological atlas of the United States* (Washington, US Geological Survey, 1894-1945) takes up 227 volumes. Many of the (UK) Institute of Geological Sciences (formerly Geological Survey) maps appear in parallel editions: 'drift', showing the superficial deposits, ie what is actually visible, and 'solid' showing the extent and nature of the solid rocks if all the superficial deposits were removed.

In many disciplines questions of geographical distribution arise, of plants, for instance, in botany, of animals in zoology, of fossils in palaeontology, of diseases in medicine, and so maps (and atlases) become indispensable tools, eg F H Perring and S M Walters *Atlas of the British flora* (Wakefield, EP Publishing, second edition 1976) and
360

Supplement (1978), J G Bartholomew *Atlas of zoogeography* (Edinburgh, 1911), L J Wills *A palaeogeographical atlas of the British Isles* (Blackie, 1951), Royal Geographical Society *National atlas of disease mortality in the United Kingdom* (Nelson, second edition 1970). Weather maps we are all familiar with nowadays, eg J G Bartholomew and A J Herbertson *Atlas of meteorology* (Edinburgh, 1899), S S Visher *Climatic atlas of the United States* (Cambridge, Harvard UP, 1954); and the heavens too have been mapped, eg *Norton's star atlas* (Edinburgh, Gall and Inglis, seventeenth edition 1978), World Meteorological Organization *International cloud atlas* (Geneva, 1956), the *Times atlas of the moon* (1969). Such thematic atlases, as they are called, can be found for ocean currents, the races of mankind, ice, marine life, oil and natural gas, and many other subjects.

In *Samson Wright's Applied physiology*, a standard medical textbook (see chapter 7) we read: 'Anatomy is to physiology as geography is to history: it describes the scene of the action'. This is probably the explanation for the hundreds of illustrated medical works calling themselves atlases, eg J C B Grant *An atlas of anatomy* (Balliere, third edition 1951), D H Ford and J P Schade *Atlas of the human brain* (Elsevier, third edition 1978), Perry Hudson and A P Short *Atlas of prostatic surgery* (Saunders, 1962). Such works have a long history: the first edition of the classic Spalteholz-Spanner *Atlas of human anatomy* (Butterworths, sixteenth edition 1967) appeared in Germany in 1895; its 1,630 illustrations 'enable the student, by comparing various views, to gain a three-dimensional impression of the subject matter'. A variation is the X-ray atlas, eg B S Epstein *The spine: a radiological text and atlas* (Lea and Febiger, fourth edition 1976); and the drug atlas, eg H G Greenish and E Colin *An anatomical atlas for vegetable powders* (Churchill, 1904). Wolfe Medical Publications specialize in such works, with something like fifty titles currently available, eg J G P Williams *A colour atlas of injury in sport* (1980).

Photographic illustrations
The half-tone block or some similar method of process illustration has been used in scientific and technical literature for many years. We all know that the camera cannot lie, and the accuracy of reproduction makes photographic illustrations (particularly in colour) of great value, for instance, in reference works designed to aid identification, eg C J Morrissey *Mineral specimens* (Iliffe, 1968) is made up of over a hundred full-page coloured illustrations; similarly B J Rendle *World timbers* (Benn, 1969-70) comprises three volumes of plates in full colour. W T Johnson and H H Lyon *Insects that feed on trees and shrubs: a practical guide* (Ithaca, NY, Cornell University Press, 1976)

has well over a thousand photographs and two hundred colour plates. Of even greater use are those photographic illustrations that provide information that no other form can, such as photomicrographs, eg British Leather Manufacturers' Research Association *Hides, skins and leather under the microscope* (Egham, [1957]), with almost a thousand illustrations; A B Wildman *Microscopy of animal textile fibres* (Leeds, Wool Industries Research Association, 1954). Rather rarer are stereoscopic photographs, which show objects three-dimensionally when viewed with the apparatus supplied, eg F C Blodi and Lee Allen *Stereoscopic manual of the ocular fundus in local and systemic disease* (Kimpton, second edition 1970). Even less commonly met is the illustrative matter described in *Satellite imagery* (Science Reference Library, 1980), a guide to the library's large collection of photographs and other images transmitted from space, and to other sources of such images.

Special types of visual material
Workers in different subjects have developed different methods of presenting their material visually, and this has produced a whole range of non-textual information sources. A selection only can be described here. Without *circuit diagrams* the electronics engineer would find his task virtually impossible, eg G A French *Twenty suggested circuits* (Data, 1960). The biologist uses *metabolic charts* to indicate the interrelations and correlations of biochemical reactions in metabolic sequences, eg W W Umbreit *Metabolic maps* (Minneapolis, Burgess, 1952) and *Supplement* (1960). *Silhouettes* have been successfully used for many years to aid identification of planes and ships, eg E C Talbot-Booth and D G Greenman *Ship identification* (Allen, 1968-). Chemists use *ring diagrams* as a convenient way of showing the molecular structure of organic compounds, eg *Parent compound handbook* (Columbus, Ohio, American Chemical Society, 1976-) in four volumes and supplements. The term 'atlas' is found here also, used to describe a variety of visual material, eg L Engel and H Klingele *An atlas of metal damage* (Wolfe, 1980) comprises 425 scanning electron micrographs; W C McCrone and J C Delly *The particle atlas* (Ann Arbor Science Publishers, second edition 1974) consists of photomicrographs and electron micrographs. A unique work is Gerhart Tschorn *Spark atlas of steels* (Pergamon, 1963), comprising over a hundred spark photographs used as a simple means of identification, based on the fact that different steels produce different sparks when at the grinding wheel. Another work with a very specific purpose is the ICI *Colour atlas* (1971): this comprises 1,379 different colour samples, but by using the neutral grey filters that are supplied the system can be
362

extended to illustrate 27,580 different shades out of the total of over a million that the human eye can distinguish.

Audio-visual materials
In the form of sound recordings and films, inconceivable only a hundred years ago, these are familiar to everyone. Very widely available too now are video recordings: they have been taken up extensively in the education world and seem destined to figure largely in the libraries of the future.

a *Sound recordings* (disc and tape – open reel, cassette, and cartridge): these are not unknown in science and technology, but have limited scope, eg Myles North and Eric Simms *Witherby' sound guide to British birds* (1969) comprises a book and two long-playing discs. Very similar in its aims is M P Fish and W H Mowbray *Sounds of Western North Atlantic fishes: a reference file of biological underwater sounds* (Johns Hopkins Press, 1970). Those interested can also acquire items like *Heart recordings*, or *North American frogs*, or *London's last trams*, but it is clear that the potential of this particular medium for the scientist and technologist is slight, except in a small number of specialized fields eg Cornell University has established a Library of Natural Sounds at its Laboratory of Ornithology.

More useful are those recordings presenting information that could also be set out in print, but which it is more convenient or agreeable to hear spoken. Discussions, lectures, courses, conference proceedings are obvious examples that come to mind. Experts speaking about their subjects are found in the Royal Society of Chemistry series Chemistry cassettes, eg Derek Bryce-Smith *Heavy metals as contaminants of the human environment* (1975) which runs for 100 minutes. Normally each cassette is accompanied by a workbook containing diagrams, equations, and often problems for the student to tackle. The American Mathematical Society has issued nearly a hundred of its series Audio recordings of mathematical lectures. The American Chemical Society has 'published' over twenty 'audio courses', eg *Principles of heterocyclic chemistry* (1975), 'taught by Professor Edward Taylor' in twelve cassettes (playing time 11.4 hours) and a '223 page integrated reference manual'. *Fuels for the next fifty years* is an example of a panel discussion issued by ACS as a cassette, accompanies by printed copies of slides illustrating the main points. The proceedings of some of the sessions of the American Association for the Advancement of Science meetings can be purchased on audiotapes, either open reel or cassette. Adding a new twist are the experimental sound journals, eg *Chemical executives audionews*, a weekly news service on cassettes initiated in 1970 by the American Chemical Society Somewhat different are the

363

series of 'Science reports' prepared for local radio stations by the American Institute of Physics: these take the form of four-and-a-half minute programmes (ten on one sound disc) with experts discussing topics in physical science of interest to the general public.

One particular application borrowed by science and technology from other fields of study is the use of audio-tapes for recording oral history, as exemplified by the tape archive at the Center for the History of Physics. This comprises recorded interviews with over sixty astronomers, and some six thousand pages of transcripts: already these constitute the most complete history we have of twentieth century astrophysics.

b *Films*: these, on the other hand, are widespread. Filmstrips (comprising a succession of still photographs) can convey little more information than a good illustration; cine-film on the other hand adds another dimension. The remarkable cart-wheeling method of getting from place to place that *Hydra* uses, as well as its less spectacular looping and shuffling movements, can only be fully appreciated by actually seeing it, preferably *in vivo*, but if not, on film. Refinements like slow-motion or time-lapse photography extend the range of applications in science and technology of cine even further, and almost all cine-films now have added sound.

Students should be familiar with the major bibliographical aids to such material: the bi-monthly *British national film catalogue* with its arrangement by the Dewey Decimal Classification contrasts usefully with the quarterly Library of Congress *Catalog: motion pictures and filmstrips* which is arranged alphabetically by titles with a subject index. Both are cumulated annually. There are a variety of other guides and indexes in specific subject fields, eg Royal Institute of Chemistry *Index of chemistry films* (sixth edition 1970) with over 2,000 titles of films, filmstrips, and film loops; American Geological Institute *Selected films on geology* (Falls Church, Va, 1978); British Industrial and Scientific Film Association *Film guide for the construction industry* (Construction Press, 1979); the loose-leaf *Film catalogue* (1974-) of the Institute of Metallurgists. By far the largest is A Seltz-Petrash *AAAS science film catalog* (New York, Bowker, 1975), claimed as the most comprehensive annotated listing of its kind ever published, with nearly six thousand films included. It can usefully be supplemented currently by the quarterly *Science books and films*, published by the American Society for the Advancement of Science, which reviews perhaps twenty new science films per issue.

Two points need emphasizing about the films in such lists: many of them are available for hire, rather than purchase; and the overwhelming majority are for teaching purposes. They could be said to correspond to the textbook in the printed literature, eg *Modern methods of*

364

underground pipe-laying (1963), *Masonry: elements of brickwork* (1976), *The story of basic slag* (1962). Some are quite technical and at an advanced level, eg *Prediction of wind loads* (1978), *Surface coating in engineering* (1971), *Quantum theory* (1971). The fact that the last mentioned is an Open University film serves to remind us of another extensive source of first-class science and technology films. As a study of the various Open University catalogues of audio-visual materials will reveal, a feature of their provision is the integration of more than one kind in one course or programme: their *History of mathematics* for example, comprises eleven films and four tapes.

c *Video recordings* (disc and tape — open reel, cassette and cartridge): for some years widely used for teaching purposes, they have now spread very rapidly into the entertainment and domestic markets in the form of videocassettes. In the form of videodiscs they seem further destined to make a major impact on the world of scientific and technical information. For the viewer they possess one incalculable advantage over sound cine-film: their simplicity of operation permits individual use without the need for blackout or a projectionist. More and more filmed material is now becoming available in videocassette form as an alternative. Indeed it has been predicted that videorecordings will eventually replace films in most institutional collections, at least as the direct viewing medium.

Like films, they are often available either for purchase or hire, eg *Introduction to sequential logic* (1977), *Lime kilns at Heilam Ferry* (1975), *Astronomy in prehistoric Britain* (1972), and the series issued by the Society of Manufacturing Engineers, such as *Metal forming, Numerical control*. Aimed more at the domestic market are videocassettes made (or at least presented) by experts in the field on home energy conversion, intensive gardening, rediscovering herbs, and many others. It is interesting to note that most of them emanate from the regular commercial publishing houses, who have not been slow to diversify from print to video.

As is so often the case with new technology, there are still problems of equipment incompatibility insofar as one manufacturer's videocassette may not be usable on another manufacturer's player. Even more radical incompatibilities are found with the videodisc, just now emerging as a practical medium. The differences here are more profound, deriving from the two quite different technologies that have emerged since 1978: optical discs comprise a variable reflectant plane embedded in a plastic disc, with a low-powered laser to track the disc; the second type of disc resembles the conventional audiodisc (gramophone or phonograph record) inasmuch as a stylus traces the helical grooves on the disc surface; a variation on this second type uses grooveless discs. In the US it is expected that there will soon be over twenty different brands of videodisc player on the market using three incompatible formats. Examples of what are already available on videodisc to the domestic consumer are Dr Benjamin Spock on *Caring for your newborn child*, which runs for two hours, and a programme about birds drawing on the expertise of Roger Tory Petersen whose bird books have sold three million copies since 1947.

For our purposes, however, what is so revolutionary about the videodisc is not its thirty minutes of playing time each side but its potential capacity to store up to 54,000 individual colour frames, which is more information than can be stored on any comparable form of magnetic media or on microform. Some systems allow each frame to be identified by an individual consecutive number (like the pages in a book) that can be called up on an appropriate player virtually instantaneously and 'frozen' for as long as the viewer desires, without damaging it as happens with videotape. Martin S White has recently pointed out that 'This would enable libraries to have on file a complete set of phase diagrams of all metal systems on one 12 inch disc'. Already videodiscs are being used successfully to store large collections of visual material, eg by museums to store photographs of items in their collections; by New Scotland Yard to store the UK national fingerprint collection of 2.5 million sets.

Pergamon International Information Corporation have incorporated videodisc storage of graphics in their on-line information retrieval system for the three-quarters of a million US patents issued since 1971. The data base is accessed on-line in the usual way, but when the required patent has been located pushing a button switches from digital storage to video and the patent drawings are displayed on the screen within two or three seconds. It is significant that microform was initially intended as the storage medium, but it soon became clear that 'the videodisc was the appropriate technology because of its ability to interface with electronic control'. NASA's Jet Propulsion Laboratory is already replacing microfiche by videodisc storage for images sent by spacecraft: they have found that the 108,000 images on one twelve-inch disc can replace up to 1,500 microfiches.

It is this vast capacity of the videodisc that has encouraged its use to store *textual* as well as visual matter – a nice reversal of roles. We are told that it is possible to store on one disc alone *two* sets of *Encyclopaedia Britannica*, or the whole of the on-line version of *Chemical abstracts* several times over. A significant advantage in the context of printed material is the interactive nature of the videodisc, permitting random access to any frame, like consulting the pages of a book. McGraw-Hill has participated in the preparation of an interactive videodisc for college biology students that is being tried out during 1981 at

Brigham Young University, and Xerox Publishing are working on a videodisc encyclopedia of gardening.

A particularly sensitive issue with video recordings, still unresolved (as with sound recordings), is the ease of reproduction which has made it virtually impossible to control illegal copying in breach of copyright. This is not yet a problem with the videodisc, which is for all practical purposes a player medium only: consumer equipment to make your own recordings is not yet feasible. Indeed, like audiodiscs (and printed books for that matter) large editions of videodiscs are needed for economic reasons, whereas videocassettes can be produced virtually on demand once the master tape has been made.

d *Teaching aids*: taking such forms as slide sets, overhead projector transparencies, film loops or cartridges, they have been familiar for many years, sharing as common ancestor the magic lantern. From these have developed multi-media learning packages, typically comprising sound as well as visual material, realia (see below), and, as often as not, conventional printed materials. Such multi-media kits are increasingly available outside the purely instructional context: the American Chemical Society is currently marketing the proceedings of a number of symposia in audiocassette form accompanied by printed copies of the original slides used by the speakers, eg *Selective insect control*. As might be expected, many of these teaching materials deal with scientific and technical topics, and can be traced through a whole range of bibliographies and indexes, eg the regularly updated series compiled at the National Information Center for Educational Media (University of Southern California) *Index to overhead transparencies, Index to 8mm motion cartridges*, etc. The most interesting development in this particular field has been the spread of simple presentations combining synchronized sound with vision, usually 35mm slides and an audiocassette. These are in effect ready-made lectures. Two contrasting examples of these tape-slide programmes are *Methods of measurement* (University of Lancaster Engineering Department, 1975) and *Theodolites* (Loughborough University of Technology, 1978). An important guide to such teaching aids is the British Universities Film Council *Audio-visual materials for higher education* (fourth edition 1979-80): the 3,000 items carefully catalogued there include sound recordings, video recordings and tape-slide programmes but the overwhelming majority are films. The third and fourth volumes are devoted to science and technology, and materials are selected for inclusion on the basis of written appraisals and recommendations sent to the Council. They are accepted only if considered suitable for degree-level teaching or research.

As well as being designed mostly for educational purposes, all the materials mentioned so far in this chapter have been professionally

367

produced for distribution, frequently on a commercial basis. They correspond to 'published' literature in the printed media. But one of the characteristics of non-print media is that its production is often well within the capability of the interested amateur, especially if he is a teacher, with access to educational technology equipment and services. There has thus developed in the universities and colleges a vast reservoir of audio-visual aids, prepared by teachers to assist their own teaching. It has been realized that much of this material, carefully produced by experienced teachers and educational technologists, may be of value outside the particular classroom or lecture theatre for which it was designed. Where such material is technically competent, and provided the teacher and his institution agree, there is no reason why others should be denied access, particularly if (as is often the case) the whole enterprise has been publicly funded. Assisting the distribution of such items is the enterprising Higher Education Learning Programmes Information Service (HELPIS) *Catalogue* (British Universities Film Council, fifth edition 1978), which contains entries for material produced at universities, polytechnics and other higher education institutions in Great Britain, eg *The energy crisis* (University of Lancaster, 1974), a sound tape; *Dissection of a rat* (Aberdeen University, 1974), a video cassette; *Medical ultrasonics* (University of Surrey, 1975) a tape-slide production.

Bibliographical control
Over such a diverse range of material, emanating from sources as varied as major universities, world-renowned publishing houses, great entertainment corporations, and individual teachers, bibliographical control will always be difficult. And apart from films and certain educational media, until recently it was quite inadequate. The ideal has always been one comprehensive multi-media catalogue, perhaps as a supplement to the *British national bibliography*. While this is still some distance away, in 1979 the British Library, following a two-year research study, published the first experimental edition of the *British catalogue of audiovisual materials*, largely based on the Inner London Education Authority Central Library Resources collection, listing in classified order over five thousand items currently available for purchase or hire in the UK. The 1980 *Supplement* added a further 2,300 items. The computerized data base is available on-line via BLAISE.

An attempt at comprehensive coverage within a particular subject field is D Maslin *Biological sciences: a subject index of audio-visual materials* (Institute of Biology, 1978) which tries in its more than six hundred pages to list 'all audio-visual materials relevant to the biological sciences of which we know and that are available in the United Kingdom'. A less successful attempt is B Eastwood *Directory of audiovisual sources: history of science, medicine, and technology* (New York, Science History Publications, 1979) which consists almost entirely of films. Anyone working in the field soon encounters a number of other multi-media guides to particular categories of materials, eg *Catalog of United States government-produced audiovisual materials* (Washington, National Audiovisual Center, 1974).

Three-dimensional material
There are limits even to what non-textual 'documents' can communicate, and for years scientists and technologists have made use of models, specimens, artefacts, and other forms of 'realia' to assist. This is an old communication problem that the rapid developments in molecular biology over the last few years have highlighted. Workers in this field concern themselves with investigating the dimensional structure of very complex molecules — their most remarkable achievement was the discovery of the structure of DNA — and make great use of three-dimensional models. Their problem is communicating with each other about molecular structures *without* the aid of models. Stereoscopic photographs can help, and holography will actually provide the answer, but a laser is needed for deciphering. This is but one recent instance of a larger problem, and no doubt some solution will be found, possibly via the computer. Valiant attempts have been made with similar difficulties in the past by providing actual specimens to accompany literature in conventional form, eg Alfred Schwankl *What wood is that?* (Thames and Hudson, 1956) comprises a volume of text and 40 actual timber samples in a case; the British Colour Council *Dictionary of colours for interior decoration* (1949) is made up of one volume of text and two volumes of strips painted in matt and gloss paint and samples of pile fabric dyed in 319 colours with 6 shades of intensity; F B Smithe *Naturalist's color guide* (New York, American Museum of Natural History, 1975) contains 86 swatches, named and numbered.

Further reading
M Alcock 'The Learning Materials Recording Study: a joint British Library/Inner London Education Authority research project' *Audiovisual librarian* 8 1979 no 2 59-65.

A Horder *Video discs — their application to information storage and retrieval* (Hatfield, National Reprographic Centre for Documentation, 1979).

C F Pinion 'The interlending and availability of audiovisual materials in the United Kingdom: report of a survey conducted in 1979' *Interlending review* 8 1980 55-61.

M Savage 'Beyond film: a look at the information storage potential of videodiscs' *Bulletin of the American Society for Information Science* 7 1980 (October) 26-9.

R Barrett *Developments in optical disc technology and the implications for information storage and retrieval* (British Library Research and Development Department, 1981) [Report no 5623].

Chapter 21

MICROFORMS

Perhaps because they too require special equipment for their utilization, microforms are often grouped with audio-visual materials.
Yet they do comprise print as well as non-print sources, and differ
only in *format* from the other categories of information discussed in
this book. Nevertheless, despite their pervasive character they deserve
separate consideration because of their uniquely distinctive attributes.

It is to be hoped that the student will be familiar with the various
kinds of microforms and their uses: all that it is intended to do here is
to indicate their particular contribution to the literature of science and
technology as an alternative form of publication. The idea is by no
means new, of course. Reporting on the Great Exhibition of 1851
James Glaisher wrote of 'the promise of future publications in miniature'.
Shortly afterwards Sir John Herschel, the great astronomer, urged 'the
publication of concentrated microscopic editions of works of reference'.
Though the Library of Congress began to microfilm for its own purposes
a range of manuscripts from foreign collections as early as 1905, high
costs and the lack of proper reading and copying facilities limited the
use that could be made of them. It was not until the 35mm camera
and the 1930s that microforms began to appear in libraries in any
numbers. Their frequent appearance throughout this work may have
been noticed, for among the bibliographies of microforms can be found
examples of practically all categories of information sources. For
very many years now they have been used for *retrospective micropublication*, ie the reprinting of originals that for one reason or another
do not warrant conventional full-size republication. Sometimes the
originals are old and rare, as in the ambitious Landmarks of Science
series, comprising in Microprint form the outstanding works of more
than 3,000 scientists and several early scientific journals, but more
often they are out-of-print works of all kinds for which there is still a
demand — but not such as to justify the expense of a new edition. Indeed so extensive is the range now that it would almost be possible to
build up a respectable working library of science and technology in
microforms alone. By way of illustration, all the following works,
chosen at random from the examples given in previous chapters, are

available in microform: American Chemical Society monographs, *Engineering index*, the US Joint Publications Research Service translations, *Report on the scientific results of the voyage of HMS Challenger ... 1872-76*, the United Kingdom Atomic Energy Authority unclassified reports, *Nature, Endeavour*, all British Standards, the *Trade marks journal, Science, American machinist*, US chemical patents, the Aeronautical Research Council *Reports and memoranda*, Dean *A bibliography of fishes*, Royal Society *Catalogue of scientific papers*, US Patent Office *Official gazette*. There are thousands of others. Not all of them are replacements for out-of-print originals, for in some cases the microform edition is on sale as an alternative to the hard-copy original, for those who may prefer it; in other words, *concurrent* or *simultaneous micropublication*, eg *New Society, Physics abstracts*, the journals published by the Institution of Electrical Engineers. Usually the price is the same for the two versions so that they do not compete for the same market, eg the American Chemical Society journals; some publishers charge less for the microform version, eg Pergamon Press. In a number of cases the microform edition is restricted to those who already subscribe to the printed version, eg the various journals of the Royal Society of Chemistry. In 1973 the American Chemical Society reported that it had over 1,400 subscriptions to its microform journals. It has been discovered, indeed, that a subscription to a simultaneous microform edition can bring benefits even to the library that already receives the hard-copy issues: for the period of their heaviest use there are duplicate copies available; binding for permanent preservation is no longer strictly necessary, as a consequence long-term storage costs are reduced; with overseas journals the microfiche edition normally arrives by airmail weeks before the hard-copy by sea mail. In the case of the microfiche version of *Index medicus* the publisher promises that *all* subscribers, at home and abroad, will get their microform copies earlier than the regular printed issues. Since 1971 the University of Toronto Press has issued all its books in printed form and in microfiche simultaneously.

What has increased greatly in recent years has been *original micropublication*, ie of material not published in any other format. Both the technology and the economics of microreproduction permit very small editions, or even one-off, on-demand publishing, which is ideal for those categories of scientific and technical literature with a limited appeal, eg university theses, research reports, etc. The services of *Dissertation abstracts international* represent the best known instance of theses micropublication, and the enormous US government programme (through the National Technical Information Service) is certainly the largest in the field of report micropublication. Other kinds of material for which micropublication has been found appro-

372

priate are the card cataloguues of libraries, eg *The card catalog of the Institute of Transportation Studies* comprising 262,000 cards on 126 microfiches; a scientist's collected works, eg Albert Einstein *Complete works* (Readex Microprint, 1966); a technologist's manuscripts, eg *The papers of James Nasmyth, 1808-90* (Micro Methods, 1968); illuminated medieval manuscripts (ie in colour), eg the tenth century *Bodley herbal and bestiary* (Oxford Microform Publications, 1979). Despite the still unsolved problem of fading with age, colour microfiches indeed have made a distinct advance in recent years, not only as a publishing medium for medical illustrations, for instance, but also as teaching aids, eg C K Edwards *A survey of glassmaking from ancient Egypt to the present age* (University of Chicago Press, 1977), described as a 'textfiche program' and comprising a brief printed full-size text and 168 colour images on fiches of pieces from the Corning Museum of Glass.

One of the effects of the recent steep rises in the cost of paper, printing, and postage, has been to multiply this economic advantage of microforms, so much so that they are now challenging the conventional media on their own ground, the provision of long-run editions. The scientific journal is an obvious target here, as we have seen (chapter 9), but no category of literature is immune to financial pressure. Even the conventional wisdom that microforms are unsuited to reference works that require frequent consultation is no longer universally accepted. Increasingly available on microform are complete archives or unique research collections, often copied *in toto*, where the sheer bulk or format of the original, as well as the limited demand, precludes conventional publishing, eg *UK land well records* (HMSO, 1980), the detailed geological results of test drillings for oil and gas; *AMSOM: atlas of macromolecular structure on microfiche* (Rockville, Md, Tractor Jitco, 1976), totalling 110,000 pages of graphics and data previously available only via computer display terminal from the Brookhaven Protein Data Bank.

The cumulative effect on library collections of fifty years of microforms has been dramatic. It is no exaggeration to say that a major research library could not function adequately today without them. The British Library Lending Division has a collection approaching three million items. A hundred of the largest research libraries in North America reported in 1979 that their collections held 220 million printed volumes and 198 million units of microforms.

User resistance

But the use made of microforms nowhere nearly matches the use of printed materials because of undoubted user-resistance to microforms, a bugbear for decades. One survey, of the members of the Institute of

Electrical and Electronics Engineers and the Society of Automotive Engineers, no strangers, it can be presumed, to technical equipment, found that only one percent preferred a microfiche to a printed journal. Microfilm (ie on reels) is particularly resented, although the advent of self-threading cassettes, particularly when used in motor-driven reader-printers, has done something to dispel reluctance. Such cassettes or cartridges can be handled and shelved and kept in order in much the same way as the bound volumes of a hard-copy journal. A survey at the Masachusetts Institute of Technology among users of *Chemical abstracts* in this format found that a substantial majority preferred it. Microfiches have always been easier to handle, but progress towards what Harold Wooster called a 'cuddly' microfiche reader is depressingly slow. There is a chance that the growing market for microforms might encourage designers and manufacturers to produce hardware that will stimulate rather than hinder their use, though the ergonomics of microform reading machines is notoriously difficult. Despite many years of research, manufacturers hold out little prospect of major improvement at modest cost in the design of optical projection readers. A particular difficulty is that library use is commonly sustained reading over a period, whereas most non-library use is consultation, where the reader only has to concentrate on the screen for a few minutes or even seconds. Since the library market makes up no more than a fraction of the whole it is not surprising that despite the hundred models currently available Desmond Taylor is able to castigate viewing equipment as 'a disgrace to the profession'. Even librarians, most of whom are anxious to promote the use of microforms, cannot honestly ignore their drawbacks. An American librarian in charge of a microform collection of over half-a-million items has confessed in public 'Personally, I must admit that I like all the advantages that microform brings, except reading it'.

It must also be admitted that the location of the reading machines in many libraries appears to have been an inconvenient afterthought. And poor image quality can be a particularly annoying problem with many microforms, especially microfiches of research reports or theses where the original documents may themselves lack legibility, perhaps because they are carbon copies or third or fourth generation photocopies. Even less easy to excuse are the wide variations in density often encountered, because these are not dependent on the original. It is little comfort to read authoritative statements such as that made in 1980 by the Chief Librarian of the Consumers' Association: 'There are numerous scientific studies which have concluded that reading microfilm does not cause eyestrain or discomfort. It is all in the mind'. This may well be so, but even librarians should recognize that user reluctance is no less real for being illogical.

Perhaps the long-term solution lies in retaining microform distribution and storage but abandoning the microform reader, replacing it with instant low-cost hard copy from the high-speed 'blow-back' machines now increasingly seen in use. For library users will continue to require access to the *content* of microform collections: in the words of Richard Boss, former Director of Princeton University Library, 'More and more people are beginning to discover that while microform may not be as nice to read as hard copy, it's a choice between a five minute wait for a microform and a fifty minute hunt for a hard copy, there is no choice'.

Bibliographical control
Bibliographical control of microforms has always been something of a nightmare, because copyright legislation rarely requires deposit, and because of the comparative ease with which microform editions can be produced, virtually by anyone with a 35mm camera. Reproduction of copies is even easier; duplicates can even be produced from duplicates. The whole terrain is also a legal jungle: the US president of Pergamon Press reports that 'in this age of information explosion and technological methods of dissemination, restrictions applied to libraries handling copyrighted material do not work. In fact, when we released the only authorized microforms for the hundreds of journals under our license more than half of the libraries already had them available, through a "custom-made" program, utilizing service companies or the libraries' own cameras'. Although we have had since 1965 the cumulative Library of Congress *National register of microform masters* with over 50,000 entries a year, and 330,000 entries in the 1965-75 cumulation, and a small but increasing number of microforms appear with their *own* International Standard Book Numbers, and therefore might be found in the national bibliographies, the best sources of information are still the trade listings. The most extensive, international in scope and covering 70,000 publications, is the annual *Guide to microforms in print* (Mansell) with *Microlist* which updates it ten times a year, and its companion *Subject guide to microforms in print*. We now also have, in *Microform market place*, an international directory of publishers of microforms; and in *Micropublishers' trade list annual* the actual catalogues on microfiche of over two hundred such publishers. Nevertheless, bibliographical control continues to be a weak point; there is still no single comprehensive listing, and Felix Reichmann's 1972 complaint remains valid: 'the Procrustean bed of existing bibliographies cannot accommodate microforms'. In 1978 W J Myrick reported that 'Although automation has made possible greater effectiveness in the sharing of bibliographic information, microforms have received almost no attention

375

in the development of these techniques'. Help may be at hand, however: in 1980 in the US the Association of Research Libraries was granted $50,000 to examine the matter and devise a plan for building a machine-readable data base of catalogue records of materials in microform. And in the UK, though the prospect is less hopeful, the Microfilm Association continues to urge that microform publication should be afforded the same treatment under copyright law as other printed publications, namely legal deposit and inclusion in the *British national bibliography*; in its 1981 consultative document on copyright the government says more public debate is needed.

'Package' collections

A recent feature of microforms in science and technology has been the emergence of microform 'systems', based on the miniaturization of a complete collection rather than individual documents. Some examples of these 'package' or 'desk-top' libraries in the field of trade literature were quoted in chapter 18 but there are subject-oriented systems also. One of the best known is *Chemical abstracts* on microfilm – over 70 years of abstracts totalling more than seven million entries housed in over 200 cassettes of 16mm film.

A particular impetus has been given to the provision of complete collections of this kind by the introduction of the high-reduction microfiche, or 'ultrafiche', giving up to 3,000 pages (equivalent to perhaps eight or ten normal volumes) on a single 6 inch by 4 inch fiche, in effect, microphotographs of microphotographs. In the 1970 Science and Technology ultrafiche collection issued by NCR in its Library Collections Program there are, for instance, 294 titles in 561 columns on 100 fiches, issued with a complete set of catalogue cards. The rival Microbook Library Series of Library Resources Inc adopted the older 5in by 3in size fiche; among the packages offered is the Library of the history of science and technology. Yet another variation, this time on standard 6in by 4in ultrafiche, are the packages offered by Information Handling Services (formerly Microcard Editions), for instance. These are linked to a standard bibliography, in this case *Books for college libraries* (Chicago, American Library Association, 1975), a core collection of some 40,000 titles. The full texts of the 203 science and technology titles listed there are available as a microform package, complete with their own individual catalogue cards. Eclectic collections of this kind sometimes offer advantages even over the originals, if, for example, the microform republisher has furnished them with an analytical index. Some such collections are in highly specialized research areas, eg *Cambridge texts in the history of Chinese science* (Cambridge University Press), with 25 Chinese titles on microfiche; *Classic literature*

376

on invertebrate palaeontology (Oxford Microform Publications).

But librarians face two problems with such collections. There is now clear evidence that ultrafiche has not proved a success, to some extent because the technique of ultra-reduction is expensive but also because it requires 'dedicated' viewing equipment with 200x magnification. Of course ultrafiche collections still form part of library stocks but as a current publishing medium it is obsolete, a good instance of how the undoubted marvels of technology frequently fail to achieve general acceptance. A number of serial publications that used to appear in ultrafiche have changed to standard microfiche format, eg *Books in English*.

The second problem is one that afflicts all 'package' collections, even those on microfiche. If the microform republisher has supplied them neither with an index nor with catalogue cards, librarians face the burden of cataloguing the individual items themselves. Sad to relate, many have decided not to do so. However, it has come to be realized, as should have been obvious from the start, that inadequate bibliographical control and low use of microform are to some extent cause and effect. C E Carroll makes the point sharply: 'cataloged materials are used; non-cataloged materials are not'. There are those who fear that 'micropublishing is creating a whole class of little used research materials'. Many librarians believe, indeed, that even where a hard copy index is provided, access to the individual titles in the collection is still hindered until each of them has been included in the library's main catalogue. Richard Boss maintains that 'microforms should be treated exactly the same as hard copy materials with every skill that we librarians have developed'. We do now have stop-gap bibliographical assistance in the shape of S C Dodson *Microform research collections: a guide* (Westport, Conn, Microform Review, 1978), a detailed analytical index to two hundred such collections.

Computer-output-microfilm

Where the material to be published is already in machine-readable form, eg magnetic tape, a most powerful tool is now available that can print out directly on to microfilm at up to twenty times the speed of a normal computer line-printer on paper. With such computer-output-microfilm there is no original document to be photographically copied: the microform is produced directly from the digital data, in a variety of ways. In one method the recording unit operates by displaying the text page by page on a cathode ray tube, and then photographing it frame by frame on to film, at perhaps three hundred frames per minute. An alternative method uses an electron beam recorder to transfer the digital data directly from the computer on to film. Output can be either roll-film (usually in cassettes or cartiidges) or fiches, though the greater reduction ratio of 1:48 or 1:42 means that a special COM reader is required.

COM is especially useful where large amounts of data are involved, for instance numerical data from experiments, or detailed bibliographical entries, particularly if it has to be regularly updated. Storage capacity is enormous: it is currently claimed by Europe's largest COM bureau that 207,000 pages of computer printout in microfiche form can fit into a standard A4 size folder. And where multiple copies of the printout are required the ease and cheapness of microform reproduction methods provide a further bonus. Some COM systems can also use the power of the computer to furnish indexes to the information they store and print out. The most obvious library application so far has been catalogues, eg of the Science Reference Library, and bibliographies, eg *British books in print*, but there is clearly much more scope.

The challenge to microforms
Despite this strikingly happy marriage between microforms and the computer, increasingly to be heard are whispers that microforms are being by-passed by the new technology. Having had the field to themselves for so long as the most obvious instance in the library of the exploitation of technical advance, they may well represent only an interim technology, like the steam car and the cylinder gramophone record. Some have given microforms no more than a decade of further useful development, and we have already witnessed the demise of ultrafiche. The future, we are told, lies increasingly with digital storage and transmission of data via the computer and telecommunications. Videodisc too, in the words of A Wight of ASLIB, is 'beginning to cast a shadow over the future prospects of microfilm'. There have been gloomy murmurings of 'the failure of microform'.

User resistance has obviously something to do with this feeling, and as has been admitted by Andrew Braid, head of reprographic services at the British Library Lending Division, 'The promised boom in micropublishing has never arrived'. Even the micrographics industry's own leaders accept that 'In some cases the new technology will compete and replace certain micrographic systems'. But others have not been slow to rise to the challenge: J W Shepard of 3M has pointed out that a rotary camera's capture of textual information is about five hundred times more cost-effective than keying-in and verifying data; filming a document takes only milliseconds, compared with the two or three seconds required for a facsimile-type scan to 'read' a page of text into a storage device. There is perhaps hope in the confluence of technologies
378

referred to in chapter 1: in a paper read at a Micrographics and Publishing seminar at Oxford in 1980 it was surmised that 'the digitalizing
of microfilm for display on a computer VDU will be in much greater
use by the 1990s'. Nevertheless an increasing number are coming
round to the view that there may not be room in the iinformation
world of the future for both computerized data and microforms.

Further reading
C W Christ 'Microfiche: a study of user attitudes and reader habits'
Journal of the American Society for Information Science 23 1972
30-5.

J Becker 'Computer output microfilm (COM) for libraries' *UNESCO
bulletin for libraries* 28 1974 242-4, 248.

S R Salmon 'User resistance to microforms in the research library'
Microform review 2 1974 194-9.

'Microimagery in the library' *Drexel library quarterly* 11 1975 no 4.

C E Carroll 'Bibliographical control of microforms: where do we go
from here?' *Microform review* 7 1978 321-6.

W J Myrick 'Access to microforms: a survey of failed efforts' *Library
journal* 103 1978 2301-4.

W Saffady *Computer-output microfilm: its library applications*
(Chicago, American Library Association, 1978).

W Saffady *Micrographics* (Littleton, Colo, Libraries Unlimited,
1978).

P Ashby and R Campbell *Microform publishing* (Butterworths,
1979).

R Campbell 'Will microforms be overtaken? a publisher's view'
Journal of micrographics 12 1979 291-2.

S J Teague *Microform librarianship* (Butterworths, second edition
1979).

S C Dodson 'Bibliographical control' *Microform review* 9 1980
145-53.

M Gabriel and D P Ladd *The microform revolution in libraries*
(Greenwich, Conn, JAI Press, 1980).

M J Gunn 'User resistance to microforms' *Microdoc* 19 1980 50-8.

W C Jackson 'Bibliographical access to microforms: on the threshold?'
Microform review 9 1980 28-31.

S Sharrock 'Microform publishing – a publisher's point of view'
Microdoc 19 1980 4-15.

N Hyde 'Bibliographical control: a British view' *Microform review*
10 1981 166-9.

Chapter 22

BIOGRAPHICAL SOURCES

It is no more possible in science than in any other area of intellectual endeavour to separate a man's life from his work. So intimately bound up with his personal life is the work of a research scientist that a full understanding of his achievements often demands close study of his biography. Perhaps this is less true of the technologist, and even of the scientist in many humble and mundane circumstances, but there is no doubt that biographical information is frequently sought in libraries about scientists and technologists of all kinds. The deep enquiry into the personal background of an individual corresponds to the 'exhaustive' approach to the literature described in chapter 1. Far more common are the 'everyday' biographical enquiries, where the searcher needs either one specific fact, eg the full name, or the address, or the present post of the man he is interested in; or in some cases rather more detail to enable him to come to some kind of judgment as to his status, eg academic qualifications, previous experience, age, publications, etc.

The literature resources that a searcher has at his disposal are not in fact scientific and technical information sources at all, as described in chapter 1. They are *biographical* sources, and even where they specialize in scientists and technologists, they are not part of the literature of science and technology. This is a point of major significance for the searcher, particularly if he is more used to investigating scientific and technical sources of information. Because of the nature of the literature, ie because it is about people rather than about science and technology, searches in the biographical sources have their own rules: it is far more common, for instance, for a biographical search to end in failure. In this science is no different from any other subject: most people are not renowned; the published biographical sources concentrate on the renowned: therefore, there is no published information available about most people. There is, on the other hand, an overwhelming amount of detail about the important figures, repeated time and again in the various sources. The student will find, for example, that some of the dictionaries and encyclopedias of science and technology contain biographical entries, eg *Hackh's chemical dictionary*, H J Gray *Dictionary of physics*. But these are invariably confined
380

to the renowned names. On the less famous, the best sources of biographical data are the directories of individual scientists and technologists, as described in chapter 6. Within their limits these are non-selective, inasmuch as a society membership list, for instance, will include all members in good standing. For the simpler everyday enquiries as to name, location, degrees, etc, such lists will often serve, but very few include real biographical information on matters such as place of birth, education, previous posts held, etc. For these one turns to the overtly biographical sources, particularly the biographical dictionaries.

Biographical dictionaries
It is frequently drawn to our attention that over 90% of all the scientists who have ever existed are alive today; not surprisingly, therefore, there is a wide range of publications providing biographical information about them. Mostly of the 'who's who' type, they are obviously selective, but do aim to include as large a proportion of the scientific and technical community as is feasible. An example of international scope is the four-volumed *Who's who in science in Europe* (Hodgson, third edition 1978), listing 50,000 names from over 20 countries. As is commonly the case with such publications, the entries are based on the responses made by the scientists themselves to questionnaires circulated by the publishers. The word 'science' in the title includes technology and medicine, as it also does in a contrasting dictionary of similar international scope but very different approach: *McGraw-Hill modern men of sicence* (1966-8). Here attention is concentrated on 'leading contemporary scientists', amounting to no more than 850 in the two volumes. Such works are frequently compiled on a national basis: by far the largest with 131,000 biographies is *American men and women of science* (Bowker, fourteenth edition 1979) in eight volumes including *Discipline and geographic indexes*. Computerization over the last few years has not only permitted the publishers to offer on-line access to the data base but has also produced as a 'spin-off' from the main work a series of 'specialty volumes' in which the identical entries are grouped by subject, eg *Chemistry* (1977) with 30,000 entries. Recently made available is a consolidated index listing all the 297,000 names from the first to the fourteenth editions. *Who's who of British scientists* (Dorking, Simon Books, third edition 1980) is much more selective, with no more than six thousand entries. The editor is fully aware of gaps, but as he candidly admits, 'each scientist of however great distinction, has the inalienable right not to submit an entry'. An unusual example of a list compiled in English but covering foreign scientists is J Turkevich *Soviet men of science* (Princeton, NJ, Van Nostrand, 1963), describing 400 of the academicians and corresponding members of the

381

USSR Academy of Sciences. By contrast, I Telberg *Who's who in Soviet science and technology* (New York, Telberg, second edition 1964) contains translations of about 1,000 entries selected from a biographical dictionary published in Moscow.

There are many 'who's who' compilations in specific subject fields in science and technology, eg *Who's who in world oil and gas* (Financial Times, fourth edition 1978), with over four thousand entries. There are two distinct compilations called *Who's who in atoms*: the larger, in two volumes (Harrap, sixth edition 1974) has over 20,000 entries from 76 countries; the smaller (Guernsey, Hodgson, sixth edition 1977) has biographies for some 7,000 nuclear scientists.

A feature of biographical dictionaries in many subjects is the division found between those listing living persons and those which only include names after death. All the titles so far mentioned in this chapter are of the first kind: an example of the second is T I Williams *A biographical dictionary of scientists* (Black, second edition 1974), which ranges from the earliest times to the present, includes technologists as well as scientists, and covers all countries. By way of contrast the 3,000 entries in the two volumes by E G R Taylor *The mathematical practitioners of Tudor and Stuart England* (CUP, 1954) and *The mathematical practitioners of Hanoverian England* (1966) are confined to a particular period, a particular discipline, and a particular country. Unusual in that it is a derivative compilation, reprinting the relevant entries from the six-volume *Who was who in America* is *Who was who in American history: science and technology* (Chicago, Marquis, 1976). Not surprisingly, perhaps, many compilers and publishers find this dichotomy in print between the living and the dead to be too restricting, and some major biographical dictionaries ignore the distinction, eg *World's who's who in science* (Chicago, Marquis, 1968), the largest one-volume compilation with 32,000 names from 1700 BC to 1968 AD; *Chambers's Dictionary of scientists* (1951) a much more selective list of 1,400 leading figures; and the great scholarly standard work, *Poggendorffs Biographisch-literarisches Handwörterbuch* (Leipzig, Barth, 1963-). Within particular subject fields too, it is usual to find past and present figures intermingled, eg J H Barnhart *Biographical notes on botanists* (Boston, Hall, 1965), comprising in three volumes reproductions of over 40,000 cards from a file maintained by the New York Botanical Garden Library.

Until recently the field lacked a scholarly multi-volumed selective work of the *Dictionary of national biography* type, but the American Council of Learned Societies *Dictionary of scientific biography* (New York, Scribners, 1970-6) in fourteen volumes, with *Supplements* (1978-) and a separate index volume, now includes lengthy biographical entries for some 5,000 of the world's major contributors to science.

It is confined to scientists no longer living, and excludes technologists. The well-nigh universal arrangement of such biographical dictionaries is alphabetically by biographee. It is likely that this suits the majority of searchers who set out with a name. Other arrangements are occasionally found, however, as in Isaac Asimov *Biographical encyclopedia of science and technology* (Allen and Unwin, 1966), which is subtitled 'The living stories of more than 1000 great scientists from the age of Greece to the space age chronologically arranged'. More common are supplementary indexes to a main alphabetical sequence to permit an alternative approach: probably the most extensive example is *Who knows – and what* (Chicago, Marquis, revised edition 1954), 'keying 12 selected knowers to 35,000 subjects'. Similar subject approaches are provided on a smaller scale by the indexes in *McGraw-Hill Modern men of science, Chambers's Dictionary of scientists*, and many others. Less frequently encountered is the geographical index, eg in W S Downs and W Haynes *Chemical who's who* (New York, Lewis, fourth edition 1956). An even rarer type of index is the chronological table of births and deaths in T I Williams' work noted above.

Individual and collected biographies

For those scientists and technologists who have made a substantial mark there is always the possibility that they may have had a whole book devoted to them, eg D McKie *Antoine Lavoisier* (Constable, 1952), R Pilkington *Robert Boyle* (Murray, 1959), F E Manuel *A portrait of Isaac Newton* (OUP, 1968), S R Ranganathan *Ramanujan: the man and the mathematician* (Asia Publishing House, 1967), and some of course have written their own stories, eg P Scott *The eye of the wind* (Brockhampton Press, revised edition 1968). Students should be aware of a particular difficulty that often accounts for the lack of a satisfactory biography for many important scientists. While a historian may by training be the best person to write the *life* of a scientist, he may lack the necessary background to do justice to the man's *work*. A number of otherwise scholarly biographies have been severely criticized on that score, eg P Williams *Michael Faraday* (Chapman and Hall, 1965). Tracing these is no problem for they appear in the standard bibliographies and library catalogues. The British Museum *General catalogue of printed books* in particular is a convenient source because of its practice of listing books about an individual under his name as well as under the author. At least one bibliography of such works has appeared; T J Higgins *Biographies of engineers and scientists* (Chicago, Illinois Institute of Technology, 1949).

Many more scientists and technologists figure as the subjects of individual chapters in works of collected biography. Samuel Smiles

Lives of the engineers (1861-2) is a classic example, but there are countless others, eg E T Bell *Men of mathematics* (Penguin, 1953) in two volumes; A Findlay and W H Mills *British chemists* (Chemical Society, 1947); L C Miall *The early naturalists* (Macmillan, 1912). In some cases such collected works are not original contributions but take the convenient form of reprints or translations of articles or extracts previously published elsewhere, eg E Farber *Great chemists* (Wiley, 1961). The tracing of the accounts of individual workers in such volumes is made easier by the existence of a number of invaluable analytical compilations such as N O Ireland *Index to scientists of the world from ancient to modern times* (Boston, Faxon, 1962) which covers nearly 7,500 scientists and technologists in 338 collections. An interesting example of a similar index in a specific subject field is R Desmond *Dictionary of British and Irish botanists and horticulturalists* (Taylor and Francis, 1977) listing some eight thousand, all deceased, with only brief biographical notes but detailed references to biographical information in books and periodicals.

Journals too are often valuable sources of biographical information. As well as overtly biographical articles, they often feature short biographical notes about their contributors, eg *Vacuum, Chartered engineer*. More useful still are obituary notices: outstanding here are the series that used to appear in the *Proceedings of the Royal Society*. So valuable a source were they that since 1932 they have been published separately, and appear currently as the annual *Biographical memoirs of Fellows of the Royal Society*. A similar series is the US National Academy of Sciences *Biographical memoirs* which started in 1877. Both these major series have indexes to speed the location of an individual obituary, eg the index to volumes 1-35 of the latter appeared in volume 36 (1962). Some of the better journals have cumulated indexes, eg the American Society of Mechanical Engineers *Transactions* index covering 1889-1923 lists some 1,400 obituaries, but with E S Ferguson *Bibliography of the history of technology* (MIT Press, 1968) 'One is inclined merely to wag one's head sadly at the voluminous but largely unrecoverable biographical information in periodicals'. A recent attempt to eliminate at least some of what its compilers call tedious page-by-page searching is C Roysdon and L A Khatri *American engineers of the nineteenth century: a biographical index* (New York, Garland, 1978), which indexes biographical citations from some thirty major journals in the field. It was preceded by the similar work S P Bell *A biographical index of British engineers in the 19th century* (New York, Garland, 1975), which locates biographical information on some three thousand engineers in 24 journals from 1823 to 1900. An index of more general scope is E S Barr *An index to biographical fragments in unspecialized scientific journals* (University of Alabama Press, 1973).

Parallel sources

For the major figures there are sometimes other sources of information productive of biographical details that are not themselves primarily biographical, eg

a *Correspondence*: individuals vary in the amount of self-revelation they indulge in in their letters, but it is common for published editions to include background biographical details as editorial matter, in an introduction, for instance, or in the notes, eg *The correspondence of Isaac Newton* (CUP, 1958-). In any case, correspondence almost invariably sheds some illumination on a scientist's work.

b *Collected works*: these are even less patently biographical, but the very fact of gathering a man's books and papers together gives them some biographical interest immediately. Like good editions of correspondence, they often include information of an introductory or explanatory nature on the author's life, eg *The collected papers of Lord Rutherford of Nelson* (Allen and Unwin, 1962-); *John von Neumann: collected works* (Pergamon, 1961-3).

c *Archives*: by definition, personal papers such as diaries, most correspondence, laboratory notebooks, etc, are unpublished as a rule, but a recent project sponsored by the Social Science Research Council and the Royal Society makes available details of the private archives of some 3,400 men and women working in the field of science between 1870 and 1950. Entitled *Archives of British men of science* (Mansell, 1972), it appears in microfiche only, although the index and brief descriptive guide is available in normal format as well. For some of the more famous figures separate catalogues (also on microfiche) have been prepared, eg M Gray and J Friday *A catalogue of the papers of Sir William Bragg and his son Sir Lawrence Bragg* (Mansell, 1975) lists some 20,000 items from the collection in the Royal Institution.

In 1973 under the guidance of the Royal Society the Contemporary Scientific Archives Centre was established at Oxford 'to ensure that notebooks, correspondence and papers of scientists, engineers and medical persons were not destroyed before their historical value had been assessed'. Restricted to those persons who died after 1945, the Centre issues catalogues of collections that it has assessed in this way, and arranges for the archives to be stored in an appropriate library.

d *Festschriften*: these dedicatory volumes of papers by colleagues or former students, etc, are not uncommonly met with, particularly in the field of academic science, and once again biographical information is usually included, frequently in the form of one of the papers, eg *Horizons in biochemistry: Albert Szent-Gyorgi dedicatory volume* (Academic Press, 1962), *Progress in applied mechanics: the Prager anniversary volume* (Collier-Macmillan, 1963), *Essays on the nervous system: a festschrift for Professor J Z Young* (OUP, 1974). The student should

385

know, however, that festschriften as a class are often criticized: Robert P Armstrong sees them as 'in general mundane and without interest to any save the specialist with a marked tolerance for the dull, the inane, and the inconsequential'. According to Max Delbrück, the chief function of unrefereed festschriften is to provide a decent cemetery for oft-rejected manuscripts.

e *Bibliographies*: in a similar way bibliographies of individual scientists and technologists can also furnish very substantial amounts of biographical data in the form of introduction or notes, eg W R LeFanu *A bio-bibliography of Edward Jenner, 1749-1823* (Harvey and Blythe, 1951), Sir Geoffrey Keynes *A bibliography of the writings of Dr William Harvey, 1578-1657* (CUP, second edition 1953). And as was noted above (chapter 8) many bibliographies list books and papers *about* as well as *by* their man. Even *subject* bibliographies may contain biographical accounts of authors, eg J Ferguson *Bibliotheca chemica* (Glasgow, Maclehose, 1906). A somewhat special bibliography which often has considerable biographical relevance is the sale catalogue of an individual scientist's library. Four examples have recently appeared in facsimile reprint as the eleventh volume in the series *Sale catalogues of libraries of eminent persons: Scientists: Elias Ashmole, Edmund Halley, Robert Hooke, John Ray* (Mansell, 1975).

f *Lists of award winners*: these vary considerably in the amount of biographical detail included. It is sparse in the Special Libraries Association *Handbook of scientific and technical awards in the United States & Canada, 1900-1952* (New York, 1956), but more extensive in E Farber *Nobel Prize winners in chemistry, 1901-1961* (Abelard-Schumann revised edition 1963) and N H de V Heathcote *Nobel Prize winners in physics, 1901-1950* (Abelard-Schumann, 1953). Most recently we have had R L Weber and J M A Lenihan *Pioneers of science: prize winners in physics* (Bristol, Institute of Physics, 1980).

Portraits

A portrait of a scientist or a technologist is an obvious example in the biographical field of a non-textual information source as described in an earlier chapter. For the information a portrait conveys there is no substitute, short of an actual face to face meeting, and a number of the biographical sources that have been mentioned above contain portraits with the text, eg Farber *Great chemists*, Farber *Nobel Prize winners in chemistry, Biographical memoirs of Fellows of the Royal Society, Hackh's Chemical dictionary*, and most of the individual biographies, correspondence, collected works, festschriften, etc. Ireland *Index to scientists* and Barr *An index to biographical fragments*, both mentioned above, also index portraits. There are examples of

386

sources that make a prime feature of portraits: H M Smith *Torchbearers of chemistry: portraits and brief biographies* (New York, Academic Press, 1949) which reproduces a selection of those collected by the Massachusetts Institute of Technology, and the Royal College of Physicians of London *Portraits . . . described by David Piper* (Churchill, 1964) are each collections of over 200 portraits with very brief biographical sketches. Most unusually, *Who's who in steel* (Metal Bulletin, 1980) has a photograph accompanying almost every entry. Some tools, while not containing portraits themselves, do indicate where they can be found, eg Williams *A biographical dictionary of scientists*. The field of medicine is especially favoured: over 4,500 portraits are recorded in A H Driver *Catalogue of engraved portraits in the Royal College of Physicians of London* (1952), and the unpublished Index of portraits in books maintained in the College is available on microfilm. The outstanding source here is the New York Academy of Medicine Library *Portrait catalog* (Boston, Hall, 1960) in five volumes and two supplements, comprising reproductions by photo-litho offset of catalogue entries not only for over 150,000 portraits from books and journals, but also for over 10,000 separate portraits (photographs, paintings, etc) in the Academy's collections. And probably unique is *Images of Einstein: a catalog* (American Institute of Physics, 1979) with miniature reproductions of over six hundred photographs of Einstein from many sources.

INDEX

No attempt has been made to index the many hundreds of titles mentioned in the text. As explained in the Introduction, these are merely examples chosen from the many possible alternatives as representatives of their class.

389

Magnetic tape *see* Digital storage of information

Manual searching 22, 210, 212, 223, 224, 225, 227, 229, 299

Manuals *see* Handbooks

Manuscripts 125, 171, 269, 290, 373, 386

Maps 360

Mathematical formulae 78, 83

Mathematical tables 82-3

Mechanization 24, 80-3, 305-6
  see also New technology; Printing; Punched cards

Media *see* Non-print media

Meetings *see* Conferences; Minutes of meetings

Membership directories 89-90, 381

Memoranda 15

Metabolic charts

Method *see* Scientific method

Metrication 84-5

Microcomputers 223, 229

Microforms 19, 30, 130, 145, 168, 225, 371-9
  abstracting services in 299
  archives in 385
  bibliographies and indexes in 178, 309, 331
  conference proceedings in 266, 268
  data in 81
  library catalogues in 125, 387
  patents in 306
  periodicals in 174
  preprints in 162
  research reports in 280, 285, 287
  synopses journals and 164, 165, 166
  theses in 350, 352, 353, 354, 355
  trade catalogues in 347-8
  videodiscs and 366
  see also Computer-output-microfilm

Microorganisms, Collections of 95

Miniprints 164, 165, 174, 348

Minutes of meetings 15

Models 359

Monographs 9, 11, 16, 20, 22, 46, 47, 98, 99, 100-4, 105, 108, 109, 112, 113, 125, 250, 251, 254, 272, 274, 289, 334, 339, 341, 350, 351, 372

Monographs (biological) 104

Multilingual dictionaries *see* Translating dictionaries

Multimedia *see* Non-print media

Museums 95, 287

Names 55, 66-7
  see also Nomenclature; Trade names

'Narrative' bibliographies 127, 251

Neglect of the literature 25-7, 207, 233-4, 293, 294, 335, 351

Networks *see* Information networks

New technology 8, 19, 30, 83, 123, 152, 154, 174-8, 183, 215, 235, 236, 365, 378
  see also Computers, etc; Digital storage of information; Micro-computers; Microforms; Mini-prints; Photocopies; Telecom-munications

Newsletters, Research 173, 182

Nomenclature 58, 359-60

Non-book materials *see* Non-print media

Non-documentary sources of information 10, 16-8, 19, 20, 23, 25, 26, 29, 30, 89-95, 135, 162, 253, 263-4, 270, 351

Non-print media 37-8, 359-70, 371

Notebooks, Laboratory 15, 385

Numerical data *see* Data

Obituaries. 135, 142, 384

Observatories 95

Obsolescence 157, 158, 158-9

OCR *see* Optical character recognition

Official publications *see* Government publications

Offprints *see* Reprints

On-line searching 10, 22, 56, 59, 81, 176, 177, 182, 186, 215-48, 275, 285, 287, 306, 331, 353, 357, 366, 368, 381

Opinion surveys 23

Optical character recognition 175, 186, 241, 378

Oral history 364

Oral sources of information *see* Non-documentary sources of information

399